Volume 2 Part G

Chemometrics in Environmental Chemistry - Statistical Methods

Volume Editor: J. Einax

With contributions by
T. E. Barnard, K. S. Booksh, R. G. Brereton,
D. H. Coomans, S. N. Deming, Y. Hayashi, Y. L. Mallet,
R. Matsuda, J. M. Nocerino, R. A. Olivero, A. Singh,
H. C. Smit, O. Y. de Vel, Z. Wang

With 88 Figures and 40 Tables

Springer

Editor-in-Chief:
Professor Dr. Otto Hutzinger
University of Bayreuth
Chair of Ecological Chemistry and Geochemistry
P.O. Box 101251
D-95440 Bayreuth, Germany

Volume Editor:
Professor Dr. Jürgen Einax
Friedrich Schiller University
Institute of Inorganic and Analytical Chemistry
Lessingstraße 8, D-07743 Jena, Germany

ISBN 978-3-662-14885-3 ISBN 978-3-540-49148-4 (eBook)
DOI 10.1007/978-3-540-49148-4

CIP data applied for

Originally published by Springer-Verlag Berlin Heidelberg New York in 1995
Softcover reprint of the hardcover 1st edition 1995

Typesetting: Macmillan India Ltd., Bangalore-25
SPIN: 10123038 52/3020 - 5 4 3 2 1 0 - Printed on acid-free paper

The Handbook of Environmental Chemistry

Edited by O. Hutzinger

Springer-Verlag Berlin Heidelberg GmbH

Preface

Environmental Chemistry is a relatively young science. Interest in this subject, however, is growing very rapidly and, although no agreement has been reached as yet about the exact content and limits of this interdisciplinary subject, there appears to be increasing interest in seeing environmental topics which are based on chemistry embodied in this subject. One of the first objectives of Environmental Chemistry must be the study of the environment and of natural chemical processes which occur in the environment. A major purpose of this series on Environmental Chemistry, therefore, is to present a reasonably uniform view of various aspects of the chemistry of the environment and chemical reactions occuring in the environment.

The industrial activities of man have given a new dimension to Environmental Chemistry. We have now synthesized and described over five million chemical compounds and chemical industry produces about one hundered and fifty million tons of synthetic chemicals anually. We ship billions of tons of oil per year and through mining operations and other geophysical modifications, large quantities of inorganic and organic materials are released from their natural deposits. Cities and metropolitan areas of up to 15 million inhabitants produce large quantities of waste in relatively small and confined areas. Much of the chemical products and waste products of modern society are relased into the environment either during production, storage, transport, use or ultimate disposal. These released materials participate in natural cycles and reactions and frequently lead to interference and disturbance of natural systems.

Environmental Chemistry is concerned with *reactions in the environment.* It is about distribution and equilibria between environmental compartments. It is about reactions, pathways, thermodynamics and kinetics. An important purpose of this Handbook is to aid understanding of the basic distribution and chemical reaction processes which occur in the environment.

Laws regulating toxic substances in various countries are designed to assess and control risk of chemicals to man and his environment. Science can contribute in two areas to this assessment: firstly in the area of toxicology and secondly in the area of chemical exposure. The available concentration ("environmental exposure concentration") depends on the fate of chemical compounds in the environment and thus their distribution and reaction behaviour in the environment. One very important contribution of Environmental Chemistry to the above mentioned toxic substances laws is to develop laboratory test methods, or mathematical correlations and models that predict the environmental fate of new chemical compounds. The third purpose of this Handbook is to help in the basic understanding and development of such test methods and models.

The last explicit purpose of the handbook is to present, in a concise form, the most important properties relating to environmental chemistry and hazard assessment for the most important series of chemical compounds.

At the moment three volumes of the Handbook are planned. Volume 1 deals with the natural environment and the biogeochemical cycles therein, including some background information such as energetics and ecology. Volume 2 is concerned with reactions and processes in the environment and deals with physical factors such as transport and adsorption, and chemical, photochemical and biochemical reactions in the environment, as well as some aspects of pharmacokinetics and metabolism within organisms. Volume 3 deals with anthropogenic compounds, their chemical backgrounds, production methods and information about their use, their environmental behaviour, analytical methodology and some important aspects of their toxic effects. The material for volumes 1, 2 and 3 was more than could easily be fitted into a single volume, and for this reason, as well as for the purpose of rapid publication of available manuscripts, all three volumes are published as a volume series (e.g. Vol. 1; A, B, C). Publisher and editor hope to keep the material of the volumes 1 to 3 up to date and to extend coverage in the subject areas by publishing further parts in the future. Readers are encouraged to offer suggestions and advice as to future editions of "The Handbook of Environmental Chemistry".

Most chapters in the Handbook are written to a fairly advanced level and should be of interest to the graduate student and practising scientist. I also hope that the subject matter treated will be of interest to people outside chemistry and to scientists in industry as well as government and regulatory bodies. It would be very satisfying for me to see the books used as a basis for developing graduate courses on Environmental Chemistry.

Due to the breadth of the subject matter, it was not easy to edit this Handbook. Specialists had to be found in quite different areas of science who were willing to contribute a chapter within the prescribed schedule. It is with great satisfaction that I thank all authors for their understanding and for devoting their time to this effort. Special thanks are due to the Springer publishing house and finally I would like to thank my family, students and colleagues for being so patient with me during several critical phases of preparation for the Handbook, and also to some colleagues and the secretaries for their technical help.

I consider it a privilege to see my chosen subject grow. My interest in Environmental Chemistry dates back to my early college days in Vienna. I received significant impulses during my postdoctoral period at the University of California and my interest slowly developed during my time with the National Research Council of Canada, before I was able to devote my full time to Environmental Chemistry in Amsterdam. I hope this Handbook will help deepen the interest of other scientists in this subject.

This preface was written in 1980. Since then publisher and editor have agreed to expand the Handbook by two new open-ended volume series: Air Pollution and Water Pollution. These broad topics could not be fitted easily into the headings of the first three volumes.

All five volume series will be integrated through the choice of topics covered and by a system of cross referencing.

The outline of the Handbook is thus as follows:

1. The Natural Environment and the Biogeochemical Cycles,
2. Reactions and Processes,
3. Anthropogenic Compounds,
4. Air Pollution,
5. Water Pollution.

Bayreuth, June 1991 Otto Hutzinger

Fifteen years have passed since the appearance of the first volumes of the solid scientific information in Environmental Chemistry has been well received, and with the help of many authors and volume-editors we have published a total of 24 books.

Although recent emphasis on chemical contaminants and industrial processes has broadened to include toxicological evaluation, risk assessment, life cycle analysis and similar approaches there is still a need for presentation of chemical and related facts pertaining to the environment. The publisher and editor therefore decided to continue our five volume series.

Bayreuth, March 1995 Otto Hutzinger

Contents of Volume 2 G: Statistical Methods

Contents of Volume 2 H: Applications

Introduction

In the last two decades mankind has become increasingly aware of environmental problems, and particularly chemical aspects of the environment. The main reasons for this trend are not only objective causes such as apparent massive environmental pollution and resulting damage to the ecosphere and mankind, but also the fact that human beings have become more environmentally aware.

Our environment is increasingly endangered by a growing population, constantly developing industrialization combined with increasing consumption, reinforced traffic growth and environmental catastrophes resulting from accidents. The potential and actual pollution of all areas of the environment, the possible serious consequences for the ecosphere and, last but not least, for mankind, require the study of all environmental media and processes.

In general, chemical processes in the environment are very complicated because they occur in open systems, are often irreversible and heterogeneous and these processes are in most cases essentially influenced by biological and physical activities. The effects of pollutants and their chemistry in the environment, on the ecosphere and human beings are very complex:

– There are many noxious substances in the environment.
– The pathways of these pollutants in the environment to where they actually do damage are essentially unknown.
– The mechanism of the reactions and the efforts of pollutants in combination and with components of the other parts of the ecosphere are mostly unknown.
– The environmental concentrations of pollutants are often below or in the order of their effect thresholds. The time of latency between emission or discharge and the obvious beginning of a change or an accident is sometimes very long.

Therefore the present level of knowledge of environmental processes and reactions is not very high and there is an essential need to obtain more information on environmental chemistry and the possible consequences for mankind.

One of the aims of modern environmental research is to obtain more and more objective information on the very complex chemical processes in the environment. However, the enormous development of chemometrics over the last two decades has given us a powerful tool for solving this task.

Chemometrics is a relatively young chemical subdiscipline. Its general purpose is the application of statistical and mathematical methods, as well as methods based on mathematical logic with the objective of finding an optimum way for solving chemical problems and for extracting the maximum of information from experimental data.

The application of chemometric methods in the field of environmental chemistry becomes necessary and useful because of the following main reasons:

– Both natural and anthropogenic processes in the environment are mostly characterized by multidimensional changes of compounds and/or pollutants in different environmental compartments.

Anthropogenic emissions are often typically characterized by their multicomponent or multielement character. Typical examples are emissions of airborne particles from heating plants into the atmosphere or discharges of refinery effluents into rivers or lakes.

These substances can react synergically or competitively. It means the interactions of pollutants with one another and with natural components of environmental media also have to be considered.

– Questions in the field of environmental chemistry are often connected with problems of analytical chemistry. Hence, the experimental data obtained contain information on different environmental processes and also on the variance caused by analytical error. An important purpose of applying chemometric methods has to be the minimization or if possible the elimination of the analytical error.

– The last twenty years have been characterized by an explosive development of information processing and also by rapid development of instrumental analytical chemistry. These changes have given us the means for more detailed and precise monitoring of the environment and the reactions of chemical substances in various media.

The necessity of mananging and handling the flood of data obtained from environmental studies and monitoring is another reason for the application of chemometric methods. The main aim is the extraction of relevant information concerning the pollution state, spatial and/or temporal changes of pollution, and the identification of emittents or dischargers.

The modelling of the behaviour of pollutants in and between environmental areas, their impact on human beings and the assessment of environmental and toxicological risks also requires the application of chemometric methods.

In other words, chemometrics in environmental chemistry should be a tool for a deeper and more objective study of the complex processes.

This book describes important basic principles of chemometric methods together with typical examples in the field of environmental chemistry and gives an overall view on the function and the power of chemometrics as an instrument for the explanation of environmental problems.

The topics of the book are the application of chemometric methods to the whole process (this means all of the steps) of environmental chemical investigations, modern methods of data analysis for the interpretation and modelling of environmental data, and the application of chemometric methods for quality assurance and control.

Starting with focusing of readers interest on the topics of chemometrics today the first part of the book consists of a discussion of the very important problems of environmental sampling and in the description of basic principles of modern chemometric methods for experimental design and optimization. Subsequently, analytical measurement combined with the application of modern chemometric methods such as correlation techniques and information theory for signal processing and resolution is discussed.

The next sections describe the very important problem of calibration in the field of environmental analytical chemistry. Newer parametric and robust methods of multiple linear regression are applied for modelling of environmental systems. Because more multielement and multicomponent analytical methods are available for solving environmental questions the extension and application of univariate figures of merit to multivariate calibration is described in detail.

Investigations in environmental chemistry are often connected with problems caused by the scattering character of pollutants in the environment. Therefore one section deals with the important problem of the application of robust procedures to identify multiple outliers.

As discussed above, methods of data analysis, mostly methods of multivariate statistics, are definitely necessary to extract the latent information contents of environmental data. The section on pattern analysis and classification is followed by a description of principles of library search and expert systems. In the section on "Empirical pattern recognition/Expert system approach for classification and identification of toxic organic compounds from low resolution mass spectra" powerful combined application of modern principles to data analysis is demonstrated.

The next sections deal with the problem of modelling processes in important environmental areas such as the atmosphere and geosphere, and modelling quantitative structure-activity and structure-property relationships as the basis for the assessment of potential damage to the ecosphere and/or human beings.

How can the applied analytical method be validated? How can the laboratory working for solving environmental problems be evaluated? How can we manage and handle data and laboratory information? Answers to these topical questions with particular consideration of the aspects of quality assurance and control are given in the subsequent three sections.

The last section of the book "Automated techniques for the monitoring of water quality" describes, as an example for the very important environmental compartment the "hydrosphere", the instruments as the "hardware" which are available for environmental monitoring. This equipment is the technical basis for the application of chemometric methods.

The purpose of the book "Chemometrics in environmental chemistry" is the description of basic principles of modern chemometric methods applied to representative problems of environmental chemistry. In addition, the contributions demonstrate that, in each environmental problem under investigation, constructive cooperation between the chemometrician and the environmental specialist is necessary for obtaining concrete and objective answers which are close to reality.

The general aim of this book is to arouse the interest of the environmental scientist who has not yet concerned himself with statistical and mathematical problems when applying chemometric methods to his own field of work.

Jürgen Einax

Environmental Sampling

Thomas E. Barnard

Dept. Geoenvironmental Science & Engineering, Wilkes University, Wilkes-Barre, PA 18766 USA

Summary

Environmental sampling and data collection is an activity that requires skill and expertise in many of the traditional scientific disciplines. Previous authors have addressed the subject from the point of view of a single medium, a single discipline or, compliance with a regulatory program. In this chapter, a general plan is presented which considers environmental sampling as a special case of the scientific method of investigation. Decisions regarding all aspects of sampling (statistical design, sample locations, tools, analytical procedures, etc.) must be made in consideration of the study objectives. The plan presents a framework for developing clear study objectives and collection of data that will lead to resolution of the objectives. Examples are presented in surface water, shallow soil, deep soil, and groundwater.

Overview of the Sampling Process

Most readers are aware of cases where many years and millions of dollars have been spent studying a hazardous waste site and scientists still cannot agree what needs to be done or even if there is a quantifiable human health hazard. Politicians and the general public are increasingly frustrated with such "studies."

There are many reasons why studies are costly and fail to deliver on promises. First, environmental science is an evolving field. Standard procedures do not exist for many aspects of sampling and data analysis. Where standards exist, they are often blindly applied to a situation where they are inappropriate. Second, the environmental field is interdisciplinary. Working professionals are required to communicate across traditional fields. This requires specialized practitioners to integrate the principles of the more traditional disciplines such as chemistry and biology. Third, the institutional and legal framework under which the studies are conducted do not understand and at times seem to be completely incompatible with the basic principles of environmental practice. As a result work plans are often a series of checklists designed to meet an endless list of requirements rather than a well thought out game plan for conducting a scientific inquiry into a specific problem. These factors and others lead to a situation where environmental investigations appear to be a group of scientists contradicting each other and making up the rules as they go along.

Although there is tremendous variability in the size, scope, and purpose of environmental investigations, there are some general principles that can be developed. These principles have been formulated into a general plan that is applicable to most investigations. This plan is based on the supposition that environmental sample collection, measurements, and data analysis are all pieces of the scientific investigation process. The plan describes how each of the basic elements of a conventional work plan: experimental design, sample collection, analytical and other measurement procedures, data analysis, and quality control are conceived, developed and, implemented in order to reach the ultimate objectives of the investigation. In addition, four controlling forces: environmental science, statistics, practical constraints, and regulatory requirements serve as guides which define the limits of the investigation. This plan provides a general framework for the development of the sampling and analysis work plan and serves as the basis of rest of the chapter.

Many other references describing the environmental sampling and data analysis process are available. These tend to present the subject from the point of view of a particular discipline. For example Keith [1, 2] discusses sampling from the point of view of analytical chemists. Barcelona [3] discusses sampling for groundwater geohydrologists, while Ward et al. [4] is for surface water professionals. Gilbert [5] has written one of the more popular books on environmental statistics. Schreuder and Czaplewski [6] present a conceptual framework for forest ecosystem monitoring. Sara [7] describes the investigation of hazardous waste sites under the Corrective Action program of the Resource Conservation and Recovery Act.

Types of Environmental Investigations

Environmental investigations are conducted for many reasons. The purposes of the investigation are formally presented as a series of study objectives. These objectives are then used to define all other aspects of the investigation including what is sampled, how the samples are collected, what measurements are made, and how the data are analyzed. Most problems that arise over the course of investigations can be traced to a lack of clearly defined objectives that are understood by all parties involved. Discussions concerning the appropriateness or acceptability of a particular procedure can only be conducted in terms of consistency with the study objectives. For example, a certain statistical test may be appropriate for a background characterization study but not suitable for a trend detection analysis.

Although most reviews of sampling are organized by media or the regulations under which the investigation is being conducted, a more enlightening method of classification is by the differing study objectives. Some of the common types of investigations are defined here. Also, a few examples of how the objectives will determine how environmental data are collected and analyzed are provided.

A study may be conducted in order to define the *baseline* status of an ecosystem. This is a basic requirement in conducting an Environmental Impact Statement under the National Environmental Policy Act. Often permit applications for a surface water discharge or air emission require a baseline study. These investigations may serve as the basis for other work such as risk assessment or evaluation of alternatives. Such studies can be exhaustive and include several years of fieldwork. In areas that are well studied, much of the information can be collected through a literature review.

Usually baseline studies assume steady state conditions. Accordingly variations in measurements are attributed to seasonal variation or assigned to sampling error. Some baseline studies try to quantify environmental processes such as geochemical weathering, transport of sediment, phase transitions, and biological activity. Results may be reported as sediment loss in tons per acre per year or primary production in mg carbon fixed per square meter of land or surface water per year.

Parameters that are measured during baseline studies are attempts to quantify some of the more basic environmental processes. These are quantifications that are averaged over time periods of months or sometimes years. Variables such as surface water flows, precipitation, and air temperature are frequently measured. In water quality studies pH, conductance, temperature, major cations and anions, and nitrogen and phosphorous are typical starting parameters. Other specific constituents are measured as they become relevant. In the biological world, baseline studies begin by determining the dominant species that are present and then quantifying the populations. Community diversity indices are sometimes used to indicate the general state of the ecosystem.

When evidence suggests that there has been a release to the environment of a substance, an *extent of contamination* study is conducted. An example would be a Remedial Investigation conducted under the Comprehensive Environmental Response, Compensation, and Liability Act. There may be hearsay accounts of chemical X being spilled or dumped. The soil or water may be discolored. Vegetation may be sparse. In such cases, preliminary studies are conducted to confirm the contamination and to determine its extent. Often the contaminant, or at least its class, is known so the analyte list is short. Sample locations are usually selected at the most concentrated areas and at what is believed to be the edge. This methodology is referred to as judgement sampling. Such a scheme may be appropriate for the study objective but will make various statistical tests inappropriate. For naturally occurring contaminants, it may be necessary to determine background levels.

Impact of contamination. Once the extent of contamination is known, its effect on the ecosystem is determined. These studies are required for the purposes of risk assessment and for the establishment of liability. While there is an attempt to further delineate the spatial and temporal extent of the contaminant, the focus is on the environmental fate of the contamination. The physical state of the contaminant and the potential for mobility must be determined. For example, it may be determined if metals are suspended, dissolved or complexed. The phase of organic contaminants (gaseous, dissolved in soil moisture, or adsorbed to surfaces) may be determined. A sub category of the impact study is the *risk assessment*. Here the potential for the contaminant to enter and move through the biosphere is evaluated along with its probability to impact the environment or human health.

Predictive studies are conducted to predict the response of an ecosystem to some type of natural or induced stress. A large scale example is the group of studies that predicts the response of a regional aquatic system to policies that regulate the emissions of sulfur dioxide and hence affect the acidity of the rainfall in the region that are being conducted under National Acid Precipitation Program. A small scale example would be to predict the rate of remediation of an aquifer during the operation of a pump and treat system. Almost all of these studies require the use of sophisticated mathematical models. Such models require the input of various parameters for calibration. It is critical that the planners of the investigation understand the data needs of the models. Typically a predictive study is carried out in two major steps. The first is model calibration and sensitivity analysis. The second is a verification process in which a model is shown to be capable of predicting ecosystem parameters with reasonable precision. It is important that an investigation provide two sets of data that are essentially independent.

In a *compliance monitoring* study a permitted facility that discharges to the air or surface water is required to demonstrate that it is operating within prescribed discharge limits. Another example would be personnel monitoring for health and safety considerations. Such monitoring plans typically contain details concerning the timing, locations and procedures for collection of samples, analytical procedures for quantitation, statistical tests and reporting format. The emphasis is not only on obtaining accurate data but also on obtaining consistent data that can be used for comparison to different time periods or different facilities. In order to reduce costs, the operators of the facility may wish to measure parameters that are indicative of the concentrations of a larger number of contaminants.

Process control sampling provides real time data that are used to monitor waste treatment facilities and provide the operators with the data necessary to insure that the facility is functioning properly. While the emphasis is on real time data there has been development of many sophisticated continuous monitoring equipment. The importance is placed on a short list of indicator parameters rather than a complete list of all analytes of concern.

In *surveillance and enforcement* monitoring there is usually a focus on a short list of analytes. The objective is to determine whether a concentration is above or below a standard. Therefore techniques are chosen that provide a high level of accuracy at the specified level. Because the study is generating evidence that must meet all legal challenges, significant quality control requirements are imposed on the study.

General Plan

A general plan (Fig. 1) has been developed that serves as framework for conducting environmental investigations. Although it cannot cover every detail of all sampling exercises, it serves as a guide to the thought process that must be followed with planning and implementing a sampling event or program. This plan is presented as a simplified, linear, top to bottom flow diagram. However, it allows for adjustments and refinements to be made at appropriate points. For example, as new information becomes available one must go through several iterations of the plan. Some iterations may be formally designated as phases while others may be simple adjustments to sample collection procedures. As long as the thought processes is followed, the plan will remain valid and it will function as a useful tool in planning and implementing sampling and data analysis programs.

There are five major steps or elements of the plan: 1) formulation of study objectives, 2) development of the sampling and analysis plan, 3) sample and data collection, 4) data analysis, and 5) refinement of the old and formulation of new questions. These elements are described in Table 1. In addition, there are four controlling forces that guide the entire process (Fig. 2). These are 1) environmental science, 2) statistics, 3) regulatory requirements, and 4) practical considerations. These forces will serve as guides throughout the process although their relative importance will vary from project to project. The application of these controlling forces, particularly statistics and environmental science, is the main focus of this chapter.

The first step is the development of the overall study objectives. These are formal statements describing the purpose of the investigation. Their development may seem to be an obvious step and hardly worthy of discussion. However, experience has shown that the importance of developing clear and concise study

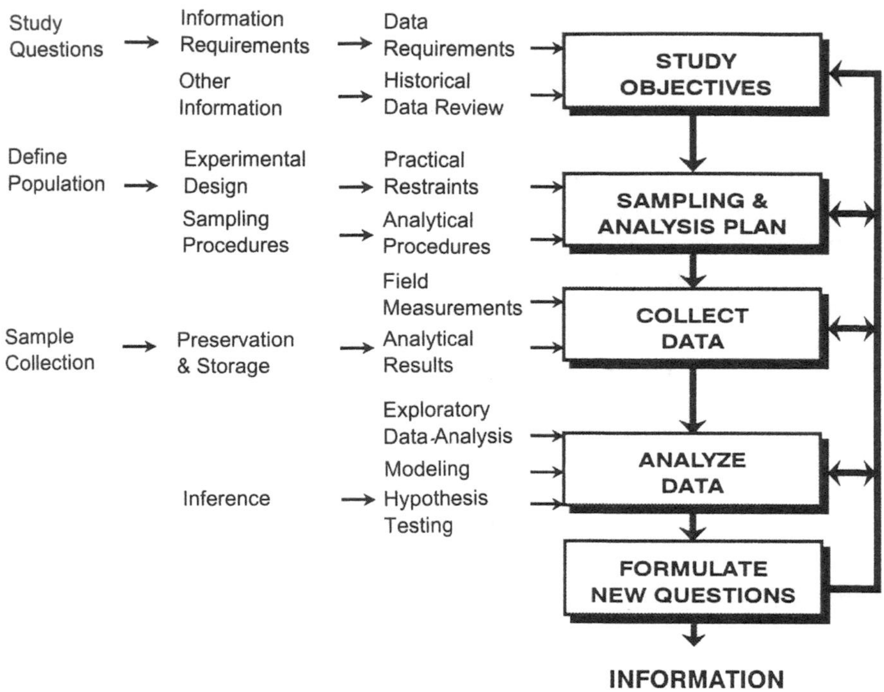

Fig. 1. General plan for environmental investigations

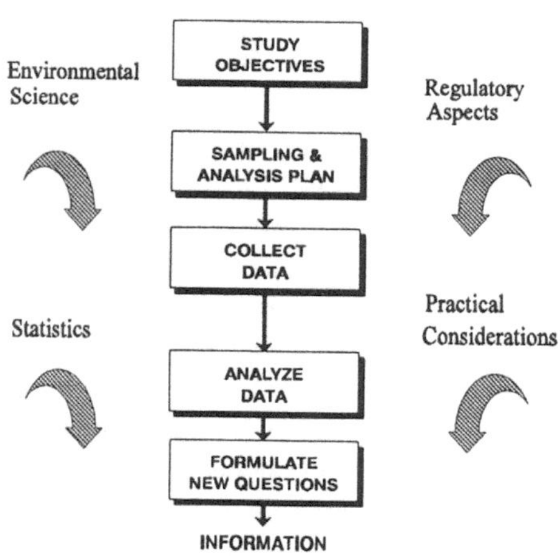

Fig. 2. Controlling forces for environmental investigations

objectives and communicating these objectives to all members of the project team cannot be understated. As previously stated, the details of the entire sampling plan are dependent on the study objects. Decisions regarding choice of sample collection procedure or analytical method can only be addressed by asking the question: "How does it provide data that can be used towards fulfilling the study

Table 1. Major elements of an environmental sampling and analysis plan

Element	Function or purpose	Examples
Study objectives	What is the purpose of the study?	Extent of contamination Impact of contamination Baseline characterization
Sampling and analysis plan, experimental design	What will be sampled? How many samples will be collected? How will individuals be selected? What statistical analysis and models will be used?	Simple random Grid Nested
Sample procedure	How will samples be collected? How will samples be preserved? How will sample be prepared for analysis?	Grab Depth integrated Sample preservation
analytical procedures	How will attribute be quantified?	Field instruments Gas chromatograph/mass spectrometer
Collect data	What are the measurements? What are the observations?	Concentration of iron Color of soil
Data analysis	What do the data say about the population?	Graphical Mapping Tabular Modeling Statistics Geostatistics
Formulate new Questions	Build on newly acquired information	Are conclusion applicable to other: sites? seasons? constiuents?
Information	Answers the study objectives	Extent of contamination Impact of contamination Baseline characterization

objectives?" For example, the analytical chemists have a choice between analyzing for total lead or extractable lead in a soil sample. If the study objective is to determine the extent of lead contamination, then total lead analysis is probably appropriate. On the other hand, if the study objective is to perform a risk assessment, then the extractable lead is probably a more appropriate measure of the potential for lead to migrate through the ecosystem.

The process of developing the objectives is outlined in Table 2. The steps are to start with the study questions and then to develop the information requirements followed by the data requirements. The distinctions between these three steps may seem vague but what is important is the thought process. The study questions lay out in plain English what the basis of the study is going to be. The information requirements define what is needed to answer the questions. Once the information requirements are established, the data required are specified. The

Table 2. Examples of the formulation of study objectives

Study questions	Information requirements	Data requirements	Study objective
Is there environmental contamination at the site?	Background and ambient levels of the constituents of concerns	Background and ambient concentrations of constituent X	Identify the constituents which originated from anthropogenic sources
What is the extent of contamination at the site?	Distribution of contaminant X in various environmental media	Concentrations of contaminant X at various points in space at various points in time	Perform an extent of contamination study
Can the contamination be treated?	Physical, chemical and biological nature of contaminant X and how it is affected by application of treatment technologies	Removal efficiencies in bench scale treatability studies as various operation parameters are adjusted	Perform a treatability study of contaminate X in media Y
What is the danger to the population human at the site due to contaminant X?	Chemical form the contaminant potential to move in the environment potential to move through the biosphere	Species distribution, retardation coefficients, diffusion and dispersion coefficients, reaction rate constants	Perform a risk assessment

scientific analysis of the study questions involves the development of information. A distinction is made here between *data* and *information*. Data are facts, quantities, or observations whereas information is defined as knowledge derived from facts. The derivation of this knowledge is what constitutes the scientific process.

A tool that is useful for developing study objectives is the *conceptual model.* This is a simplified description of the fundamental environmental processes that are controlling behavior of the constituents, organisms, and phenomenon of concern. It is developed on the basis of the findings of similar studies. Initially this model is very simple. During the course of the investigation the model becomes more sophisticated. It evolves from a conceptual model to a quantitative or predictive model. It can be a rigorous analytical or stoichastic or statistical model. As shown below, the development of the conceptual model helps the investigator to determine the data requirements.

As an example, the study question may be "What is the extent of contamination of lead in surface soils at a site?" The information required is the establishment of the distribution of lead in soil, the data requirements are the measurements of lead in mg/kg dry weight at various locations at the site and the study objectives are to perform an extent of contamination study. A conceptual model is developed that attributes the lead contamination to a single source (a pile of mine tailings) and the transport to a single process (airborne dust). This model is used to help define spatial extent of sample collection. The contamination would be expected to be greatest in the direction of the prevailing wind and extend as far as dust particles could be transported by the wind.

It is almost always less expensive to review and build on the findings of others than to implement an investigation where nothing is known. The most efficient application of environmental science maximizes the utilization of information that has been previously obtained towards answering the study objectives. This usually begins with a literature review beginning with a general overview of the subject and quickly focuses in on the specific study questions. For example if the study question is to determine the yield of an aquifer at a specific location, the review would begin with general knowledge of flow in porous media. It would then focus in on the regional geology and the information that has been developed from existing wells. It may turn out that the aquifer yield is essentially dependent on the thickness of one permeable formation. As a result, the study objective can be refined to measuring the thickness of the permeable formation.

The historical data review is a step that is often neglected in the planning process. It is too easy to discard or ignore previously collected data because of incomplete documentation, questionable procedures, or because the sample locations were less than ideal. This tragedy is referred to as "throwing away useful information." Even the best planned and most expensive investigations result in a less than ideal accumulation of data. In emphasizing the iterative nature of the environmental investigation process, it is pointed out that each study should build on the results of previous ones. A skillful environmental scientist should be able to infer useful information from all but the most sketchy data sets. The procedure for conducting the historical data review is to ask four questions. They are: 1) what data are available?, 2) how complete are the data?, 3) what is the quality of the data? and, 4) what data are missing? At a minimum, historical data should provide assistance in determining the range of sampling, the experimental design (simple, random, stratified, etc.), and the number of samples required to meet the study objectives with the desired precision. In the lead in soil example, historical data may be utilized to estimate the concentration gradient which would be helpful in planning the spacing of sample points. Alternatively, it may be useful in estimating the order of magnitude of the contamination. This will assist in determining whether the samples need to be concentrated or diluted prior to analysis.

The sampling and analysis plan is the heart of any environmental investigation. It is a basic requirement of most regulators and is typically specified as a major milestone in project planning. Too often the sampling and analysis plan is merely a laundry list of equipment and procedures for field collection and laboratory analysis. While these components are important, they do not constitute a complete plan. The plan should be thought of as a guide book for conducting sampling and analysis. In the real world, conditions are constantly changing. A well conceived plan will allow the investigators to make the proper adjustments that will assure that the data collected will answer the study objectives. Therefore, in addition to the detailed sample collection and analysis procedures, the plan must also discuss the study objectives, statistical design, and proposed data analysis procedures.

In this chapter, the term *experimental design* is defined as a statistical term. It is the processes of eliminating known sources of bias, and reducing or quantifying the sources of variation. During the design process, the investigator must determine the sample selection scheme, the number of replicates at various levels, and the procedures for sample randomization. Good experimental design leads to an efficient sampling program, i.e. the most useful information is obtained for the least cost. Many statistical tools are only valid for specific experimental designs. Once the number of samples required at each level is determined, a safety factor is included to account for units unavailable for collection, samples broken or lost during shipment, or errors in measurement or analysis.

There are two basic types of studies. In an *experimental study*, the variables of interest are fixed or controlled at predetermined values for each run of the experiment. A waste treatability study at specific concentrations of a contaminant would be an example. In an *observational study*, there are many variables of interest that cannot be controlled, but they are measured and analyzed. A background characterization would be an example. In the environmental field, both types of studies are conducted and sometimes a mixture of both is used. It will be shown that controlling variation through careful design is almost always a more cost effective technique for reducing variance than to collect more samples.

The *target population* is that about which inferences are to be made and is divided into *population units*. For some studies, applying the definitions is an obvious and straightforward exercise while in others it involves making arbitrary decisions. For a treatability study it must be determined whether the target population is the effluent over the course of a day, week, or life of the treatment process or, possibly the effluent from all wastes that are treated. Obviously the choice would affect the range of time over which samples are collected. In biological studies, the population unit is typically that of an individual of a species or a group of species while in other studies, it may be arbitrarily defined as one square meter of soil at a specified depth or one liter of surface water.

Next, the attributes to be measured are determined. The parameters about which conclusions are to be drawn are measured. Other parameters which may affect variability are measured in both experimental and observational studies. For example, parameters such as pH, temperature, and alkalinity are commonly monitored in studies involving water chemistry. However, too often little thought is given to the determination of the attribute list and the "measure everything" approach is employed yielding an analyte list numbering in the 100s. This is a very inefficient sample design. The experimenter's knowledge of environmental science is relied on to determine these attributes.

The ideal experimental design must be tempered with all of the practical constraints such as cost, duration of the project, and availability of the population for sampling. Procedures for optimizing some of the basic sample designs will be developed later in this book. At this point, tradeoffs are made between costs and the expected precision that can be attached to the study conclusions. If it becomes obvious that the study objectives cannot be met within all of the project constraints, then decisions must be made regarding revision of the objectives versus increasing the time and/or budget of the investigation.

The sample analyses are usually specified before the collection protocols are finalized. This is because most analytical methods have some sample collection requirements with them. The analytical procedures are typically selected from a library of reference methods. Again these must be selected with consideration of the requirements of the investigation. The criteria are sample matrix, detection level, precision, and accuracy. Use of "accepted" methods is typically advocated when defensibility of the study is an issue. Too often laboratory managers are asked to recommend analytical methods in a vacuum. This may result in the specification of inappropriate methods or methods that cost more than is necessary.

Specification of a reference method may not settle all of the sample analysis issues. Often a pretreatment step such as sub-sampling, phase separation, particle size separation, or sample clean-up is required. The criteria for selection of pretreatment procedures must be included in the plan. When unusual analytes or complex matrices are involved some method development may be required.

The preparation of the sample collection protocols is a straightforward, although detailed and at times tedious process. The items that need to be considered are presented in Table 3. The level of detail required for each item varies from project to project. Most plans borrow heavily from or directly reference previously developed protocols. Major environmental programs typically have a library of standard operating procedures that can be utilized. The details of the collection protocol must be determined in conjunction with the rest of the plan. For example, the sample volume, the container type, and the preservative are selected based upon the analytical procedure that will be used. In another example, a unit selected from sampling may be unavailable. The sampler needs to know if that unit can be skipped or if an alternative unit needs to be sampled. This question can only be answered in consideration of the experimental design and the proposed data analysis procedures.

Table 3. Items to be included in the sampling protocol

Pre sampling	Personnel protection
site preparation	training requirements
equipment list	protective clothing
decontamination	personnel monitoring
containers, labels, paperwork, etc.	contingency plans
Sample collection	Quality control
number of samples	custody forms
location and time	documentation
alternative locations	spikes and duplicates
field measurements	
field observations	
Sample handling	
field treatment	
preservative	
containers	
labelling	
temporary storage	
shipment	

In terms of time and manpower, data collection is the largest task of an environmental investigation. With proper planning, it should be the most straightforward. Any collection program will encounter hitches such as equipment failure, units unavailable, changing weather. With experience, an environmental scientist can prepare a sampling and analysis plan that anticipates changing conditions and allows for contingencies.

One aspect of data collection that is not adequately addressed is data management. A tremendous volume of data flows into project headquarters from a variety of electronic and hard copy sources. Issues such as input, preprocessing, storage, quality control, and providing access to the data must be addressed prior to data collection. Procedures for the development of a data management plan are discussed elsewhere [8].

Once the data has been collected, environmental scientists and statisticians begin the process of data analysis. As previously stated, the basic outline of data analysis is a requisite part of any sampling plan. Many experimental designs have specific statistical tests associated with them. However, data analysis is a creative process. It takes skill and experience to glean useful information from the large quantities of data that are collected in most studies.

The first step is *exploratory data analysis*. Summary statistics such as means, variances, and ranges are calculated for all parameters. Graphs are produced that show how parameters vary in time, space, or other parameters. Attempts are made to classify the data into groups of differing levels or trends for graphing or calculation of summary statistics. A mass balance may be performed. This exercise usually results in the formation of hypotheses (in addition to those originally developed) that can be tested by more rigorous statistical procedures.

Statistical tests consist of *planned* and *unplanned* analyses. Planned tests are those that were conceived prior to reviewing any of the data. Unplanned tests are those that are suggested by the exploratory data analysis. Some statisticians require greater levels of significance in order to reject null hypothesis that are suggested by the data. Since there is a lack of any consensus regarding significance levels in the environmental field, the distinction between the two types of analysis will not be emphasized here. Briefly, the statistical analysis process is one of formulating a hypothesis, performing an appropriate test, drawing a conclusion, and assigning a level of confidence to that conclusion.

Another type of data analysis is modelling. Environmental models can be as simple as a linear calibration curve or as complex as the global atmospheric models that are used to predict the impact of carbon dioxide emissions on global warming. Such models are referred to as mechanistic or deterministic. They are based on the premise that all processes follow the laws of chemistry, physics, and biology. If scientists were only smart enough to understand these laws and to measure all of the parameters that cause processes to occur, then the rate and extent of all processes can be predicted. Of course, the laws are not fully understood and everything cannot be measured with the required accuracy. Therefore all models are necessarily simple and the predictions are approximations. The models serve as useful tools for the understanding of the factors that control environmental process. Indeed, statistical analysis is a form of modelling. It involves models that allow for variance or that component of differences that can only be explained by chance. These are called stoichastic models.

The final step is the formulation of new questions. This is not presented as a guarantee of full employment for environmental scientists. It is merely a statement of the scientific process. No single study can result in the final definitive law that is never questioned. The new questions may be slight modifications of the initial study questions or they may be based on totally unexpected results. As shown in Fig. 1, when these questions arise they are pursued by a number of options. These are: a reanalysis of the data, collection and analysis of more samples, collection and/or analysis of essentially the same samples by alternative techniques, a revised experimental design, or the development of entirely new study questions.

Surface Water Sampling

Water is a very important medium in the environment. It commonly exists in the gaseous, liquid and solid form and readily makes transitions between these phases. In its liquid form, its unique chemical properties allow it to dissolve a tremendous diversity of ionic and nonionic constituents to varying degrees. Its unique physical properties (temperature-density relationships, surface tension, electrical conductivity, and viscosity) affect the way water moves and suspends and moves particulate matter. Indeed water is the medium of life. All living forms either exist in water or have developed a cellular structure to contain a water medium. The presence and availability of surface water (quantity) and its chemistry (water quality) are an integral component of most environmental investigations. Environmental scientists know and are constantly expanding their understanding of how these two concepts are interrelated. This section reviews the instantaneous and continuous measurement of surface water flow and the collection of samples for the determination of dissolved and suspended constituents in the following surface water bodies and flows: rivers, streams, lakes and reservoirs, surface water runoff, estuaries, liquid waste streams.

Study Objectives

In many water resources investigations quantifying the flow of water in surface streams is the primary objective. The time scale of concern for the variation in flow has a tremendous range. For storm water management and flood control projects instantaneous flow quantities and the response of a watershed to a storm event are the focus. Where water use is seasonal such as for irrigation, the variation in monthly mean flows may be the key parameter. In arid regions of the world where there is considerable year to year variation in weather, the time scale may extend to a decade or longer.

In water quality investigations, quantifying the spacial and temporal variation of parameters is the primary objective. In rivers the spacial variation is typically described along the dimension parallel to the river bed. In simple lake systems depth is used as the dimension of variability. Studies frequently are conducted to investigate how the parameters of concern vary with time or with flow. For example, in areas of contaminated runoff surface waters are more polluted during high flow conditions while in pristine regions the highest flows are associated with snowmelt which tend to dilute surface waters.

The mass transport of a contaminant through a system is frequently the focus of an environmental investigations. Because mass is the product of flow and concentration and, there is variability inherent in both of these parameters, extreme caution must be taken in acquiring and interpreting this attribute. First the significance of a statistic that is the product of two factors, each with their own error must be determined. Secondly, because the concentration is often a function of flow, the two parameters are rarely independent. Thirdly, most investigations are often unable to collect all of the desired measurements. As a result, a mass balance analysis is often used to calculate a flow or concentration value by difference. In light of these factors, it is often difficult to perform a rigorous statistical analysis on mass transport data.

In water quality investigations, the use of indicator parameters is frequently employed as a means to reduce the cost of sample analysis. Sanitary engineers have employed the concept of Biochemical Oxygen Demand (BOD) and Chemical Oxygen Demand (COD) as single parameters that measure the biodegradability or the total organic contamination of a water body or a waste. Public health professionals have used counts of *Escherichia coli* as a measure of the potential for water to transmit diseases. The simple measurement of electrical conductivity is used to estimate the concentration of dissolved ionic species in water. Although the analytical procedures for the determination of these and other indictor parameters are well established, care must be taken in their interpretation. Each dissolved ion has a different electrical conductivity and therefore make the conductivity of a high sodium water not comparable with that of a high calcium water. Between waters that have the same relative ratios of major ions the conductivity may be a very useful means of comparison. In the early phases of an investigation, the establishment of useful indicator parameters is a recommended objective.

Background Information

An effective and efficient surface water sampling program can be designed and implemented only after a review of the general knowledge of the behavior of surface waters and the constituents contained within has been conducted. This information is then augmented with site specific data obtained through field measurement and sample collection. As discussed in the first section, this interactive process continually builds on previously acquired knowledge.

Figures 3 and 4 are graphical presentations of stream flow data showing the range of time scales that are typically encountered in the environmental field. Figure 3 shows the response of a stream to a storm event. These data are used for designing flood control structures and for evaluating the impact of storm water runoff on water quality. It is obtained from field measurements is total flow and hydrologic analysis is used to separate the total flow into the direct runoff and the base flow (that resulting from groundwater inputs) components. In small to medium size streams it is not uncommon for the peak flow to increase more than two orders of magnitude during a storm event. A review of Fig. 3 might suggest that either continuous measurements or those taken at very short time steps are required to obtain this data for a storm event. However, the general shape of the direct runoff curve can be estimated from a knowledge of the basin morphology. Alternatively, if the peak flow and the time to peak are known, the direct runoff curve can be approximated with a triangle.

Figure 4 shows annual flow of the Colorado River at Less Ferry, Arizona. Note that even though the data are annual mean flows there is still a tremendous variability. The range is from a minimum of 5.64 in 1933 to a maximum of 24.0 in 1917. Also note that periods of low flows such as in the early 1930s and the early 1950s tend to be grouped together. A casual inspection of Fig. 4 would suggest that a record as long as 20 years may not be long enough to obtain an accurate estimate of the average flow. In order to be comparable, data obtained after 1956 would have to be adjusted for the effects of major upstream water projects such as the Glen Canyon Dam.

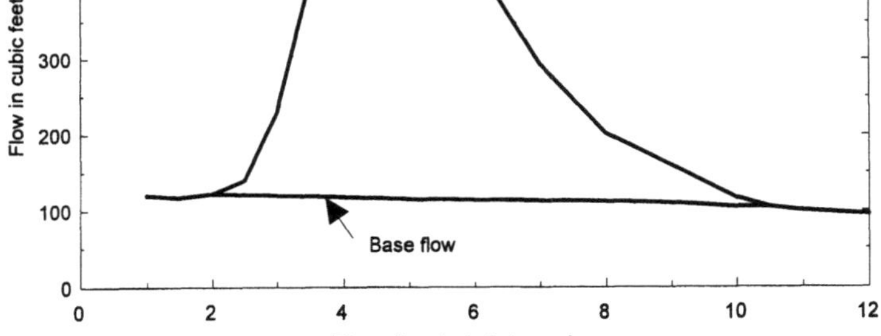

Fig. 3. Storm event hydrograph

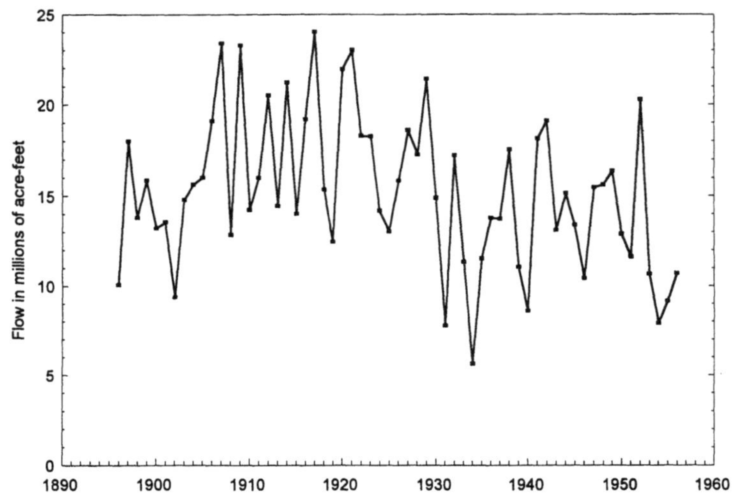

Fig. 4. Annual flow of the Colorado River at Lees Ferry, Arizona 1896–1956

In the United States, stream flow is monitored by the Geological Survey (USGS). This organization maintains a network of approximately 4000 stations where stream flow is recorded. The data are published in annual reports (Water Supply papers) that are organized by watershed. These data are also available in electronic format. It is rare that a USGS station with a sufficient length of record is located at the study site. A variety of hydrologic procedures are available [9] to apply the flow data of a nearby station to the study site. The simplest approach would be to install a gauging station at the study site and develop a short term record (1–5 years). A correlation relation could be developed between the flow at the study site and that of a nearby station with a longer record.

The typical distribution of instantaneous velocities and suspended sediment in a natural stream channel at a given cross section are shown in Fig. 5. These diagrams were developed for a regular section (i.e. that stream bed is relatively straight and there are no flow obstructions). At any point in the distribution of velocity as a function of depth may be approximated by a parabola with velocity of zero at the bottom and a maximum at approximately 20% of the total depth when measured from the surface. The distribution of suspended sediment is also complex. Generally, a stream is mixed just below a hydraulic jump, an outlet of a flume, or the nappe of a weir. As the water flows downstream, the sediment is settling downward with a velocity proportional to the square of the particle diameter and is being resuspended by the scouring action of the flow.

In temperate regions of the world, freshwater lakes of moderate depth or greater (5 meters) exhibit seasonal patterns of thermal stratification [11, 12]. This stratification has tremendous implications on the water quality and the biology of lakes. Figure 6 shows the typical distributions of temperate and dissolved oxygen during the period of summer stratification for a oligotrophic and a

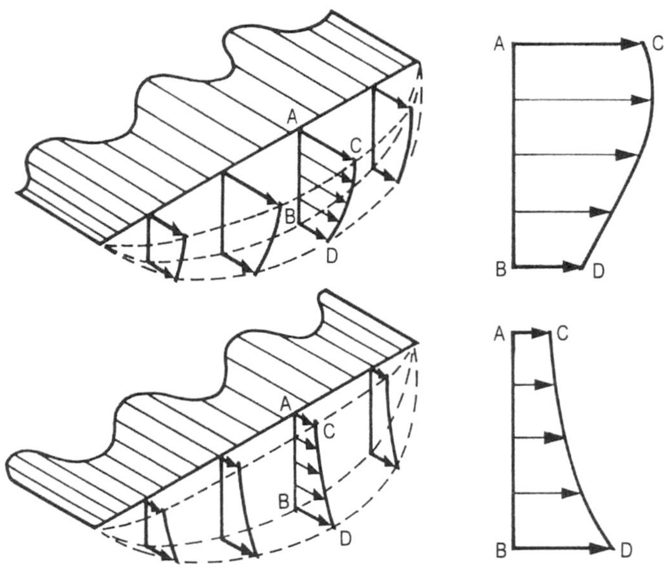

Fig. 5. Typical distributions of velocities (*top*) and suspended sediment (*bottom*) in surface water streams [10]

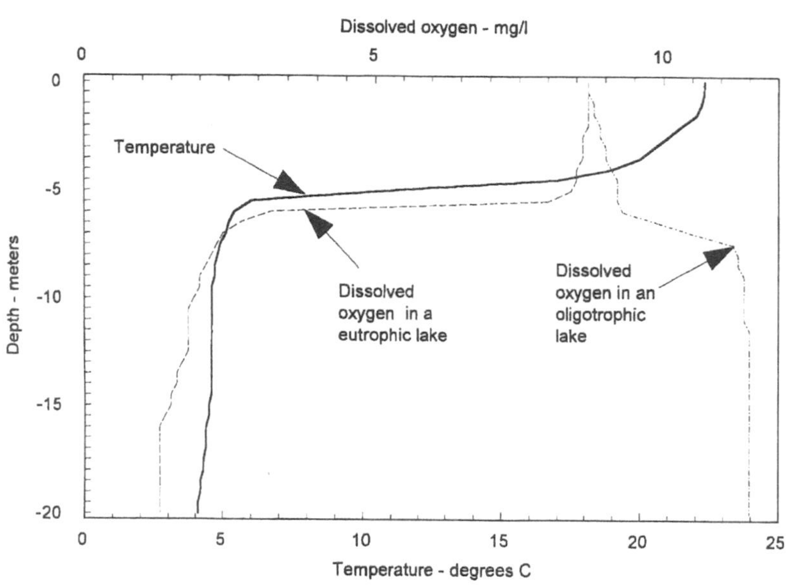

Fig. 6. Temperature and dissolved oxygen profiles in a stratified lake

eutrophic lake. Because of differences in density, the warmer water rests on top of the colder bottom layers. This may result in layers with radically different water quality. The warmer upper layer is referred to as the epilimnion while the colder lower layer is called the hypolimnion. In the oligotrophic lake the increase in dissolved oxygen with depth is due to the higher saturation values in the higher temperatures. In the eutrophic lake, oxygen is consumed by biological activity. In the epilimnion the oxygen is replenished by diffusion from the atmosphere, while in the hypolimnion, with no external supply oxygen, levels are diminished over the course of the summer. The complete depletion of oxygen in the hypolimnion is common in eutrophic lakes. This in turn affects the biology as well as reducing the redox state of elements such as nitrogen, sulfur, and iron. Use of field instruments such as temperature probes and meters for dissolved oxygen, pH, and conductance to record measurements at various depths at monthly intervals are extremely useful for preliminary studies that indicate the seasonal behavior of a lake system. Other factors such as basin morphology, inflows, outflows, and weather will have significant impacts on vertical and horizontal mixing in lakes as well as sediment deposition patterns. See Lehrman [13] for a more detailed discussion.

The role of sediments in aquatic systems is frequently inadequately addressed in environmental investigations. The subject has been reviewed on a fundamental basis by Stumm [14] and on a practical basis by Horowitz [15]. Many environmental processes such as adsorption, ion exchange, precipitation and even biological uptake occur at interfaces. Because of their large surface to volume ratios suspended particulates serve as the focus of these processes. Trace metals and hydrophobic organic substances tend to have low solubility and their transport in aquatic systems is governed by their ability to interact with surfaces. Also, the chemical form of trace metals gives clues to its origin or its bioavailability. For example, a common technique is to fractionate the particulate trace metals into the following partitions: crystalline, bound to oxides, and bound to organic matter. The crystalline fraction may be considered as having a geologic origin and generally unavailable for biological uptake. The fraction bound to oxides may have been adsorbed from the solution and hence recently mobilized and more likely to be from an anthropogenic source. The fraction associated with the organic coatings of the particulates may be biologically active. Particulates can serve as scavengers of contaminants by removing material from the dissolved phase. Scientists can exploit this property in using deposits of sediments as integrators of environmental quality.

The physical behavior of particulates in aquatic systems is most strongly governed by the particle size. The Stokes settling velocity of a 0.5 mm sand grain is 0.21 m/s while that of a 5 μm clay particle 2.1×10^{-5} m/s. The distribution of suspended sediment shown in Fig. 5 is obviously a simplification. In a fluvial system with moderate turbulence, the sand sized particulates settle rapidly and only become resuspended during periods of high flow while the clay sized particulates may never settle. In lakes lower levels of turbulence and long residence times result in a general downward flux of particulate matter and a buildup of sediments at the bottom. In all but the most extremely oligotrophic lakes, significant amounts of organic matter would be expected in these sediments. Biological decomposition and the lack of an external source of oxygen cause the reduction of major elements such as nitrogen, sulfur, and iron. This in turn results in significant variations in the ecosystem over a very short change in depth. Seasonal turnover may bring oxygen to the upper layers of bottom sediment and cause other transformations. See Wetzel (1983) for further discussion. In estuaries, flowing fresh waters tend to have a higher sediment load than ocean water. As the current fresh water decreases, sediment is deposited. Tidal induced currents play a major role in the deposition of these sediments.

Flow Measurement

The velocity-area method for the measurement of discharge in a natural stream channel is a basic tool of hydrology [16]. The ideal point at which the measurement is to be taken should be a straight section of channel with no irregular obstructions to flow. The section should be free of eddies, reversed currents, or other irregular flow patterns. Shallow streams can be easily accessed by wading across the stream. Deeper streams may be measured from a bridge. Permanent stations have a cable car assembly to convey the personnel and equipment across the stream. Flow is measured by dividing the cross section into segments and measuring the average velocity in each section. A rectangle is used to approximate the shape of each segment. Total flow is calculated by Eq. (1).

$$Q = \sum \bar{v}_i d_i w_i \tag{1}$$

where:

$Q =$ total stream discharge, ft^3/s or m^3/s

$\bar{v}_i =$ average velocity in subsection, ft/s or m/s

$d_i =$ depth of subsection, ft or m

$w_i =$ width of section ft or m

A variety of instruments are available for velocity measurement (Table 4). The so called Price meter consists of a wheel of cone-shaped buckets which rotate in a horizontal plane. The rate of rotation is proportional to flow velocity. The Pigmy meter is a miniaturized version of the Price meter. Another variation is the propeller meter. These devices are designed with a combination of fins and counterweights to assure that they are properly aligned with the flow of water. As the bucket assembly or propeller rotates it emits a small electric signal. The rate of rotation is detected either by a set of headphones or an electronic module which translates the rotation directly into velocity readings in a digital display. A more modern current meter design is the electromagnetic current meter in which a capacitor is contained in a probe. The flow of water past the probe induces an electric current. An electronic device converts the current to a velocity. All

Table 4. Velocity meters

Name	Supplier	Measurement principle	Notes
Model 2100	Swoffer	Propeller	Electronic readout 0.1–25 ft/s
Price meter	Teledyne Gurley	Rotating bucket	0.2–25 ft/s
Pygmy meter	Teledyne Gurley	Rotating bucket	0.05–3.0 ft/s
4150 Flow Logger	Isco	Depth – pressure transducer velocity – Doppler ultrasonic	Calculates discharge based on readings
Model 260	Marsh-McBirney	Electromagnetic	–5 to 20 ft/s

of these meters can be used to measure velocities between 0.5 and 20 ft/s. The current meter is attached to a calibrated wading rod. This rod has a sliding scale which assists the operator in determining depth and in setting the meter to the proper depth. For deep depths where measures are taken from a bridge or a cable car, the meter is suspended from a cable.

Chemical tracers are used for both velocity measurements and for analysis of flow paths in complex flow patterns such as seepage from waste piles or foundations drains. Tracers are ideal where for small flows and where access is difficult. The selection of a tracer is based on two criteria. First they should not interact with the water solution and secondly they should be easily detectable over large concentration rages and at low levels. Materials used as tracers include sodium or potassium which are detected with ion selective electrodes, radioactive substances such as ^{18}O or deuterium (^{2}H) and, fluorescent dyes such as Rhodamine. The tracer is introduced into the stream in a concentrated form. The measuring point should be located at point far enough downstream to insure adequate mixing in the plane perpendicular to the flow direction.

The stream discharge is calculated using the mass balance analysis in Eqs. (2) and (3).

$$M_{\mathrm{in}} = M_{\mathrm{dn}} = C \times V = \sum c_{\mathrm{dn,i}} \Delta t \qquad (2)$$

where

M_{in} = mass of tracer placed in stream, µg
M_{dn} = mass of tracer detected at point downstream, µg
C = average concentration, µg/l
V = volume, l
$c_{\mathrm{dn,i}}$ = concentration of tracer at point downstream at time i, µg/l
Δt = time step between concentration measurements, s

since:

$$Q = V/t$$

$$Q = \frac{M_{\mathrm{in}}}{\Sigma c_{\mathrm{dn,i}} \Delta t} \qquad (3)$$

where:

Q = discharge in l/s

The concentration of the tracer as it moves downstream is shown in Fig. 7. Note that as the trace moves downstream, the shape of the distribution is distorted by the longitudinal dispersion in the stream. The total area under the curve (in units of μg-h/l) is diluted by an increase in discharge. An alternative method is to pump a concentrated solution of the tracer into the stream at a constant rate. The concentration is measured downstream until it reaches an equilibrium level. The discharge is calculated by Eq. (4).

$$Q = q\left\{\frac{c_t - c_{eq}}{c_{eq} - c_b}\right\} \tag{4}$$

where:

- q = flow of tracer
- c_t = concentration of tracer in injection
- c_{eq} = equilibrium concentration of tracer at sample point
- c_b = background concentration of tracer

See Kilpatrick and Cobb [17] for further discussions on the use of chemical tracers for flow measurement.

For improved accuracy and ease of measurement engineered control sections are usually place in locations where long term flow measurements are desired. Control sections employ the concept of critical flow. In open channels the flow of water through a given cross section is either controlled by conditions occurring

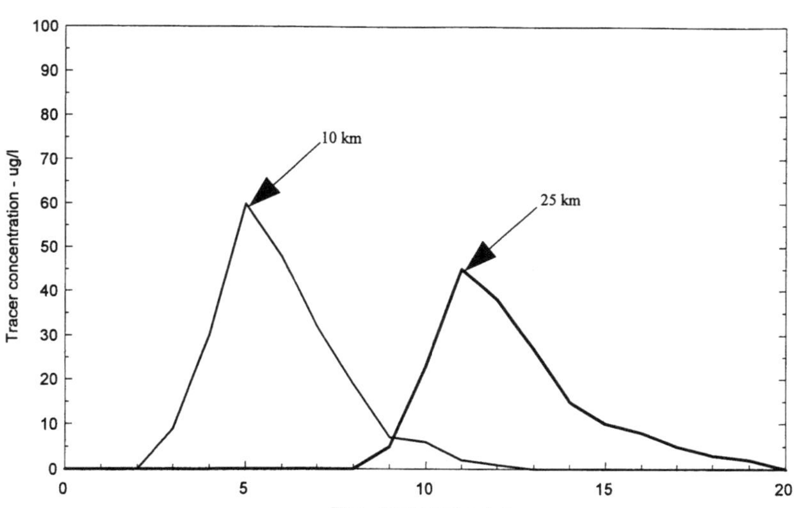

Fig. 7. Concentration of tracer downstream of injection point

upstream or downstream. When flow is controlled by downstream conditions, it is referred to as subcritical flow. Supercritical flow is flow controlled by upstream conditions. The transition between these states is referred to as critical flow. It occurs when the minimum amount of hydraulic energy is moving through a given cross section and is noted by observing a standing wave. Devices that cause flow to achieve critical flow are very accurate tools for measuring flow. In all cases a relationship is developed between the upstream elevation of the surface of the water and the discharge through the control section. Two common types of control sections that are used in open channel flow measuring are weirs and flumes.

Sharp crested or thin-plate weirs are structures where the nappe separates from the weir. Rectangular weirs can extend across the full channel width or may be contracted. V-notch weirs can be constructed at any angle but 45, 60, and 90° are the most common. The selection of the type of weir is based upon the range on flows that are anticipated as well as other factors such as the channel geometry. The basic equations describing the flow over weirs are presented in Eqs. (5)–(7). The terms are defined in Fig. 8. The weir coefficient, C_w includes factors that account for friction losses and correct for units. Values for C_w are list in Table 5. For low flows C_w is a function of H. More accurate determinations of C_w should be obtained by calibration [16]. H, the elevation of the surface of the water above the crest of the weir is measured in feet for English units and meters for SI units. The minimum flows are based on a value of H of 0.2 ft. At flows below these values the nappe begins to cling to the crest and the flow equations are no longer valid. The maximum flows are usually determined by the maximum depth of flow that can be allowed in the stream channel. See Grant [19] for a more detailed discussion on the use of weirs in flow measurement.

$$Q = C_w B H^{3/2} \quad (5)$$

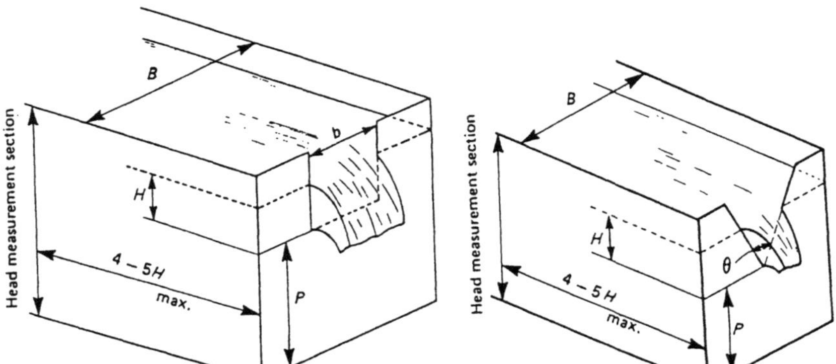

Fig. 8. Sharp crested weirs showing dimensions used in Eqs. (5)–(7). Contracted sharp crested weir is on *left*, V-notch weir is on *right* [18]

Table 5. Discharge coefficients for sharp crested weirs

Type	Equation number	Discharge in ft^3/s		Discharge in m^3/s	
		Weir coefficient	Minimum discharge[1]	Weir coefficient	Minimum discharge[2]
Rectangular					
full width	11.5	3.33	0.298	1.84	0.0277
contracted	11.6	3.33	0.286	1.84	0.0266
V-notch					
45°	11.7	1.035	0.019	0.57	5.38×10^{-4}
60°	11.7	1.443	0.026	0.80	7.36×10^{-4}
90°	11.7	2.50	0.077	1.38	2.18×10^{-3}

[1] ft^3/s per foot of width for rectangular, ft^3/s for V-notch
[2] m^3/s per meter of width for rectangular, m^3/s for V-notch

$$Q = C_w(b - H)H^{3/2} \tag{6}$$

$$Q = C_w H^{5/2} \tag{7}$$

For larger channels an natural rivers, broad crested weirs are often constructed with concrete. Because there is no separation of the flow from the surface of the control section, placement of a weir results in less alteration of natural flow conditions. The geometry can have a rectangular profile or it can have a curved upstream or downstream edge. The general relationship between surface elevation and flow is described by Eq. (5) although C_w is a function of the smoothness of the weir and the overall channel geometry. Variations on the basic broad crested weir have been developed in order to insure similar flow conditions over a larger range of flows. These include the Crump weir and flat-V weir. See Shaw [18] for further discussions.

The Parshall flume has been used extensively for flow measurement. It has the advantage of requiring a low drop in surface elevation and most suspended sediment will pass through the control section. In addition, floating objects will readily pass through the flume. It is constructed so that the critical flow is achieved through the converging section. Large Parshall flumes can be constructed of cast in place or precast concrete. Smaller flumes are available from a variety of suppliers and can be obtained as portable models. There are 22 standard designs with a flow range from 0.091/s to 93 m^3/s. They are used in locations such as headworks to treatment plants, irrigation channels, and inside of gravity sewers. Parshall flumes are rated by the throat width. Minimum and maximum flows for each of the standard designs are available from references such as Herschy [16]. As long as free flow is maintained, the discharge can be accurately measured as a function of h_1 the upstream elevation given in Eq. (8)

$$Q = Kh_1^u \tag{8}$$

where

Q = discharge m^3/s

K = coefficient

h_1 = water surface elevation in converging section, m

u = coefficient

A variety of devices are available for measuring and recording water level. The simplest is the simple staff gauge. These tools are calibrated by feet (0.01 ft increments) or by meter (0.01 m increments) and are designed to be easily read when a portion of the scale is submerged. The gauges may be coated with a porcelain or plastic material for protection against weather. Gauges are attached to permanent structures where they may be read from a convenient location.

Recording gauges are usually located in a stilling well. The well serves two functions. First it protects the recording equipment from the physical forces of moving ice and water and secondly it moderates short term fluctuations in water level caused by turbulence and wave action. In older installations, a floatation device is connected via a pulley and counterweight system to a recording device. The device can be a strip chart, punched tape, or magnetic tape recorder. Such devices require frequent servicing and are prone to mechanical breakdown. Modern equipment utilize pressure transducers to measure depth or ultrasonic sensors placed directly over the water to measure the distance. These devices can be attached to data loggers to electronically record water levels at preprogrammed time increments. Note that in using engineered control sections for discharge measurement, the location of the water level reading is critical to the accuracy of the measurement.

Permanent flow stations utilize a rating curve to convert water level readings to discharge. For natural channels the velocity area method is utilized to generate discharge measurements. Because of the geometry of most channels is irregular and the hydraulic properties of the floodplain are usually different from that of the channel rating curves tend to not be regular. Ideally, measurements taken at the extreme low and high flows and at regular intervals in between would be taken. However this is not practical. High flow readings are especially difficult to obtain as they may present a safety concern to field personnel. An additional problem is that sediment deposition, scouring and ice flows change the channel geometry and hence the rating curve. In order to enable the computerized use of the rating curve an empirical equation is fitted to the data.

When engineered control sections are utilized, care must be taken to insure that the textbook equations adequately describe the flow behavior. Dingman [9] recommends field calibration whenever possible. The most common problem is extreme flows that are beyond the calibration range. Other problems include uneven settlement taking water level readings at the wrong locations.

Surface Water Sample Collection

It is difficult to make any general statements concerning the design of a network of surface sample stations in the absence of consideration of study objectives. Ward and Loftis [20] presented a five step approach for the design of a water quality monitoring systems. This approach is consistent with the general model that was developed in the first section. The study objectives are the basis for determining the location of sampling points, schedule for sample collection, parameters to be qualified and the analytical procedures to be used. The number of samples (number of stations × sampling frequency) must be based on a valid statistical design as discussed by Ward and Loftis [21]. The population unit is usually arbitrarily defined as a standard volume (liter, gallon, etc) of water. The determination of whether dissolved or total (including suspended) concentrations should be measured as well as the use of indication parameters must be based on the knowledge of the environmental behavior of the constituents of concern and the study objectives. To overcome the temporal variability of water quality the use of organisms and sediment as indicators of longer term concentrations is encouraged.

Sample locations are selected based on a knowledge of the general behavior of body of water. In fluvial systems a common location is downstream of the point of mixing of a side stream or a discharge. This point may be several miles downstream in larger streams where laminar flow condition occur. In lakes typical sample locations may be the midpoint of the epilimnion and hypolimnion in the deepest part of the lake. Using the outflow of a lake to characterize the general chemistry is often a poor choice because this flow generally reflects only the epilimnion. Sampling frequencies are again determined on the basis of the study objectives and the environmental behavior the constituents of concern. In streams typical quarterly or monthly intervals that are commonly used to meet regulatory requirements are probably inadequate to define trends or cycles. A better approach may be to have defined time of sampling on the basis of flows. Such a scheme would sample at peak flows and during the low dry weather flows and would probably be a better technique for making mass balance determinations. In lakes a common technique is to sample during fall and/or spring turnover and during well stratified conditions.

Table 6 lists equipment that is commonly used in surface water sampling. The table is not meant to be comprehensive. It gives one example of a piece of equipment that is used for common sampling activities. Small streams and shallow flat waters are sampled with simple hand equipment. Dippers are available with handles up to 12 feet for the collection of 500 to 1000 ml for the collection of samples at the surface or at a free fall point such as weir. The dipper is filled and the sample is poured into a sample bottle. The device is usually made of polyethylene for easy cleaning. Shallow depths can be sampled with the subsurface sampler. The sample bottle is clamped to the end of this device and the screw top is attached to a suction cup. A rod extends from the suction cup to the opposite end of the device. The bottle is lowered to the desired depth and the top is unscrewed by twisting the rod. After the bottle is filled the top is screwed back on and the bottle is lifted. The buoyancy of the empty bottle and the weight of the filled bottle limit the depth of collection to a few feet and the total length of the device to 12 feet.

Table 6. Surface water sample collection devices

Name	Supplier	Matrix	Sample size	Materials	Comments
Hand samplers					
Dipper	Wheaton	Surface water	500–1000 ml	Polyethylene	Handles up to 12 feet
Sub surface grab	Wheaton	Shallow depths	1000 ml	Depends on bottle	With handle for remote opening of sample bottle
Sediment sampler	Scientific Instruments	Suspendend sediment in flowing streams	500 ml	Brass	Nozzle designed to collect with flow proportional to velocity
Coliwasa	Wheaton	Drums and tanks – sludges and viscous liquids	50 ml	Glass	Retractable inner rod is withdrawn to allow sample intake
Dipstick	Cole-Palmer	Drums and tanks – sludges and viscous liquids	50 ml	Telflon	Thumb or stopper placed on upper end to hold sample as device is withdrawn
Used from boats					
Kemmerer	Cole-Palmer	Deep water	0.4–2.21	Acrylic, stainless steel	Messenger is dropped to close ends at desired depth
Bomb	Wheaton	Deep waters	250–1000 ml	Telflon, stainless steel	Messenger is dropped to seal device at desired depth
Plankton net	Cole-Palmer	Plankton, suspended matter	Depends on concentration	Nylon mesh	Net and filter lowered for vertical or horizontal sampling
Ekman dredge	Cole-Palmer	Bottom sediment	6″ × 6″	Brass	Lowered to bottom, messenger releases spring loaded jaws
Sediment core sampler	Forestry Supplies	Bottom sediment	2″ dia	Stainless steel with plastic sample liner	Dropped to bottom, penetrates by gravity 4
Automatic Samplers					
Compact sampler	Isco	Stream, waste flows	24–500 bottles or 1–2.5 gal bottle	Glass, polypropylene	Programable for time or flow integration, refrigerated

The collection of samples of suspended sediment can be accomplished with a flow proportional sampler. The device is lowered into the stream with a rod or cable. Stabilizing fins assure that the assembly is pointed into the flow. As this device is lowered into a stream, water enters the nozzle at a rate propor tional to the stream velocity and enters the collection bottle. The sampler travels a vertical path from the surface to the bottom and returns to the surface while maintaining a constant velocity. The bottle is removed. The normal procedure is to collect samples in this manner at evenly spaced intervals across the stream. At each segment, the bottle is removed and emptied into a larger composite bottle. This procedure results in a flow weighted composite sample. Chapter 3 of the water data acquisitions manual (Interagency Advisory Committee on Water Data [22]) contains more details on procedures and equipment for suspended sediment sampling.

Deeper samples are collected with devices such as the Kemmerer sampler. This device is an open cylinder with spring loaded end caps. It is lowered by rope or cable to the desired depth and a weighted messenger is allowed to drop down the line. The messenger trips a release and causes the end caps to snap shut. The Kemmerer sampler is used to obtain large volume samples of dissolved and suspended matter as well as biological samples. Although larger faster swimming organisms will escape collection. Bomb samplers are equipped with seals that keep the sample chamber empty until a plunger is activated. The device is weighted to overcome the buoyancy effects of the empty chamber.

Biological samples are collected from lakes with plankton nets. The apparatus consists of cone shaped nylon net with a detachable filter bucket at the end. It is lowered to a desired depth and towed horizontally. For quantitative analysis the amount of water entering the net is metered. Nets are available with mesh openings ranging from 60–1000 μm. The use of this and other zooplankton collection devices is discussed by Wetzel and Likens [23].

A variety of grab and coring devices have been used for the collection of bottom sediments. The simple core sampler consist of a stainless steel tube (typically 2″ dia.), a sharpened nosepiece, plastic liners, and a core catcher. In shallow waters, the rod is attached to a rod and can be pushed into the bottom sediment. In deeper waters, the device is attached to a cable and dropped. Stabilizing fins maintain a vertical orientation and attached weights can increase the penetration into the sediment. The core catcher prevents sediment from escaping the sampler when it is raised. Because considerable force may be required to raise the device, a winch may be necessary. The Ekman dredge is one of the many grab samplers that are available. This device has spring loaded jaws that are retracted when the device is lowered to the bottom. A weighted messenger is dropped and it trips the jaw release. Sediment samplers generally do an adequate job of collecting organic and fine inorganic sediments. All have some degree of problems associated with sample disturbance. See Mudroch and McKnight [24] for a more detailed discussion bottom sediment collection.

Automatic samplers are frequently used for the collection of time or flow weighted composite samples in small streams and in waste flows. These devices consist of a peristaltic pump either a circular array of small bottles or a large bottle. The bottles can be pre-filled with the appropriate preservative. Advanced models store the sample bottles in a refrigerated chamber. The devices are programmed to collect samples at timed intervals. Advanced models are connected to a flow measuring device and a computer sets the duration of sample and hence the volume of sample, at a value proportional to the flow. For each sample cycle, the line is rinsed with source liquid and then the appropriate sample volume is pumped to the collection bottle. When operating in discrete sample mode, the bottle array is then rotated one step in preparation of the next sample cycle.

While automatic samplers offer convenience, errors are associated with the sample intake and with contamination. The intake should be located downstream of a zone of natural mixing and the pump should be operated so that intake velocities are the same as those in the stream channel. As shown in Fig. 5 in flowing water suspended sediment concentration is a function of depth. Intakes placed too shallow will cause low measurements while intakes placed at the bottom of the channel with high intake velocities will scour settled sediment and cause erroneously high measurements. The most common cause of leaching organics from sampling equipment is the addition of plasticizers that are added to make plastics flexible. Hence the tubing of the peristalic pumps is the most likely source of sample contamination. Tygon and Teflon are the most commonly used tubing materials. See Newburn [10] and references cited therein for further discussions on automatic sampling equipment.

Surface Soil Sampling

Soil is a medium that serves a large number of physical, chemical and biological functions in the environment. It provides a base for building foundations, roads and slopes. It temporarily stores water in surface depressions and in pores and then conveys it across the surface or downward to plants and the groundwater table. It provides a surface for a variety of chemical reactions that may retain or alter nutrients and contaminants before they move in the biosphere and the groundwater table. Soil provides the habitat for a tremendous assortment of microbes, plants and animals. Finally dissolution of soil minerals through weathering reactions provides the source of many constituents of ground and surface water.

Surface soil is the upper layer of the earth where parent geologic material are transformed into a series of layers or horizons. Environmental scientists must understand the nature of the soil system and how soil processes affect environmental quality. This section discusses investigations into the physical, chemical, and biological properties of shallow surface soils; defined as the soil that can be sampled with hand tools. The sampling for and the determination of geotechnical properties of soils is not covered here. Investigations into deeper geologic materials that require power equipment are covered in the section on deep soil and groundwater sampling.

Study Objectives

The study objectives serve as the basis for selection of specific parameters through the process described earlier. Environmental investigations in soil are generally concerned with determining how soil affects some process such as the movement or the transformation of a constituent. For the purposes of discussion the parameters of soil that are typically quantified in environmental investigations are organized in Table 7. Because of the heterogeneous nature of soil most of the parameters provide only indirect information concerning the environmental processes under investigation. Studies that utilize the standard sampling and analysis procedures without a clear understanding of what they are measuring will have a difficult time resolving the study objectives.

Siting studies are frequently conducted to determine the suitability of the soil to support some function. These studies focus on a particular soil property listed in Tables 7 and 8. For example, if a site is to be used as leachfield for the subsurface disposal of effluent from a septic tank, then the hydraulic conductivity of the soil is the most important parameter. Additional parameters would be the surface slope and the distance between the proposed disposal zone and the groundwater table. If a soil was being investigated as a barrier to contain waste then in addition to the hydraulic conductivity, a description of the ability of

Table 7. Typical soil properties that are utilized in environmental investigations

Category	Parameter	Environmental significance
Particle size	Particle size distribution, texture class, uniformity coefficient, effective size	Used to describe distribution of particle sizes
Bulk	Porosity, void ratio, bulk density, moisture content	Ratio moisture and gas to total soil volume
Hydraulic	Slope, infiltration capacity, permeability, distance to groundwater, effective porosity, specific yield	Defines ability of water to move into and through soil matrix
Geotechnical	Compaction ratio, Atterberg limits	Defines response of soil to mechanical stress
Visual	Color, mottling, facies codes	Indicates mineralogy, degree of oxidation, structure of pores
Soil profile	Soil horizons	Define extent of pedogeneic processes
Chemical	Mineralogy, nutrient level, cation exchange capacity, fraction of organic matter, contamination concentration	Describes chemical makeup of soil solids and the extent of interaction between solid and liquid phase
Biological	Vegetation cover, micro fauna, bacterial activity	Describe the extent and rate of biological activity in soil matrix

Table 8. Site use, soil functions and sampling objectives

		Soil properties to be evaluated	
Facility	Soil Function	Field	Laboratory
Building, roadway	Provide base	Slope, distance to groundwater	Texture, geotechnical properties
Leachfield	Liquid waste disposal	Slope, distance to groundwater	Moisture level, permeability
Landfill	Contain contamination	Distance to groundwater	Texture, mineralogy, exchange/adsorption capacity, permeability
Park	Support vegetation	Slope, profile	Fertility, texture

the soil to chemically interact with the waste would become part of the study objectives. Parameters such as surface mineralogy, cation exchange capacity and fraction of organic matter would be measured. The study could be conducted on the native soil or on soil that will be imported to the site.

In extent of contamination studies the total concentration of a constituent is usually measured first. Additionally two other issues must be addressed. The first is the determination of which phase (gas, liquid, or solid) the contaminant resides. The solid phase may be further divided into the fractions that are absorbed, precipitated and in the primary minerals. The second issue is the background level of the constituent. For synthetic organic chemicals the presence of any amount is usually regarded as contamination. However, for metals and naturally occurring radionuclides, the need to determine background levels is also a data requirement. This is accomplished by an appropriate statistical design of the sampling and analysis program.

The study of the potential for migration of contamination for migration is based on the potential transport pathways. If wind is the mechanism then only the contamination attached to the surface of the smallest soil particles will move. If groundwater is the mechanism then only the fraction of the contaminant that is soluble may move. The rate of movement of groundwater must be determined. In addition the ability of soil solids to adsorb or otherwise interact with the dissolved contaminant must be quantified. If the transport mechanism is biological uptake then the ability of indigenous plants or other organisms to uptake the contaminant is an information requirement.

Background Information

The analysis of soil properties can be divided into three scales as shown in Table 9. On a micro scale soil is a heterogeneous matrix that contains multiple phases as shown in Fig. 9. Solid particles of different chemical structure, size and shapes are interspaced with voids that contain water and gas. Properties are expressed on an unit expressed on a unit volume (cm^3), unit mass (kg on a dry

Table 9. Scales of soil properties

Scale	Study dimensions	Unit name	Typical unit size	Example parameters
Micro	3	Ped	1 kg 1 cm^3	Porosity Moisture content cation exchange capacity
Meso	1	Pedon	0.5–2 m	Profile
Macro	2	Soil series	1–100 ha	Soil series Slope Infiltration capacity

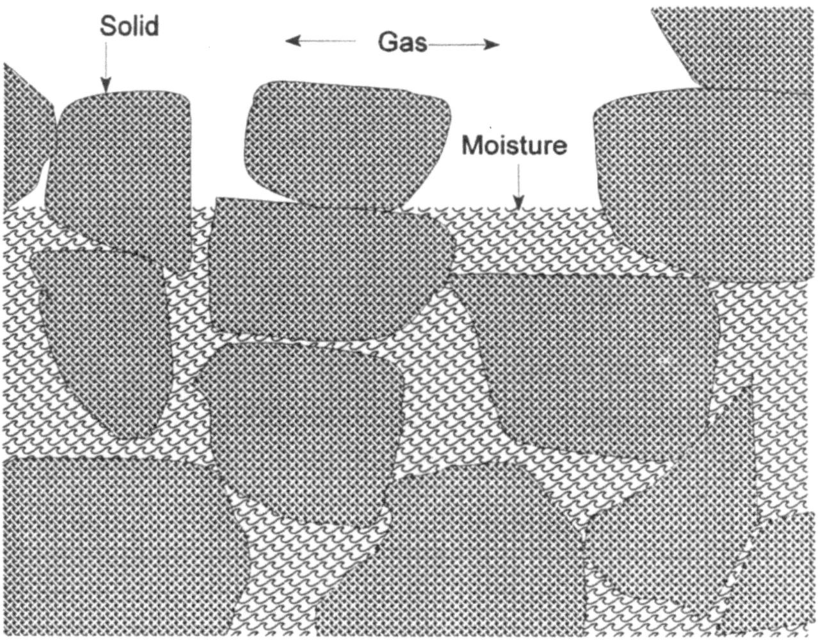

Fig. 9. Schematic diagram showing phases in soil matrix

weight basis) or may be unitless. The meso scale properties are based on the concept of soil genesis. It is concerned with the expression and properties of soil layers or horizons. The basic unit is the soil pedon which is the smallest volume that can be called a soil. It extends from the soil surface to the lower limit of pedogenically altered geologic material. The lateral dimensions are large enough to permit study of the nature of any horizons present (typically 1 m^2). The macro scale has a fixed vertical dimension equal to the depth of soil. The basic unit, the soil map unit which is the smallest area of soil with the same profile, vegetation cover and, surface slope.

As a result of its complex nature a variety of parameters have been developed to describe soils. Although these parameters provide only indirect answers to the study objectives, environmental investigators rely heavily on the sampling procedures and test methods developed in the geotechnical and soil science disciplines in conducting their studies. Geotechnical tests measure how soil responds to mechanical stress while soil science tests are concerned with the ability of soil to support vegetation under various agricultural practices. It is important that the environmental scientist understand the definitions of these parameters, how they are measured, and how they can be used towards the understanding of overall environmental processes. It is also important that environmental scientists understand the classification systems that have been developed to describe soil.

Soils are classified by their particle sizes and the distribution of sizes. Particles larger than approximately 0.074 mm can be sorted by a seive analysis. Smaller particle sizes are determined by hydrometer analysis of a dilute suspension of the soil. The terms gravel, sand, silt and, clay are applied to various ranges of particle sizes. Because natural soils contain a distribution of particle sizes, two parameters have been developed to describe the distributions. The uniformity coefficient, C_u is the ratio of the particle size that is 60% finer by weight d_{60} to the grain size that is 10% finer by weight, d_{10}:

$$C_u = d_{60}/d_{10} \tag{9}$$

As C_u approaches unity, the closer the particles are in size and the volume of voids becomes larger. Conversely, large values of C_u indicate a greater distribution of particle sizes and because the smaller particles will situate themselves in between the large ones, the overall volume of voids would be expected to be less. The USDA textural classification is based on the distribution of particles less the 2 mm in diameter. The textural name is based on the fractions of sand, slit and clay in the soil as shown in Fig. 10.

Bulk properties describe the overall density and the ratio of gas and moisture to total volume. Moisture content is often very dynamic in soil. Hydraulic proper ties describe the ability of water to move through the soil matrix. Infiltration rate is the rate at which water enters the soil from its surface. Permeability quantifies the rate at which a fluid moves through the soil matrix under pressure. Effective porosity is a measure of the pores that are connected and available for fluid flow. Specific yield is the amount of water that will drain from a saturated soil due to the force of gravity.

The geotechnical parameters describe the response of a soil to mechanical stress and changing moisture conditions. Soils in which the adsorbed water and particle attraction work together to produce a body of which holds together and deforms plastically are known as cohesive soils. The soils that do not exhibit this behavior are known as cohesiveness. These soils generally contain larger size grains and the response to stress in dependent on interactions between par ticles. The degree of cohesiveness and the extent of plastic behavior is a function of the clay and moisture content. The Atterberg limits describe the moisture content at which soil exhibits specific behavior. Environmental scientists must

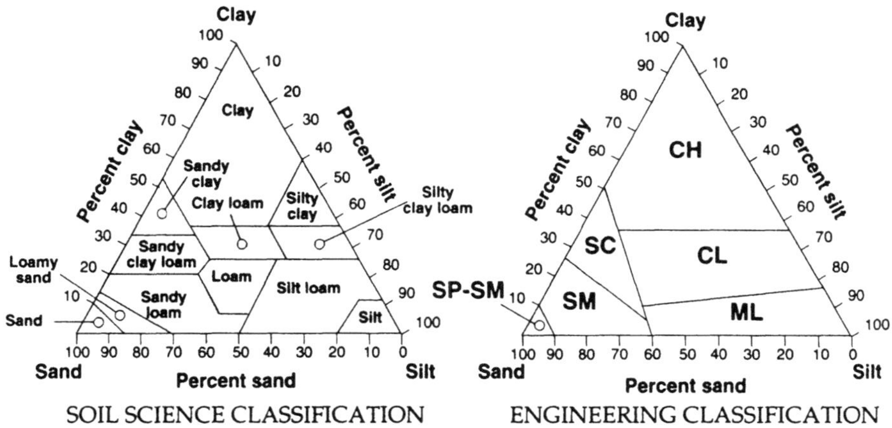

Symbols for engineering classification system

CH inorganic clays of high plasticity
CL inorganic clays of low to medium plasticity
ML inorganic silts and very fine sands
SC clayey sands, poorly graded sand-clay mixtures
SM silty sands, poorly graded sand-clay mixtures
SP poorly graded sands, gravelly sands little or no fines

Fig. 10. Soil textural classification systems. Soil science system developed by US Department of Agriculture and the engineering system is from the American Society of Testing and Materials. Adopted from [25]

appreciate the fact that as soils are subjected to stress they can release water through consolidation and the hydraulic properties will change. This in turn has impact on the transport of constituents through the soil.

Soil particles form larger aggregates with planes of weakness known as peds [26]. The study of the structure of soil at this level and smaller is called soil micromorphology. Visual observation of the soil is used to understand the chemical behavior of the soil matrix. Microscopes are used to view features smaller than 200 μ. The abundance, shape and color of features such as voids, cutans, and mottles are noted.

The mineral structure of soil particles have been divided into three broad categories in Table 10. Primary minerals were formed when the original magma, molten at very high temperatures, was cooled. Two types of magma are recognized based on their contents of silica. The first is granitic having a SiO_2 content greater than 60% and the second is basaltic with less than 50% SiO_2. Primary minerals are very stable in the environment at the surface of earth today. However, physical and chemical processes are slowly breaking them down through a process referred to as weathering in Equation (10).

$$\text{primary minerals} + \text{carbonic acid, organic acids} \rightarrow \text{secondary minerals} + \text{soluble components} \tag{10}$$

Table 10. Categories of minerals found in soil

Category	Examples	Environmental significance
Primary	Micas Feldspars Silica	Initial source for groundwater and secondary mineral constituents generally inert from environmental point of view silica is the most inert
Secondary	Silicate clays Al, Fe oxides Gypsum	Clays have high ion exchange capacity oxides have high absorption capacity swelling behavior will decrease permeability
Organic	Humic substances	Formed by incomplete degradation of organic materials acid-base, ion exchange and absorptive properties effect contaminant transport

Carbonic acid originates from the dissolution of CO_2 in water and its hydrolysis while the decomposition of organic matter serves as the source of the organic acids. The weathering process results in the release of the more solid elements such as Na and K while the less soluble elements (Si, Fe, Al) remain in the reformed three dimensional structure of the secondary minerals. The fate of elements of intermediate solubility (Ca and Mg) depends on the many factors such as temperature, water, and the structure of the primary minerals.

Secondary minerals are of a greater interest from an environmental point of view because: 1) they have a large surface area, and 2) the chemical structure of the surface is such that there is a great amount of interaction between the soil solution and the mineral surface. The specific surface area of clay size par ticles range from 1 m^2/gm for a 2 μm particle to over 760 m^2/gm for smectites and vermiculites. Isomorphous substitution of Al^{+3} for Si^{+4} and Mg^{+2} for Al^{+3} in the lattice of clay minerals and substitution of dissociable –OH for O at the edges of the particles results in a net negative charge at the surfaces. This surface charge is balanced by an excess of cations from the soil solution. These cations are readily exchangeable and hence has a great impact on the chemistry of the solid solution.

The structure of silicate clays consist of three dimensional layers of Si, Al, and O. Kaolinite is an example of a 1 : 1 clay. It is made of alternating layers of tetrahedral and octahedral sheets to from hexagonal particles. It's surface charge and hence its ability to flocculate is largely a function of the pH and concentration of cations in the soil solution. Montmorillonite is an example of a 2 :1 clay. It has an octahedral sheet coordinated with two tetrahedral sheets. Hydrated cations and water molecules can become strongly bound between layers resulting in the shrinking and swelling behavior of montmorillonite. 2 :1 clays typically have a greater charge than the 1 : 1 clays. A typical cation exchange capacity for Montmorillonite is 100 meq/gm [27].

A large number of crystalline aluminium and iron hydroxides, oxyhydroxides and oxides are found in soils. They exist in numerous chemical forms ranging from amorphous $Al(OH)_3$ and $Fe(OH)_3$ to the stable minerals such as gibbsite (Al_2O_3) and goethite (Fe_2O_3) [28, 29]. The oxides are formed by the precipitation of hydroxides followed by slow dehydration and rearrangement or polymerization reactions. Aluminium and iron oxide minerals are generally found in the clay sized fraction of soils or as coating layers of clays or primary minerals. The environmental significance of the oxides is the absorptive behavior that is a result of the surface charge. In the crystalline structure, the outermost oxygen position can be occupied by either O, $-OH$, or H_2O. These functional groups undergo ionization reactions H^+ and OH^- ions in the soil water and result in a net surface charge that is a function of pH. As a result oxides can have a positive or negative surface charge. The charged surfaces undergo adsorption/desorption reactions with nutrients (NH_4^+ and PO_4^{3-}) and many environmental contaminants that exist as charged species.

Organic materials in soil consist of a hetrogeneous mixture recognizable plant debris; macro molecules such as polysaccharides, proteins and lignins and; humic materials. Microbial action removes the decomposable portions of the material leaving behind a relatively stable substance called humic material. Environmental investigators are interested in humic materials because of their ability to interact with other soil constituents. Because its structure defies characterization, it is frequently divided into three categories with operational definitions. Fulvic acids are soluble in acid and basic solutions are have a molecular weight less that 10 000. Humic acids are insoluble in acid and soluble in base and have a molecular weight of 10 000–100 000. Humin is insoluble in acid or base and has a molecular weight greater than 100 000. Humic materials contain carboxylic and phenolic functional groups which provide acidity and undergo complexation reactions with cations in the soil. Parameters such as exchangeable acidity are used to quantify this behavior. A second characteristic is the ability of soil organic matter to adsorb hydrophobic substances such as those containing chlorine.

Although soil may be considered as saturated (no gas) or dry (no moisture), in the natural environment all three phases exist at measurable levels. Soil gas tends to have higher levels of carbon dioxide and lower levels of oxygen than the atmosphere. In zones of higher moisture and biological activity it is often void of oxygen. Soil water tends to have a much higher ionic strength than natural surface waters. In addition soils can have microenvironments where radically different soil properties exist.

Natural soil forms horizons as part of the pedogenic process. A soil horizon is "a distinct layer of soil, approximately parallel to the soil surface with characteristics produced by the soil forming process" [30]. The definitions of the major horizons are presented in Table 11. Subdivisions of the major horizons are designated with Arabic numerals. Transitional horizons are also defined. Mature soils generally have O, A, B and C horizons. E horizons are frequently found in forested soils. Immature soils are lacking or have incomplete expression of one or more horizons. The process of soil formation involves 1) the buildup of organic matter, 2) the weathering of the primary minerals accelerated by organic acids, and the build up of secondary minerals such as clays and oxides. The pedogenic process is affected by factors such as age of soil, primary minerals, and climate.

The basic unit of soil science is the padon. It is defined as the smallest volume that can be recognized as a soil individual. Its lower limit is the vague

Table 11. Definitions of soil horizons (adopted from [30])

Category	Name	Description
Organic	O	Organic horizons of mineral soils include horizons (1) formed or forming above the mineral part of mineral soil profiles: (2) dominated by fresh or partly decomposed organic material: and (3) containing more than 30% organic matter if the mineral fraction has no clay. Intermediate clay content requires proportional organic matter content equal to 20 + (0.2 × % clay).
Mineral	A	Mineral horizons consisting of (1) horizons or organic matter accumulation formed or forming at or adjacent to the surface; (2) horizons that have lost clay, iron or aluminum, with resultant concentrations of quartz or other resistant minerals of sand or silt size; or (3) horizons dominated by 1 or 2 above but transitional to an underlying B or C concentration
	B	Horizons which feature one or more of the following; (1) an illuvial concentration of silicate clay, iron, or aluminum alone, or in combination, (2) a residual concentration of sesquioxides or silicate clays, alone or mixed that has formed by means other that solution and removal of carbonates or more soluble salts, (3) coatings of sesquioxides adequate to give a conspicuously darker, stronger, or redder colors than the overlying and underlying horizons in the same sequm; or (4) an alteration of material from its original condition in sequm lacking conditions 1, 2, and 3 that obliterates original rock structure, that forms silicate clay, liberates oxides, or both, and that forms a granular, blocky, or prismatic structure
	C	A mineral horizon or layer excluding bedrock, that is either like or unlike the material from which the solum is presumed to have formed, relatively little affected by pedogenic processes, and lacking properties diagnostic of A or B.
	E	Mineral horizon in which the main feature is loss of silcate clay, iron, aluminum, or some combination of these, leaving a concentration os sand and silt particles of quartz or other resistant materials
	R	Hard bedrock including granite, basalt, quartzite and indurated limestone or sandstone that is sufficiently coherent to make digging impractical

and somewhat arbitrary limit between soil and "not soil". The lateral dimensions are large enough to permit study of the nature of any horizons present. The soil profile is defined as the vertical exposure of the horizons of soil. This inspection will give an indication of the heterogeneity of the soil and help to determine sample locations.

Numerous systems for classification of soils have been developed [26]. The classification system used in the United States is described in Soil Taxonomy [30]. This system has evolved under the direction of the Soil Conservation Service to a hierarchy that considers the geologic origin, mineralogy, moisture condition, vegetation, and pedogenic processes. Table 12 lists the six levels of classification used in this system. The highest level is order while the lowest is series.

Maps of soil series are available in the United States in scales ranging from 1 : 15 840 to 1 : 24 000. Modern soil maps are printed on aerial photographs and are grouped by counties. The associated text contains a description of the characteristics and suitability of each series. Soil maps are an excellent source of information for preliminary environmental studies. They can be used for an ini-

Table 12. Levels of classification in US Comprehensive Soil Classification System

Category	Examples	Interpretation	Type of classification
Order	Spodosols Vertisols	Acid ashy gray sands over a dark sandy loam shrinking and swelling dark clay soils	Soil forming process, presence/absence of major diagnostic horizons
Sub order	Alb Fluv	Presence of albic horizon fluvial deposit	Soil moisture, parent materials, organic content
Great groups	Camb Torr	presence of cambric horizon usually dry	Degree of expression, base status, temperature, moisture status
Sub group	Typic Vertic	Represents central concept of great group has properties of vertisol	Intergradations to other great groups
Family	Fine loamy Carbonatic	15 – 35% clay, > 15% sand > 40% carbonate	Textural, mineralogy and temperature classes
Series	Pocono Chippewa	Deep, well drained, gently sloping deep, poorly drained, nearly level	Arrangement of horizons, color, texture, structure

tial assessment of the suitability of the soil to support various functions such as those listed in Table 7. Karlen and Fenton [31] discuss the use of soil maps for planning of soil sampling. It is noted that soil mapping is generally focused on undisturbed soils. Areas where urban development or surface mining has radically altered the soil are identified but no other information is provided.

Soil Sample Collection

Soil sample designs must be based on the knowledge of the distribution of the parameters of concern. In most cases the designs are stratified with depth. For the initial phases of the investigation, the most obvious stratification scheme is to use the major horizons. In disturbed soils the vertical strata are selected arbitrarily. A one meter square is often selected as the soil unit. In order to reduced variance it is common to subsample with the selected unit and composite subsamples. The procedure for compositing must consider potential alteration of soil properties. The vertical extent of sampling depends on the study objectives. For example soil fertility studies deal primarily with the A horizon while studies involving water movement would focus on the B horizon where permeability would probably be reduced.

The sampling equipment and collection procedures for surface soils have been developed by the soil science and geotechnical engineering fields. Special adaptations have been made for environmental considerations. The commonly used tools are listed in Table 13. The table is not meant to be comprehensive. It gives one example of a piece of equipment that is used for common sampling activities. There are two basic approaches to soil sampling. In the excavation method,

Table 13. Tools for collecting soil and participate material samples

Name	Supplier	Material	Comments
Hand tools			
Shovel	Numerous	Iron, stainless steel	Used for preliminary excavation
Scoop	Weaton	Polyethylene	Variety of sizes available
Trier	Weaton	Stainless steel	Extract plug from pile of material
Bucker auger	AMS	Steel, hardened high carbon steel tips	$1\frac{1}{2}$–7″ dia., models for regular, sand, mud
Core	Clements	Hardened stainless steel, PETG tubes	May be pushed with T-handle, driven with mallet or pushed with foot pedal
Sludege sampler	Weaton	Stainless steel	1000 ml capacity, 6′ long handle
Power tools			
Backhoe	Numerous	Hardened steel	Provide access for observation or hand sampling
Flight auger	AMS	Steel, hardened tungsten carbon steel tips	$1\frac{1}{2}$–2″ dia. solid, 3″ hollow,

a trench is dug with a shovel or a backhoe in order to provide across to the sample depth. Discrete samples are collected with hand tools after scraping away disturbed soil. In the probe method, an auger or coring tube is pushed, screwed or driven into the soil to the prescribed depth. The tool retains the sample as it is withdrawn. The probe method is faster and results in less disturbance of the sample as well as the site. The excavation method is required for a visual analysis of the meso scale properties of the soil such as delineation of horizons. Also, wedges of sand or clay that constitute only a small portion of the total soil volume yet may have a great environmental significance are often missed with probing.

The shovel is a basic tool for soil sampling. Manufacturers have made special adaptations to the standard shovel. They are available in shapes that facilitate digging a deep, narrow hole. Sampling shovels are made of stainless steel and can be coated with rubber. Scoops are available in a variety of shapes, sizes and materials. The size should be selected so that one scoop full will contain more than one sample volume. Typical materials are stainless steel and polyethylene.

When granular material is stockpiled it forms a cone shaped pile. Large particles tend to role to the bottom of the pile. Collection of a representative sample requires special consideration. The first step is to have a clear definition as to what constitutes the population and population unit as discussed in the first section of this chapter. If the unit to be sampled is large and the particle sizes are larger than 1″ then sub-sampling procedures should be implemented. If the material is relatively homogeneous and the particle sizes are small then a device such as the trier can be used. This device was developed for sampling of grains and requires a small amount of cohesiveness. It is pushed into the material, rotated 180° and, retracted. This results in the collection of a plug of material.

A large variety of augers and coring tools are used to collect soil samples. Augers are attached to a T-shaped handle to facilitate screwing the device into the soil. They are typically made of stainless steel with tungsten coated high carbon bits. Outside diameters range from $1\frac{1}{2}$ to 7 inches. Models with larger openings are used for mud and other cohesive materials. Extension rods with threaded ends can be added to allow the device to be screwed any depth although the required effort increases significantly beyond 6 feet. Flighted augers in diameters of $1\frac{1}{2}$ and 2 inches can be driven vertically or horizontally. The most common application is to prepared a hole for the collection of soil gas or moister or the placement of an instrument. Collection of soil samples with a hollow stem auger is more appropriate. Coring tubes can be pushed, screwed or, driven into the soil.

Sample disturbance is always a concern in soil sampling. While disturbance can be minimized, it can never be eliminated. Because removing a sample involves release of compressive forces and motion, there will be some alteration to the void structure. Properties that involve the void structure such as porosity and permeability and constituents that are partitioned in the void spaces such as moisture and volatile organcis are suspectable to alteration. Some laboratory procedures specify that a sample maintain its original orientation from the point of collection until the analysis is completed. The level of effort required to reduced disturbance depends on the parameters being measured and the laboratory procedure that is used. In the excavation method, disturbance is reduced by removing material that has been scraped, compacted or otherwise effected by the excavation. Small hand tools are then used to select the sample from the unaffected zone and quickly placing the sample in its container. In the probing method most of the disturbance occurs when the same is extruded from the core or auger. Most collection procedures specify that only material from the inner zone that has not touched the surfaces of the device be placed in the sample container. Samples can be removed from coring tubes with plungers. While this action maintains the core as a single unit, the removal compresses the core and will distort the micro structure. An alternative technique is to collect the sample in an internal sleeve. This sleeve is removed from the device and the ends are immediately capped. The assembly is left intact until it undergoes analysis. Subsamples are obtained by cutting perpendicular to the sleeve. This technique is commonly used when volatile organic compounds are being measured.

Contamination is always a concern in soil sampling. Samples can be contaminated by the collection tool or by carryover from a previous sample. Tools are suspectable to scraping by the sharp edges of soil particles. Many tools have replicable tips. Others are coated with plastic or rubberized material. However, these coatings tend to be even more suspectable to abrasion. The general cleaning procedure is to first remove all visible traces of soil with a dry brush. Next the tool should be washed in either in tap water or a dilute solution of a mild detergent. The final rinse should be with distilled or deionized water. Extensive field decontamination procedures with solutions of acid or organic solvents are generally not recommended. Tools showing extensive surface abrasion should be replaced or used only for preliminary excavations.

Authors of sampling plans need to be aware of the subsampling and processing that samples undergo prior to the measurement of the desired parameter. While not always explicitly stated most soil sample designs involve either the compositing or multi level sampling at some point. Laboratory procedures almost always require some subsampling prior to analysis. A typical mass of a soil sample is 500 gm while a typical aliquot for chemical analysis may be 5–50 mg. Some analytical procedures specify that this aliquot be selected randomly while others specify that organic debris or particles greater than a certain size not be analyzed. Samples analyzed for total concentration require extensive grinding in order to get the constituent into solution prior to measurement. In many cases the soil property associated with a certain size fraction is the parameter of interest. Common soil sample handling procedures are listed in Table 14. In addition to physical processing, a large number of chemical fractionation schemes have been utilized to extraction certain minerals. See Murdroch and MacNight [24] for details on extraction procedures for core samples and Klute [32] for specific methods of soil analysis.

Deep Soil and Groundwater Sampling

In this section deep soils are loosely defined as the unweathered geologic materials below the developed soil that was discussed in the section on surface soil sampling. In the environmental field deep soils are investigated for their role in the transformations and transport of constituents. Biogeochemical transformations typically take place at the surface of the soil particles. Transport can occur in saturated groundwater, in the unsaturated zone, as a nonaqueous liquid or in the gaseous phase. The sampling of deep soils is typically the most expensive part of environmental investigations. There are several reasons for this. First, the probing and drilling of this deep involves the collection of relatively small samples from a complex three dimensional matrix. The nature of the medium between locations can only be inferred. Second, once sample locations for groundwater are set they cannot be moved. Only additional wells can be installed. Thirdly, the most important parameter in groundwater investigations, the velocity is rarely measured directly. It is calculated from the aquifer properties and the measured hydraulic gradient.

Table 14. Soil and sediment sampling handling procedures

Process	Equipment or procedure	Applications
Splitting	Cone and quarter, riffle	Subsampling, aliquot selection
Mixing	V-shaped cone	Compositing, batch preparation
Drying	Air, oven, freeze	Moisture content, pretreatment
Grinding	Mortar and pestles, ball and pebbl mills	Total constituent analysis
Sieving	Dry, wet	Texture classification, fractionation
Extraction	Chemical solven	Mineralogy, constituent analysis

As is common in the environmental field, tools developed in other disciplines are utilized. In deep soils, procedures developed in geology, hydrogeology, oil exploration, and geotechnical engineering are all utilized to gain an understanding of the behavior of constituents in deep soils. However, care must be taken not to become too indoctrinated in specific procedures, as these procedures must be selected and adapted in environment investigations.

Study Objectives

The development of a clearly defined set of study objectives serves as the initial step towards conducting an efficient and effective deep soil investigation. The study question are first defined and then translated to information requirements and then data requirements. These requirements are then combined with other information, through the process described in the first section into the study objectives.

In the early phases of the investigations, study questions focus is on the existence and availability of groundwater. The direction and magnitude of flows must also be addressed. The amount of water in an aquifer, the rate at which it can pumped and the change in water table elevation during pumping are typical study questions. Groundwater quality investigations analyze the chemistry and the spacial and temporal components of its variability. A study question may concern the existence of contamination. As discussed in the surface water section, these studies may require extensive sampling and analysis of the background or ambient groundwater quality. Once identified, the spacial extent of contamination needs to be defined. The natural processes that affect contaminated groundwater: convection, diffusion and, dilution are then investigated.

Not all transport in deep soil is via saturated groundwater. Many constituents enter the soil at the surface and are transported downward in the vadose zone. Others move through the soil as gases. The distribution of constituents within three phase system (Fig. 9) is more complex that the two phase system of saturated flow. The study question may be: What is the phase distribution of contaminates and what are the factors that effect that distribution? The data requirements would then be the concentration of that constituent in each phase and a determination of the environmental factors that correlate to the distribution. Further study questions would involve the rate of movement of unsaturated water and gases through the vadose zone. The information requirements would be the infiltration capacity and soil moisture potential as well as the rate of gas flux.

Geologic materials are studied from the point of view of how they affect the movement of groundwater and contaminants. The stratigraphy or macro scale variation is examined in order to yield information on the existence water bearing zones. The chemical properties are measured in order to gain an understanding of the interaction between the groundwater, contamination and deep soils. As the investigation progresses predictive studies are performed to determine the behavior of the contamination under various scenarios.

Background Information

An effective and efficient surface deep soil and groundwater investigation can be designed and implemented only after a review of the general knowledge of the behavior of groundwater and the constituents contained within has been conducted. This information is then integrated with information on the regional geology and whatever data can be acquired from previous borings and existing wells at or near the study site. The physical and chemical properties of the soil matrix were discussed previously. The major difference is in deep soils the minerals tend to be less oxidized and contain very little organic material. Therefore less chemical interaction is expected between the solid and liquid phases.

The background review begins with the regional topography. In the United States topographic maps on scales ranging from 1 : 24 000 to 1 : 100 000 are available for a minimal charge. Topographic and hydrologic features as well as land use can easily be identified with these maps. Regional geologic maps define the extent of the uppermost formation. Cross sections and stratigraphic columns provide general information on each of the formations in the area such as age, thickness, rock characteristics, and water bearing properties. Drilling or boring of deep soils is often done for water supply, oil extraction or geotechnical sampling. Logs of these drillings are often filed with public agencies.

A survey of existing wells can be utilized to provide a preliminary understanding of groundwater flow and quality in an area. Although significant differences exist between supply and monitoring wells, the water supply wells should provide good data on piezometric elevations and qualitative information on aquifer yield and water quality. Typically, the survey is limited by incomplete information on well completion information and wells being located outside the study area.

Data Collection and Sampling Techniques

Geophysical techniques are measurements of the earth's ability to transmit, reflect or refract some portion of the electromagnetic spectrum. These techniques provide indirect measures of density, conductivity, clay content, moisture content and, other properties of the geology and aquifer. They utilize differences in the measurements can be taken from the air, on the surface or, downhole. Measurements can be made can be taken quickly and readings can be taken continuously and are therefore use for finding anomalies such as contamination or buried structures. The principal advantage of these procedures is that they provide a 2 or 3 dimensional picture of an attribute of the geology. The major disadvantage is that the measurements are indirect and must be calibrated with site specific field observations before they can provide useful scientific information. See Benson [33] for a more complete discussion on this subject.

There are two techniques for accessing deep soils and acquiring samples of soil and groundwater: boring and punching. Drilling is used here as a general term for the variety of coring, augering and hammering techniques that have been developed. In the environmental field, the method of choice is the continuous flight hallow stem auger. In this procedure, a drill advances ahead of 5 foot lengths of screw type augers with outside diameters ranging from 2.5 to 10 inches. As the assembly rotates cuttings are conveyed up the hole. Periodically, the drilling is stopped and undisturbed soil samples are obtained with a split-barrel or thin-walled tube samplers. These devices are driven into the hole ahead of the drilling head. The sampling device is withdrawn from the hole and the sample is extracted. Augering works well in depths less than 150 feet in granular materials. Circulation fluids (water, mud, air or foam) are used to reduce resistance to drilling and to help convey cuttings to the top of the whole. In deeper holes or in rock formations, rotary drilling is required.

Groundwater samples are obtained by means of a properly designed, constructed, and developed monitoring well. This subject has been extensively reviewed by others [34, 35] and will only be briefly summarized here. The basic components of well are defined in Fig. 11. The purpose of the structure is to allow for the collection of an undisturbed sample of water from a specific point in the aquifer. Monitoring wells are designed around the processes that will alter the sample as it is removed from the well. The most significant consideration

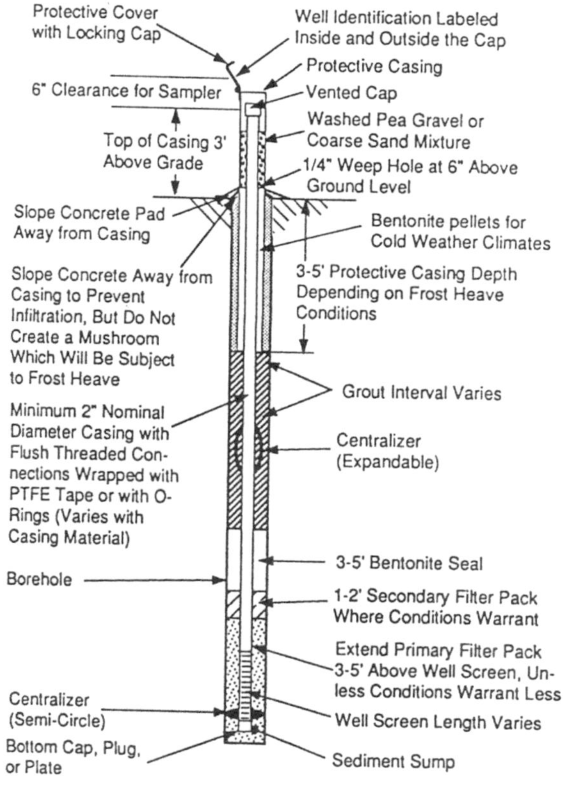

Fig. 11. Components of a typical groundwater monitoring well [35]

is alteration of groundwater during sample collection. The filter pack and the well screen are designed so that fines will not be carried from the aquifer into the well. The geometry of the well screen is a compromise between collection of sample from a discrete point in space as possible, providing a large enough sample volume, and not inducing large flow velocities during sample collection.

Groundwater samples are obtained with hand bailers, syringes or pumps [36]. Bailers probably are the most versatile tool and result in less sample alteration during purging and sample collection. They are available in diameters ranging from less than 1 inch to 3 inches. The disadvantages of bailers are the small sample volumes, labor intensive efforts required for they operation, and the potential for contamination from the line. Syringe devices are used for small sample volumes and where transfer of analyte in of the gas phase is a concern. A variety of pumps are available for sample collection including suction lift, gas driven, submersible, and bladder. Suction pumps are limited to approximately 26 feet or shallower. Other factors include flow rates, material of construction and power requirements. Due to the labor intensive decontamination requirements many long term monitoring programs utilize permanently dedicated pumps for each well.

It is difficult to discuss groundwater sampling without addressing alteration of the sample during the collection procedure [37]. Removal of water from a well induces abnormal velocities in the aquifer which may alter the movement of constituents. As groundwater is removed it is subjected to atmospheric pressures and gas composition quite different from that in the geosphere. Movement of O_2 and CO_2 in to or out of the sample will effect the redox potential, pH, and alkalinity of the sample. This in turn will affect the speciation and solubility of the constituents. In addition, volatile hydrocarbons are easily lost from the sample. Alteration is reduced by design of the monitoring well, pumping selection and purging prior to sample collection. Samplers have an almost unlimited number of options available with preparing the sampling protocol. Decisions regarding pump selection, pumping rates, purge volumes, and sample handling must be made in accordance with the study objectives and site specific conditions.

The measurement of the piezometric surface in monitoring wells or simple piezometers can be done with manual or automated methods [38]. Manual procedures for measuring static water levels include the wetted tape and the electronic probe. The bottom two or three feet of a steel tape is marked with carpenters chalk and it is placed below the water level in the well. The distance between the water mark and the top of the casing is noted. Modern battery operated electric probes have a tip that sense water. The presence of water is noted with a light and/or audible signal. The tip is attached to a calibrated electric cable. Both methods require that the top of casing elevation be accurately determined with a survey. Aquifer tests require multiple measurements over a short time steps. Use of pressurized air lines with gauges or pressure transducers is common. Since aquifer tests require only the change in water level in response to water withdrawal or addition, absolute calibration of these devices is not required.

Because many environmental investigations focus on a relatively shallow zone (less than 100 feet) and because of the large expense involved in traditional drilling and monitoring well installations, vendors have developed a variety of probing or punching equipment [39]. These tools use low power hydraulic jacks attached to small motorized vehicles to push small diameter probes into the soil as it is displaced. Tips are available for collection of soil, groundwater or soil gas samples at desired depths. Multiple samples can be obtained from each hole. After sample collection, the tool is withdrawn from the hole. Probing offers advantages. It is much faster than traditional drilling techniques. It can be used in applications where smart sampling techniques are appropriate such as following a plume of contaminated groundwater. The technique produces essentially no spoils that typically constitute a waste disposal problem.

Conclusion

The lead in soil example will be used to demonstrate the iterative nature of environmental investigations. Figure 12 shows three successive iterations. The initial study objective is to perform an extent of contamination study in surface soils. Based upon a conceptual model with a single source and a single transport mechanism, a sampling plan is developed and implemented. Core samples extending from the surface of the soil down to 10 cm are collected and analyzed for total lead. Data analysis reveals that there is a bimodal distribution to the lead contamination. This suggests that there is at least one additional source of lead

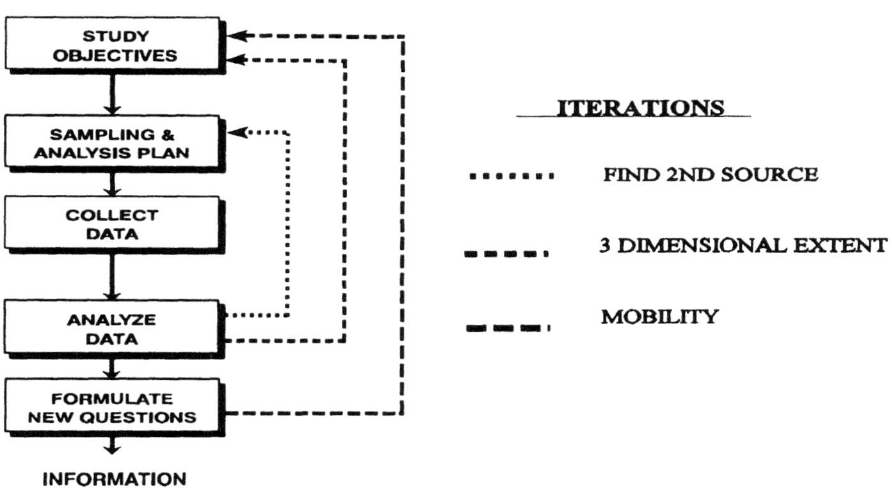

Fig. 12. Iterations in example investigation

contamination. It is hypothesized that a second tailings pile that was removed several years ago is that source. A second round of sampling is proposed with the same sampling procedures except that the spacial extent of the sampling is extended to include the area believed to be contaminated by the second tailings pile. This logic is indicated by the lines with the shortest dashes in Fig. 12.

Upon further review it is discovered that lead concentration varies with the length of each core sample. The study question is revised to "What is the extent of contamination of lead in surface soils at a site and how does it vary with depth?" The information required is the establishment of the three dimensional distribution of lead in soil, the data requirements are the measurements of lead in mg/kg dry weight at various locations and at various depths at the site and the study objectives to perform a three dimensional extent of contamination study. Samples are collected at the same locations and are extended to a depth of 25 cm. The variation of total lead along each core is analyzed. This third iteration is indicated by the medium length dashes on Fig. 12.

After the extent of contamination has been quantified, questions are raised regarding the potential for the lead to contaminate an underlying aquifer. The study questions is revised to "How mobile is the lead?" The information required is the chemical form of the lead and the data required becomes fractionation of total lead (soluble, exchangeable, mineral, etc). In addition, data regarding the movement of water through the soil column would also be required. This fourth phase of the study is indicated by the longest dashed in Fig. 12.

The simple collection of data does constitute a scientific investigation. In fact, automated equipment for sample collected and data logging may lead to the collection of an overwhelming amount of data of questionable quality and useability. The skilled environment scientist must utilize all available information and resources into an organized plan. This plan must be flexible enough to adjust for changing conditions and new information while at the same time provide a framework for overall logic.

References

1. Keith LH (1988) Principles of Environmental Sampling, ACS Professional Reference Book
2. Keith LH (1991) Environmental Sampling and Analysis. Lewis Publishers, Inc. Boca Raton, Fl
3. Barcelona MJ (1988) Overview of the sampling process, In: Keith LH (ed) Principles of environmental sampling ACS Professional Reference Book
4. Ward RC, Loftis JC, McBride GB (1990) Design of Water Quality Monitoring Systems, Van Nostrand Reinhold
5. Gilbert RO (1987) Statistical Methods for Environmental Pollution Monitoring, Van Nostrand Reinhold Company
6. Schreuder HT, Czaplewski RL (1993) Long-term strategy for the statistical design of a forest health monitoring system, Environmental Monitoring and Assessment 27:81–94
7. Sara MN (1992) Standard Handbook of Site Assessments for Solid and Hazardous Waste Facilities, Lewis Publishers, Inc
8. Barnard TE, Weinel DH (1991) Database managment in environmental investigations, National Research and Development Conference on the Control of Hazardous Materials
9. Dingman SL (1994) Physical Hydrology, Macmillan Publishing Company, New York

10. Newburn LH (1988) Modern sampling equipment: design and application, In: Keith LH, (ed) Principles of environmental sampling, ACS Professional Reference Book
11. Wetzel RG (1983) Limnology, 2nd edn, Saunders College Publishing
12. Goldman CR, Horne AJ (1983) Limnology, McGraw-Hill, Inc
13. Lehrman A ed. (1978) Lakes Chemistry, Geology, Physics, Springer-Verlag, New York
14. Stumm W (1992) Chemistry at the solid water interface Processes at the mineral-water and particle-water interface in natural systems, John Wiley and Sons
15. Horowitz AJ (1991) A Primer on Sediment Trace Element Chemistry, 2nd edn, Lewis, Boca Raton, FL
16. Herschy RW (1985) Streamflow Measurement, Elsevier, New York
17. Kilpatrick FA, Cobb ED (1985) Measurement of discharge using tracers, In: Techniques of water resources investigations book 3 Chapter A16, US Geological Survey
18. Shaw EM (1988) Hydrology in Practice, 2nd edition, Chapman and Hall
19. Grant DM (1989) Isco Open Channel Flow Measurement Handbook, Isco, Inc. Lincoln, NE
20. Ward RC, Loftis JC (1989) Monitoring systems for water quality, CRC Critical Reviews in Environmental Control 19(2): 101–118
21. Ward RC, Loftis JC (1986) Establishing statistical design criteria for water quality monitoring systems: review and synthesis, Water Resources Bulletin 22(5): 759–769
22. Interagency Advisory Committee on Water Data (1982) National Handbook or Recommended Methods for Water-Data Aquisition, Prepared cooperatively by agencies of the United States Government
23. Wetzel G, Likens GE (1991) Limnological analyses, 2nd edn, Springer Berlin Heidelberg, New York
24. Mudroch A, Macknight SD (1991) CRC handbook of techniques for aquatic sediments sampling, CRC Press
25. Sara MN (1991) Standard handbook for solid and hazardous waste facility assessments, Lewis, Bocca Raton, FL
26. Buol SW, Hole FD, McCraken RJ (1980) Soil genisis and classification, 2nd edn, Iowa state University Press
27. Borchardt GA (1977) Montmarillonite and other smectite minerals, In: Dixon JB and Weed SB (ed) Minerals in the Soil Environment, Soil Science Society of American, Madison, WI
28. Hsu PH (1977) Aluminum hydroxides and oxyhydroxides, In: Dixon JB and Weed SB (ed) Minerals in the Soil Environment, Soil Science Society of American, Madison, WI
29. Schwertmann U, Taylor RM (1977) Iron oxides, In: Dixon JB and Weed SB (ed) Minerals in the Soil Environment, Soil Science Society of American, Madison, WI
30. Soil Survey Staff (1975) Soil Taxonomy, US Dept. of Agr. Handbook 436
31. Karlen DL, Fenton TE (1991) Soil map units, In: Nash RG and Leslie AR (ed) Groundwater residue sampling ACS Symposium Series 465
32. Klute A (ed) (1986) Methods of Soil Analysis, American Society of Agronomy, Madison, WI
33. Benson RC (1991) Remote sensing and geophysical methods forevaluation of subsurface conditions, In: Practical handbook of ground-water monitoring Nielson DM (ed) Lewis Chelsea, MI
34. Barcelona MJ, Gibb JP, Helfrinch JA, Garske EE (1985) Practical guide for ground-water sampling, EPA/600/2–85/104
35. Nielsen DM ed, (1991) Practical Handbook of Ground-water Monitoring, Lewis Publishers, Chelsea, Mi
36. Herzog B, Pennino J, Nielsen G (1991) Ground-water sampling, In: Nielsen DM (ed), Practical handbook of ground-water monitoring
37. Barcelona MJ (1990) Uncertainities in ground water chemistry and sampling procdures, In: Melchoir DC, Basset RL (eds) Chemical modeling of aqueous system II, ACS Symposium Series 416
38. Dalton MG, Huntsman BE, Bradbury K (1991) Aquisition and Interpretation of Water-Level Data, In: Nielsen (ed) Practical handbook of ground-water monitoring
39. Chalfant C (1992) HydroPunch and soil gas probes simplify ground water assessments, National Environmental Journal 2 : 24

Topics of Chemometrics Today

Richard G. Brereton,

School of Chemistry, University of Bristol, Cantock's Close, BRISTOL BS8 ITS, U.K.

An historical overview of chemometrics is presented, and the role of chemometric techniques in analytical chemistry, statistics, organic, physical and environmental chemistry and computer science evaluated. The interaction between statistics and other disciplines, including environmental chemistry, is described. The main textbooks and meetings are listed. Considerable emphasis is placed on software. The type of software used depends on the expertise and experience of the user ranging from statistical macrolanguages to microprocessor based packaged software, instrumental software and home grown programs using a variety of languages and, also, object oriented techniques. An extensive bibliography is presented.

Introduction

The name chemometrics was first proposed by the Swedish physical organic chemist, S. Wold, when submitting a grant application in the early 1970s. A few years later, he joined with the American analytical chemist B.R. Kowalski to establish the (International) Chemometrics Society (ICS). Despite this formal origin of the name, there have been many diverse strands of development relating to modern chemometrics.

Analytical Applications

Analytical chemistry applications have been most publicised. There are several reasons for this, the main being that much of the revolution in use of chemometrics has been linked to the revolution in computerised laboratory based instrumentation. Measurements are required not only in analytical chemistry, but also in geology, medicine, environmental science and so on. Many of the early users of chemometrics techniques published in journal such as *Analytical Chemistry* and *Anal. Chim. Acta (Computer Technology and Optimisation Section)*. Regular reviews, such as in *Analytical Chemistry* [1–7] and the *Analyst* [8], did much to classify the literature in the subject. Many less formal articles were written in journals such as *Trends in Analytical Chemistry* and *Analytical Proceedings*. The membership of the ICS was overwhelmingly composed of analytical chemists. Although there was a strong organic strand from the Umeå group, quite a number of their early papers related to analytical methods such as chromatography, NMR and pyrolysis.

In some regions, the term chemometrics was slow to take off. Major ideas were developed in the Benelux countries, largely under the guise of analytical chemistry. The early books on *Quality Control in Analytical Chemistry* [9] and *Evaluation and Optimisation of Laboratory Methods and Analytical Procedures* [10] were written by analytical chemists, who did not regard themselves primarily as chemometricians, during the 1970s or early 1980s. As the name chemometrics spread, the Benelux groups gradually reclassified themselves chemometricians, but brought a wider variety of methods to the subject: quality control, signal processing and information theory were particularly strong interests of the Nijmegen group, for example. This "marriage of convenience" between a US/Scandinavian tradition of multivariate pattern recognition and a Benelux tradition of mathematical analytical chemistry, formed a strong focus for the introduction of chemometrics techniques into analytical chemistry.

A final important strand of analytical chemistry is the tradition of using statistical methods to assess the accuracy, precision and quality of data [11]. Many analytical chemists felt chemometrics was simply the next step, and a logical progression from traditional analytical chemistry. Many early books were written by such major figures such as Youden [12] and Davies and Goldsmith [13], demonstrating, in the precomputer age, the importance of using statistical methods in the chemical industry. However, such people rarely regarded themselves as chemometricians.

One of the difficulties with analytical chemistry is that there is always a conflict of interests between analytical chemists and workers in more applied areas. For example, should a user of chromatography equipment in an environmental laboratory be regarded as an environmental scientist or as an analytical chemist?

Statistics

There are many well established areas of applied statistics. These include econometrics, biometrics, medical statistics and so on. Although the word chemometrics was not invented by statisticians, many statisticians regard chemometrics as a natural area.

Chemistry is the central science, spanning soft science such as biology and medicine, and hard science such as physics. Chemists require a variety of different statistical approaches. Quantum chemistry and statistical mechanics are essential tools of the physical chemist and have been developed over very many years. Although originating within mathematics and physics, chemists have now moved the subjects forward independently, so that these areas have their own notation, journals and conferences. A major feature of this type of statistics is that the sample size is extremely high and so predicted physical models can be measured with an exceptionally high degree of accuracy. On the soft side of science, environmental studies, for example, result in comparatively low levels of reproducibility, and there may be major measurement problems. Consider the example of measuring the organic compounds and productivity at sampling sites in an ocean. Each sample is extremely expensive, and the reproducibility of the data may be poor. Completely different statistical problems arise here.

Traditionally, geologists, biologists, economics, sociologists and clinicians have turned to statisticians to solve their problems. Chemists tend to be more numerate and are able to develop their own statistics. However, the problems of reproducibility and the multivariate nature of data mean that traditional statistical methods cannot always be applied to chemical problems. In chemometrics, many measurements can be made per sample. Even in the example of sampling sites in an ocean, once the samples are available, it is possible to measure the relative amounts of different organic fractions, phosphates, nitrates, algal productivity etc. (Fig. 1). Many of the chemical measurements may also be replicated. This contrasts to other areas such as econometrics, where samples cannot easily be replicated, and where there are a limited number of possible measurements.

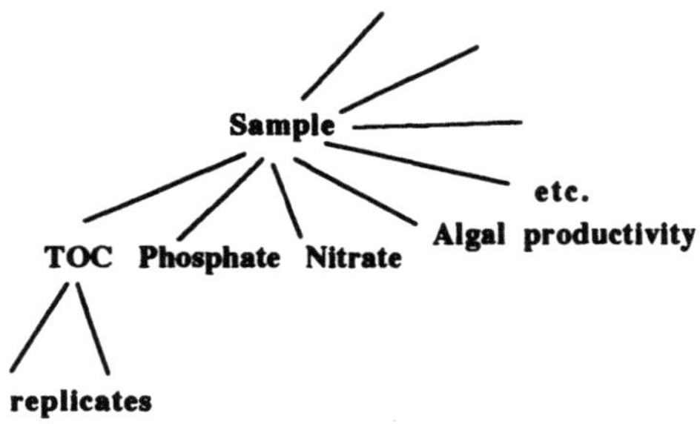

Fig. 1. Typical scheme for sampling in chemometrics as applied to looking at the chemistry of oceans: a large amount of information can be obtained from a single sample, and the information can be replicated

This ability to replicate and take several measurements distinguishes chemometrics from most other branches of applied statistics. Hence, unique statistical approaches are required.

There are many uniquely interesting features of chemometric statistics. The PLS (partial least squares) method [14–16] is very commonly used, probably the major success story of chemometrics. Factor analysis has a very unique status in chemometrics [17, 18]. In conventional statistics, factors are mainly used to simplify data: for example psychometricians determine factors from intelligent tests that signify different types of ability. It is possible to believe the results or disbelieve them. Factor analysis is often used for graphical simplification [19]. In chemometrics, factors usually correspond to real compounds, such as the spectra and elution profiles of different coeluents in DAD-HPLC (High Performance Liquid Chromatography) [20–22], and have a precise physical meaning. Conventional statisticians do not distinguish strongly between PCA and factor analysis, yet within chemometrics the two approaches have quite different implications. Another feature of chemometrics is the ability to perform PCA (principal components analysis) without mean-centring data. In conventional statistics, variation about a mean is the most significant form of variation: for example, measurements may be taken on a number of banknotes and used to determine whether a banknote is forged or not. A model of the "average" banknote is determined. Elaborate tests, such as the F-test [23], have been developed to find a variation about a mean. In chemometrics, often the most significant variation is above a baseline (e.g. in chromatography or spectroscopy) and not around a mean, so that different procedures both for analysing the data and for assessing significance are necessary [24].

One of the most used statistical package is SAS (Statistical Analysis System). Despite the very large user base, chemometricians will be disappointed by many omissions from the package. For example, there are often major limitations in the number of variables, and when many variables are used algorithms and calculations tend to be slow. Common chemometric methods such as PLS or SIMCA (soft independent modelling of class analogy) [25] are missing. Even the ability to perform principal components analysis (PCA) on uncentred data is missing. Factor rotations are restricted to abstract rotations, which are of little use to chemometricians.

Therefore, there is a very wide gap between most other branches of applied statistics and chemometrics. Statisticians tend to have an interest in a small number of chemometrics methods such as PLS which have wider applications outside chemistry, but most of the development of specific approaches (e.g. in factor analysis) occurs within the chemical community. Over the past few years, chemists have learned the basis of multivariate statistics, and are adapting methods to their own problems.

Organic Chemistry

The original organic chemistry interest in chemometrics was quite strong. However, many of the earlier work concentrated on analytical organic chemistry, involving instrumental methods such as NMR (Nuclear Magnetic Resonance) [26], pyrolysis [27] and chromatography [28].

Independently, other investigators, not calling themselves chemometricians, applied computational approaches to large chemical databases. Problems of structure representation, crystallographic databases, and spectroscopy required multivariate methods for pattern recognition. Some examples are as follows: use of pattern recognition in NMR (nuclear magnetic resonance spectroscopy) [29], crystallography [30, 31], and data storage and retrieval [32]. There is much literature on computational chemistry, much of it touching on approaches that relate strongly to chemometrics. However, this group does not call itself chemometricians.

Another major area of organic chemistry is the field of QSAR (Quantitative Structure Activity Relationships) where structural parameters are related to activities [33–36]. These activities may be dipole moments, potential toxicity of compounds or spectroscopic properties. A great deal of classical physical organic chemistry concerns QSAR studies. The Hammett relationship where structural parameters are related to reaction rates of substituted benzenes is well known [37]. Chemometricians extend this to multivariate situations where properties/activities are quite complex, such as the biological effect on an organism. Again, many applications of QSAR are not reported in the so-called chemometrics literature. QSAR has been especially well developed within the pharmaceutical industry, where the potential activity of drugs can be predicted from structural properties, and is complementary to molecular modelling. In environmental science, QSAR has a major role to play. Measured properties such as LD50 and mutagenicity can be related to the structural properties. Most chemometric approaches such as PCA, classification and PLS can be successfully employed in environmental QSAR studies. Within certain regions, such as Scandinavia, these results are reported in the chemometrics literature, and other regions such as the US tend to report the results in the environmental literature.

Physical Chemistry

Physical chemists have been using techniques related to chemometrics for many years. Classical examples are factor analysis [17, 38] and the use of maximum entropy to enhance the quality of NMR [39–41], MS (mass spectrometric) [42] and chromatographic [43] data. Problems of noise and peak resolution are as interesting to physical chemists as they are to analytical chemists. A few physical chemists, such as Malinowski, have published in the chemometrics literature, but most have published in other areas, such as specifically spectroscopic journals. One of the classical books, on deconvolution [44], is rarely cited by chemometricians, yet contains a vast amount of information and ideas relevant to chemometrics. One reason for this division is that analytical chemistry tends to be weakly organised in relation to physical chemistry in regions such as the UK, so physical chemists tend to dominate the instrumental literature in certain geographical regions.

A difficulty of this division is that many chemometricians do not know about techniques that could be very relevant to their work. Most classical chemometricians are schooled in multivariate analysis, and are reluctant to advocate approaches they do not understand or have not developed. Hence, there is a major division between analytical and physical chemists. Despite this, books on, for example, curve fitting [45] are very widespread within physical chemistry, and the discipline of data simplification in spectroscopy and deconvolution should be regarded as part of chemometrics.

Computer Science

Yet another strand comes from computer science. Many methods, such as cluster analysis, optimisation, PCA, and regression analysis, have been developed by mathematically oriented computer scientists. Some of the best books describing the theoretical basis of chemometric methods have been written by computer scientists. The "Numerical Recipes" books are widely used throughout science [46–48].

With the modern development of microprocessor based instrumentation, and the ability to acquire, and the need to analyse, large quantities of data in a short time span, an understanding of computing is essential to the modern day chemometrician. Environmental chemists, in particular, have the ability to monitor, continuously, a large number of processes, such as the quality of water, and need rapid, on-line, methods for determining if the quality is outside defined limits. On-line monitors should be inexpensive, and be able to make decisions quickly. This can be done by programming microprocessors effectively, often requiring some knowledge of memory management, and rapid algorithms.

At the other extreme, large databases may be set up, especially in the environmental area, by major bodies such as the EPA (Environmental Protection Agency). Good knowledge of database structures [49] is required for searching for information, and obtaining good environmental predictions. Often some methods familiar to chemometricians, such as cluster analysis, are part of these databases.

Environmental Chemistry

Environmental chemistry is a major application area for chemometrics. Chemometrics should be distinguished from environmetrics. The latter normally involves univariate statistics of samples collected from the field. In most cases, chemometrics implies some instrumental analysis has been performed on the chemical data, either from field studies or in the laboratory.

There is no well defined division between chemometrics and conventional statistics, but the following example illustrates a possible distinction. Consider the example of growing plants in differing levels of heavy metals such as Pb, Cd, Hg etc. [50, 51]. After a few days, the lengths of roots and shoots are measured, and a quantitative model is established between the plant growth and the heavy metal concentration. Although quite sophisticated statistical analysis is required, and well designed experiments are essential for meaningful interpretation of data, the methods required derive from conventional statistics. There are only two measurements per plant, and the replication problem is severe. Chemometrics might be used if a large number of chemical measurements were taken on the plants, e.g. concentrations of various chemicals: in such cases, there will be more possibilities of replication (the plant tissue extracts can be analyzed several times by HPLC, for example), there will be a large number of measurements per plant (hence multivariate methods can be obtained) and there may be problems of instrumental deconvolution (e.g. if spectroscopy or chromatography is employed in the measurement process).

The potential applications to environmental chemistry are vast, and overlap with applications discussed above. More details are found elsewhere in this publication, so only a few sample areas are listed. The interaction with geologists and geochemists is particularly important. The area of statistics in geology is well established, and there are many potential applications of sophisticated chemometric methods to geochemical data [52–54]. These often involve simplifying data, and looking at long term or geographical trends. Factor analysis plays a major role, for example, in looking at sources of potential pollutants [55–57]. Simple exploratory data analysis can relate chemistry to geography, and also to directions of pollutants [58]. Classification methods can be used to relate analytical data to groups of samples, e.g. polluted and unpolluted mussels [59]. Calibration has a very major role to play. Many biological parameters, such as mortalities, mutagenicity and toxicity, can be related to chemical composition [60–62]. QSAR is an important area: large databases of potentially toxic compounds can be built up. Composition activity relationships are a specific application [63, 64]. An understudied area is the potential of chemometrics to relate several blocks of measurements using calibration and Procrustean methods, and so determine the cheapest approach for measurement [65–67]. For example, if the parameter of interest is the toxic effect on mammals, simple bacterial tests might be substituted: these are cheaper and involve less legal difficulties. Clinical chemists have used these approaches for many years – e.g., bacterial tests on potentially active drugs could be substituted for tests on rats. In environmental chemistry, another major problem involves field tests. A rigorous method of monitoring the composition and potential toxicity of fumes in factories might be by using GC-MS (gas chromatography mass spectrometry) but not many factories could afford to do this routinely. Can a simpler method such as IR (infrared) or UV/VIS (ultraviolet visible) spectroscopy be substituted? Calibrating one method to another helps. Obviously, deconvolution and enhancement of sophisticated instrumental signals such as occur in GCMS, DAD-HPLC and AS (atomic spectroscopy) has major potential in environmental chemistry, where detailed information on samples is required. Chlorophyll degradation, where complex mixtures are produced when algae die in aquatic environments, is an example where quantification and detection is hard [22].

Environmental chemistry poses considerably harder problems to the chemometrician than straight analytical chemistry. Normally there are several parameters of interest. One or more non chemical parameters are usually measured. Examples are as follows. (1) Physical dimensions such as lengths of organisms, e.g. root lengths of plants or shell deformities of mussels. (2) Survival data, e.g. the time taken for 50% of a population to die. (3) Bioassays such as the change in cell counts. (4) Physiological data such as measured cardio-vasculatory variables. (5) Depth in sediments or water columns. (6) Tests on dead meat such as force and deformation of muscle. (7) Biological productivity levels such as planktonic productivity. (8) Microbiological parameters such as mutagenicity. (9) Distance along rivers or from sites of potential pollution. (10) Human mortality rates over a long period. (11) Geochemical maturity of oils.

Generally, the interest is to relate the chemistry to these response parameters. Sometimes the response parameters are only of indirect scientific interest. For example, the study effect of a potential pollutant on human beings may be the main objective of an investigation. However, it is not practicable to perform tests on humans, and more acceptable tests, such as mutagenicity of bacterial populations, may be chosen instead. In some cases it is possible to calibrate the directly measured variable to another variable of interest, e.g. in geochemistry, where depth is related, but in a non-linear fashion, to geological time: elaborate statistics (e.g. use of forminifera counts) may be required to estimate the depth/age calibration curve, but when sampling a core this curve is initially unknown.

The environmental chemist generally uses one of two types of chemical data. The first involves employing an instrumental technique to determine the relative composition of elements or organic compounds in a sample. Common methods include atomic spectroscopy, chromatography, and mass spectrometry. Quantitative data is sometimes obtained, e.g. isotope ratio mass spectrometry. Alternatively, chemical structure data is sometimes used as in QSAR studies. Sometimes an extra, signal processing step is required to handle the instrumental data before it can be interpreted by statistical methods, e.g. signal deconvolution.

In most cases of chemometrics in environmental chemistry, some relationship is obtained between the chemical and non-chemical datasets. Methods such as calibration, factor analysis, classification and response surface modelling have been developed to establish this relationship.

Environmental chemistry is an interdisciplinary subject, there being very little "pure environmental chemistry" and interfaces with many other areas. Figure 2 illustrates some of the main interfaces between different disciplines. Conventionally, only areas 7 and 9 would unambiguously be regarded as chemometrics. Historically, however, many workers in area 11 would class themselves as chemometricians, whereas only selective workers in area 14 publish in the chemometrics

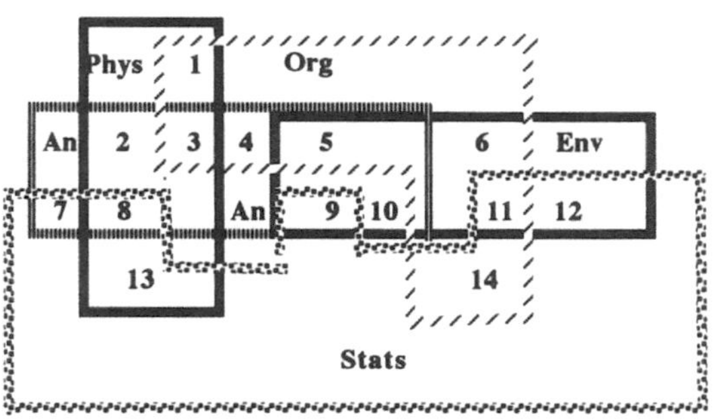

Fig. 2. Interface between major disciplines
Env = Pure environmental chemistry
An = Pure analytical chemistry
Phys = Pure physical chemistry
Org = Pure organic chemistry
Stats = Pure Statistics
1 Physical/organic chemistry: example – kinetics of organic reactions
2 Analytical/physical chemistry: example – electrochemistry
3 Organic/physical/analytical chemistry: example – developing new mass spectrometry
4 Organic/analytical chemistry: example – new assignment methods in NMR
5 Environmental/analytical/organic chemistry: example – chromatographic detection of toxins
6 Environmental/organic chemistry: example – synthesis of potential toxins
7 Analytical chemistry and statistics: examples – linear calibration, "conventional chemometrics"
8 Analytical/physical chemistry and statistics: example – maximum entropy spectral deconvolution
9 Environmental/analytical chemistry and statistics: for example – use of factor analysis to resolve HPLC peaks of mixtures of potential pollutants.
10 Environmental/analytical chemistry: example – atomic spectroscopy of heavy metals
11 Environmental/organic chemistry and statistics: example – QSAR of possible pollutants
12 Environmental chemistry and statistics: example – measuring root lengths of plants and relating to chemistry
13 Physical chemistry and statistics: example – statistical mechanics
14 Organic chemistry and statistics: example – QSAR of drugs

literature, despite a great number of methods in common with area 11. Workers in areas 8 and 13 rarely regard themselves as chemometricians. There is extensive literature in area 12, much of it relevant to chemometrics, but using conventional statistical approaches, most of which were available long before the word chemometrics was invented.

Most chemometrics in environmental chemistry is at the interface between analytical and environmental chemistry and statistics. It is hard to understand chemometrics without some appreciation of its relationship other disciplines.

Information about Chemometrics

Literature

Although there is much literature relevant to chemometrics, most of the organised literature is within the area of analytical chemistry. The regular reviews in the journal *Anal. Chem.* [1–7] concentrate primarily on analytical applications. There are two main chemometrics journals, *J. Chemometrics* and *Chemometrics Intell. Lab. Systems.* The former is more theoretical, whereas the latter is more applied and concentrates on conference proceedings and analytical applications. The tutorial articles from *Chemometrics. Intell. Lab. Systems* have been reprinted in two volumes [68, 69].

There are several texts on chemometrics. The book by Sharaf et al. [70] is somewhat theoretical, but was the first comprehensive text with chemometrics in its title. The book by Massart et al. [71] is an excellent general text, but oriented strongly towards the analytical chemistry laboratory manager. The book by Brereton [72] is a more low level and introductory text. However, all three are strongly oriented towards analytical chemistry applications. A series of monographs by Research Studies Press in collaboration with Wiley on Chemometrics presents a broader range of topics [73–77]. Some of the many other books are referenced [78–82].

Specialised areas of chemometrics are covered in great detail in certain books. Notably these are pattern recognition [83–87], calibration [88], experimental design [89–91] and factor analysis [17].

Numerous journals publish articles relating to chemometrics.

Meetings and Courses

The original group of chemometricians was quite small, consisting mainly of specialist computer programmers and statisticians applying their work to chemical data, hence meetings in the 1970s were informal in nature. As awareness of chemometrics increased during the 1980s, the number of meetings expanded substantially.

A notable early meeting that brought many international experts on chemometrics together was held at the NATO Advanced Study Institute in Cosenza in 1983 [79]. The US NBS (National Bureau of Standards) organised a meeting where chemists and statisticians met together [92].

Within the area of analytical chemistry, there are several ongoing series of meetings, namely CAC (Chemometrics in Analytical Chemistry), COMPANA (organised by the University of Jena) and COBAC (Computer Based Methods in Analytical Chemistry). Most of these meetings have resulted in proceedings being published, mainly in the journals *Chemometrics Int. Lab. Systems* and *Anal. Chim. Acta.* The COBAC meetings have now largely been subsumed within major conferences.

There are a vast number of regional meetings, but one of the most international is the Scandinavian Symposium in Chemometrics (SSC) series, whose substantial proceedings have been published in the journal *Chemometrics Int. Lab. Systems* [93, 94]. Informal meetings have also been arranged by the Umeå group over a number of years.

A predominantly statistical series of meetings entitled "Statistics in Chemistry and Chemical Engineering" has been held as part of the Gordon conferences in the US every year since 1951. More applied meetings, emphasising the application to spectroscopy, are the Snowbird meetings on "Computer Enhanced Spectroscopy" held in the US.

There have been a number of conferences specifically oriented towards environmental chemists. These include meetings organised by the EPA (Environmental Protection Agency) on "Progress in Chemometrics" and a meeting organised by the ISI (International Statistical Institute) on "Chemometrics and Environmetrics" in Bologna in 1993.

It is not possible to enumerate all the meetings organised in the area of chemometrics, but one of the misfortunes of the rapid and highly diverse development of the subject is that there is no general meeting where experts from many different disciplines meet together.

Major courses were slow to evolve, but one of the first was organised by Scientific Symposia in the UK. Since 1987, interdisciplinary workshops in chemometrics have been organised by the University of Bristol. A large number of courses have been organised in Europe under the umbrella of the COMETT scheme between 1990 and 1994. With the advent of good texts, numerous courses have evolved over the last few years, but most are heavily oriented towards certain well defined application areas. Various manufacturers of chemometrics software (see below) have also developed courseware, most notable being InfoMetrix in Seattle, US, CAMO A/S in Trondheim, Norway and Umetri in Sweden.

There is no overall international group representing all strands of chemometrics. Active statistical organisations include the ASA (American Statistical

Association) and ISI (International Statistical Institute). Analytical chemists have formed several national groups. Pharmaceutical and organic chemists tend to regard chemometrics as part of computational chemistry, although there are several organisations involved in chemical structure/QSAR/molecular modelling studies. The EPA in the US has an active interest in chemometrics. Surprisingly, NASA is also an active supporter of chemometrics activities.

Software

Introduction

It is impossible to discuss modern chemometrics without an appreciation of new software developments. The general acceptance of chemometrics depends on the wide availability of modern, user-friendly software and cheap, powerful microprocessors.

There is a wide variety of software packages and methods, and the best approach depends greatly on the background of the user. In computing, a wide variety of ranges of sophistication exist side-by-side. In a typical department, there may be one or two systems managers who understand technical manuals and documentation, set up new systems, interface with manufacturers and train other users. A few programmers may exist. It is not necessary to understand how computers work, or even set up new systems, to be effective programmers. Many programmers rely on other people, with a greater technical knowledge of computers, to maintain systems. At a less computationally sophisticated level are users of packaged software, either developed in-house or else commercially purchased. Many such people will be laboratory based scientists. Finally, there will be secretarial and related staff whose main use of computers is wordprocessing and graphics. Different packages exist for each group. All uses of computers are legitimate and equally important. Hence, in chemometrics, likewise, there has emerged a wide variety of software.

Early Chemometrics Software

In the 1970s, most computer users developed programs for mainframes. Punched card and paper tape input, off-line printers and job submissions on shared mainframes were typical for this era.

Many of the early chemometricians were also strong programmers. Very few scientists used computers unless they, too, were good at programming. Hence rather cumbersome, off-line, software was developed during these early days. The main emphasis was on producing libraries of subroutines, often in FORTRAN. A very large number of chemometric approaches were coded into these early packages. Notable packages are as follows. ARTHUR is a fairly comprehensive package of FORTRAN subroutines, mainly for pattern recognition, developed by the Seattle group [95]. Despite their vintage, these routines span a wide variety of methods, and the code is still useful nowadays. Developments by InfoMetrix still make use of these early routines. The SIMCA package [96] was developed by the Umeå group, mainly for PCA and classification purposes, although PLS was added later. This package should be distinguished from the method of soft independent modelling of class analogy, although in the early days the two were almost indistinguishable, the package promoting the method and vice versa. Malinowski developed software for factor analysis called TARGET [97], again, largely in the form of subroutines. User-friendliness was not a main issue. A very wide number of "indicator functions" for determining the number of significant components in mixtures by factor analysis were included. Some chemometrics routines were deposited with the QCPE (Quantum Chemistry Program Exchange) in the US [98]. Hopke's group developed a package FANTASIA [99].

Most of these early packages have been incorporated into more modern microprocessor based software as discussed below.

Statistical Macrolanguages

Parallel to this development of software specifically for chemometrics, were large, international, developments of packages for statisticians. Originally intended for mainframes, most packages now run on microcomputers.

Probably the largest statistical package is SAS (Statistical Analysis System) [100–102]. Originally mainly for statistics, it has been substantially expanded to include very extensive graphics, matrix routines, econometrics and time series analysis, database, interactive access, quality control and so on. SAS is an industry standard. It comes with its own language, making it easy to perform tasks such as PCA in a few lines of code, and very powerful macros for handling complex datasets. Many statisticians find this language easy to use, but, on the whole, chemists dislike it, as it requires a certain amount of programming and thinking in a statistical manner. Despite this, around 700 people are employed by the company, there are innumerable conferences, manuals, newsletters and books, and SAS is implemented on most common computer systems. Having started as a statistical language primarily for off-line use on IBM mainframes, it now is available under Windows on PCs. A weakness, alluded to above, is that SAS does not contain some common chemometric methods such as PLS, many types of factor rotations and SIMCA. Using the SAS IML (Interactive Matrix Language), it is, though, possible to code these approaches in. Once into the SAS system, the graphics and data handling are excellent.

BMDP (Biomedical Data Processing) [103] and SPSS (Statistical Packages for Social Sciences) [104] are well established packages that contain a large number of routines for pattern recognition. Although the graphical and general facilities are much less than with SAS, there are, nevertheless, well supported packages, and are available on PCs. Some statisticians, especially in the food industry, prefer the GENSTAT package [105]. This is, indeed, very elegant, but is maintained only by a small number of people, and cannot possibly match the facilities of large commercial concerns. Despite this, regular conferences are held, and books written, on GENSTAT, and some mathematically minded chemometricians find it a useful development tool. CLUSTAN [106] is a package primarily for cluster analysis, used mainly by biometricians.

A recent development is the S-plus package for statisticians [107, 108]. This runs on both Unix and DOS systems, and is strongly oriented towards interactive graphics. It contains its own macrolanguage, which allows users to develop new methods and then create new commands which involve these new approaches. For the statistically oriented user, this is a very powerful approach. A statistician would regard algorithms such as PCA or PLS as easy to program and understand. For the novice environmental chemist, however, S-plus is not suitable.

There are several other statistical packages, many now running on micros, such as SYSTAT [109, 110] and Minitab [111]. These are limited in capabilities, but very easy to use, and good, general, introductory packages that can be purchased cheaply and used for simple problems without much difficulty.

The major advantage of the large statistical packages is that there is a very big user base. This means substantial support in terms of newsletters, conferences, books, courses and help. It also means that most packages are regularly updated and take into account new developments in hardware. A final, and extremely important, factor is that most of these packages are "industry standard". This particularly applies to SAS – the routines are very thoroughly validated and can, therefore, be used as benchmarks. The difficulties are that none of these packages are oriented towards chemometrics, and that some statistical expertise is required to use the packages and understand the manuals. Although most professional statisticians will find the packages easy to use, this does not mean that environmental scientists will necessarily like them. Software is normally designed with the user's needs uppermost in mind, and an approach that is favoured by a statistician may not be favoured by a laboratory based scientist and vice versa. It is, however, useful that active chemometrics research groups have at least one of these packages available for benchmarking and for testing out new methodology.

Early Microprocessor Based Packages

In the early to mid 1980s, there was a flood of new microprocessor based chemometrics software. The majority worked on PCs under DOS. This phase also resulted in the development of efficient new algorithms. Early microprocessors were often limited in memory, and also fairly slow. PCA algorithms such as SVD (Singular Value Decomposition) work on an entire data matrix. If 20 variables are measured, this involves setting up 20×20 matrices, and performing operations such as inversion to calculate 20 principal components. For the users of off-line mainframes, it did not matter if the size of such problems became quite large. A job was submitted in the afternoon and the next morning a printout arrived in a pigeon hole. If the job took a very long time, applications would be made for extra funds for computer time, or, in the long term, even a new computer. When the first microprocessor applications were developed, there emerged serious difficulties resulting in the need to develop better code. Calculations could be seen to happen interactively and it was not acceptable to have to wait several hours before an answer became available. The NIPALS algorithm and associated cross-validation [112–115] was developed with efficient computing needs in mind. Instead of calculating all the principal components, NIPALS extracts one component at a time, and cross-validation being used to determine whether sufficient components had been calculated. If 20 environmental variables are measured, there may only be 3 or 4 significant factors, so no need to calculate all 20 components.

Many microprocessor based package of that vintage are still available, and are often quite cheap. The microprocessor based SIMCA package continued its development over the 1980s, continually evolving. Ein*Sight [116] is a microprocessor based evolution of ARTHUR.

SPECTRAMAP [117] was developed by Lewi and colleagues, primarily for pattern recognition, mainly PCA with associated scaling. The scope of this package is limited, but within its limitations it provides excellent graphics and is widely used by pharmaceutical statisticians. This package is written in APL [118]. SIRIUS [119] was developed by the University of Bergen. Tutorial versions of both these packages are available for small datasets, with the text *Multivariate Pattern Recognition in Chemometrics* [87]. Both packages are DOS based, and are compiled, so that source code is not available.

PARVUS [120] is a large collection of routines for pattern recognition written by the group in Genoa. Although very interesting for chemometricians, such a package is written in BASIC and is rather slow with fairly limited graphics. However, source code is available, and the package is a useful exploratory tool. Some knowledge of programming is useful. During the early 1980s, various students in the Nijmegen group wrote routines in BASIC, primarily for teaching chemometrics, which have been bundled into CLEOPATRA [121]. The difficulty is that this package is not very user-friendly and quite expensive, but was an important historical landmark for computer based teaching of chemometrics. A difficulty with academic groups building up routines over the years is that microprocessor technology is likely to move quite fast, so the software rapidly becomes out of date.

SPIDA [122] is a set of statistical routines, many of which are interesting to chemometricians. It also contains a small language rather like S-plus, but is not so large or elaborate. It is, however, an affordable DOS based package that includes a variety of programs for clustering, PCA, factor rotations etc. and it is not too difficult to incorporate new macros such as PLS.

Academic groups in particular continue to contribute a wide variety of chemometrics packages, many of which are available openly on networks. Most packages, though, are maintained and marketed by small groups, and many, because they have taken several years to build up, are DOS based and written in languages such as BASIC or FORTRAN and are fairly limited in capacity.

Historically, the widespread distribution of microprocessor based packages in the mid 1980s did a great deal to catalyse the use of chemometrics methods.

Modern Developments in Packaged Chemometrics Software

Although there are a vast number of packages on the market, the most fruitful ones are those maintained by large groups of people or commercial organisations. Often these packages are quite specialised and expensive, but there is good user support, newsletters, updates and maintenance.

UNSCRAMBLER [123–125] has been developed over 10 years, and is marketed by CAMO A/S. Many of the principles arise from the work of H. Martens in multivariate calibration. A very large number of workshops are organised by this group, in addition to newsletters and associated literature. The original package was primarily concerned with multivariate calibration, although now there are modules for experimental design and calibration. Great effort has been put into good graphics and help facilities. It is strongly oriented to the processing of instrumental data such as NIR (Near Infrared) spectroscopy. A weakness at present is that the package is DOS based rather than Windows based, and requires a maths coprocessor, although this limitation may change shortly. It is very widespread within chemical laboratories.

Umetri have developed interactive Windows based versions of their packages SIMCA3B and MODDE. InfoMetrix continue with a large number of interactive packages, notable of which is Pirouette [126]. These are major developments from the original SIMCA and ARTHUR software discussed above SCAN, produced by Minitab, is a recently available package.

Many chemometricians like to think in terms of matrices, being able to manipulate them and visualise data. MatLab [127, 128] is widespread. It requires some understanding of algorithms and is not packaged software in the same way as UNSCRAMBLER or SIMCA3B. However, it does run on both Macintosh and PC compatible computer. There are many MatLab macros for PCA, PLS, factor analysis and so on, and so this language can be considered a valuable development tool for the chemometrician who has some understanding of the basics of algorithms and does not wish to program from scratch. Because it is used by scientists for many disciplines, not just in chemometrics, it is likely to be available for some years, and to be properly supported.

Instrumental and Applications Software

Over the last decade, many instrument manufacturers have incorporated some chemometrics routines into their software. Particularly important have been many standard methods for signal enhancement, such as digital filters, smoothing functions etc. Optimisation methods such as simplex for tuning signals are very common. Outside the direct realm of chemometrics, a large amount of Fourier transform software has been developed. Several instrument manufacturers have incorporated PCA, PLS, PCR and some simple factor analysis approaches into their software, especially in the area of NIR spectroscopy. Maximum entropy has been widely applied to a variety of instrumental datasets, and some instrument manufacturers market the package with their operating systems; MS is a good example here.

In the area of QSAR and molecular modelling, there is also a very large number of commercial software approaches, most including some form of PCA, multivariate calibration and clustering/classification facilities. Many database packages contain clustering algorithms.

One weakness of commercial packages is that there is rarely any interpretative facility. The methods are presented as options for data processing, and it is often possible to obtain completely meaningless answers if the methods are misapplied. Often various options such as scaling methods are included "by default" and it may require a great deal of technical knowledge to change these. A good example involves determining the number of significant components in IR spectroscopy. A large number of methods can be employed [129, 130], and it is not untypical for answers to vary between 2 and 20 for identical datasets. It is unwise to use black boxes from instrument manufacturers, or packaged software, unless the background to the methods is first understood.

A major difficulty is that instrument manufacturers are very conservative as to the amount of effort they are prepared to put into implementing chemometric software. Most instrumental software must be user friendly, and great emphasis is placed on good graphics, windows, mouse and icon control, menus etc. The algorithms are only a very small part of a commercial package. Instrument manufacturers do not like releasing packages with a poor user interface. The choice of system is often made by technicians who are more attracted to multicolour, interactive graphics than to advanced statistical output. Therefore, many instrument manufacturers prefer to invest resources on the interface, and many state-of-the-art chemometrics methods are not implemented on commercial software. It takes several years for a method to become widely accepted before commercial companies invest time and funds in widespread implementation.

One exception involves development of chemometric sensors to perform very specific and limited tasks. For example, a device is available that will monitor water quality by electronic absorption spectroscopy using chemometric methods to deconvolute the spectrum. This device can then be used to warn if nitrates and other potential pollutants are above a threshold, and so check the water purification methods. The package is sold as a "black box" and is only useful in one context.

Another common problem is that instrument manufacturers very rarely release source listings, so the user is absolutely dependent on the implementation.

Programming Environments

Many chemometricians develop their own methods by writing specific programs.

There are a large number of languages and environments. The first problem is to choose a hardware configuration. Until recently many numerically intensive programs were written on powerful minicomputers such as a VAX (under VMS) or a SUN (under Unix). However, the power of microprocessors has increased dramatically recently. A 486 PC with 16 Mbyte memory, 500 Mbyte disc space, the Windows operating system, colour Super VGA is routinely available, and has the capacity of a departmental minicomputer of a few years ago. Hence many new developments in chemometrics software are microprocessor based.

Traditionally, a great deal of numerical software has been written using the FORTRAN language [131–133]. This is one of the original programming languages, specifically developed for numerical work. It has evolved a great deal from the original specifications, with FORTRAN-77 and FORTRAN-90 appearing. The original language was fairly unstructured, concentrating largely on numerical methods, with very few facilities for input/output or structured loops such as "if .. then .. else" facilities. The present specifications are somewhat hybrid, and most microprocessor implementations are very awkward to use if good, modern, Windows based graphics and mouse control are required. The main advantage is that there is a great deal of historical continuity. In chemometrics, both the ARTHUR and TARGET packages were written in FORTRAN. Large numerical analysis libraries, such as the NAG (Numerical Algorithms Group) [134] library have been developed over many decades and have been adapted to microprocessor applications.

C [135–137] is an attempt to improve on FORTRAN. It has been much used by computer scientists over the last decade. A specific advantage is the ability to manage memory, which is automatically handled by the compiler in FORTRAN. For microprocessor based implementations this is a major advantage, as it allows the best use of the available memory. This language is very flexible and structured, but has several disadvantages. First, strong knowledge of computing is required for use of this language, meaning that it is not easy to develop numerical software. Second, because so many facilities are under the control of the programmer, it is much easier to make mistakes: the compiler will not automatically correct these errors or optimise the code. Whereas the programmer can do almost anything, he/she has to think more carefully about the consequences of his/her actions.

Quite a lot of early microprocessor based software was written in BASIC. There are a large number of implementations of BASIC, many of the early ones having poor structure. Early editions of BASIC were also interpretative, meaning that programs ran slowly. However, this language was included with most microprocessor based systems in the early 1980s. FORTRAN was originally developed as a mainframe programming language, so early microprocessor implementations of FORTRAN were not very successful. BASIC served a useful purpose for microprocessor based numerical routines. Early editions of the SIMCA package were written in BASIC. There were innumerable early "dialects" of BASIC, causing a great deal of confusion when reading the literature.

Another language sometimes used was PASCAL [138] which derived from Algol. This was originally developed as a mathematician's language, good on symbolic programming, but its numerical facilities are fairly limited. For example, there is no direct facility for powers: they have to be calculated via logarithm functions.

A revolution happened in the late 1980s with the introduction of OOP (Object Oriented Programming) [139–141] and the Windows operating system [142]. An object is something that has both functionality and properties. A good example is a square: it has an area, a position, a circumference, a diagonal etc. A generic class of squares could be developed, each member with specific properties. The programmer defines a number of objects, each with their own properties. In the Windows operating system objects may be scroll bars, menus, icons etc. Events activate objects. For example, clicking a mouse on an icon may activate a program; clicking a menu item selects certain actions associated with the menu icon. Programs are no longer "linear", but the software consists of a number of objects. Another feature of OOP is that it is possible to define a hierarchy of objects. In chemometrics, it may be possible to define a class of "graph" objects The graph objects may be subdivided, lower down the hierarchy, into scores, loadings, eigenvalues etc. Matrix objects may be subdivided into vectors, square matrices etc. all with their own, specific, properties.

OOP changes the way programmers work. One of the most successful OOP environments is C++ [143–144]. This is an extension of C including object oriented methods, among other features. Vast class libraries have been developed for C++. Microsoft link C++ to the Windows operating system, so classes of objects include scroll bars, menus, windows and so on. This programming environment is exceptionally flexible, but not best suited for most chemometrics programmers, the difficulty being that it takes a great deal of time to set up class libraries and define basic properties of objects. It can be useful for large team efforts, where one type of operation is going to be performed repeatedly.

VISUAL BASIC [145–147] is a relatively recent OOP environment of increasing popularity. Although it includes "BASIC" in its name, and has a certain degree of compatibility with previous versions of this language, it has evolved very far from BASIC. A very flexible object oriented environment is provided, but an additional advantage is that conventional code, such as straight numerical routines, can be attached to objects. Scientific programming should not be entirely within an object oriented environment: some algorithms work best in a linear fashion, so a combination of both traditional and object oriented methods is favoured. OOP methods are very useful for control of programs and good user interfaces, whereas conventional methods are best for traditional algorithms. One disadvantage of VISUAL BASIC is that the compiled code is quite slow, although this may change as implementations improve.

Other OOP environments worth mentioning are as follows. VISUAL C++ is an extension to C++ with some of the facilities of VISUAL BASIC. Toolbook [148] is an excellent environment for a Windows interface, but is not very useful for numerical programming. Linking Toolbook to a numerical package such as VISUAL BASIC or C would seem an optimal solution.

In conclusion, the chemometrician has a very wide variety of programming environments to choose from. The choice must depend on many factors as follows. The technical ability of the programming has to be taken into account. The hardware and operating system should be considered. The importance of a user interface must be considered at the beginning of a project. The time scale of the project, and whether it is necessary to interface to other people's programs are important. There is no one single optimal language or approach, but the ability to develop "homegrown" software is important for development of state-of-the-art methods.

Conclusion

We have provided an overview of modern chemometrics. One of the difficulties in this field is that there is a very wide variety of groups of people, all with different backgrounds and expectations. Some investigators do not class themselves as chemometricians, despite the relevance of their work to chemometrics. Other investigators, such as mainstream analytical chemists, are often very keen to be classed as chemometricians.

A wide variety of people will want to know about and use chemometrics methods in view of the diversity of approaches to this very broadly based subject.

References

1. Kowalski BR (1980) Chemometrics. Anal Chem 52:112R–122R
2. Frank IE, Kowalski BR (1982) Chemometrics. Anal Chem 54:232R–243R
3. Delaney MF (1984) Chemometrics. Anal Chem 56:261R–277R
4. Ramos LS, Beebe KR, Carey WP, Sanchez E, Erickson BC, Wilson BR, Wangen LE, Kowalski BR (1986) Chemometrics. Anal Chem 58:294R–314R
5. Brown SD, Barker TQ, Larivee RJ, Monfre SL, Wilk HR (1988) Chemometrics. Anal Chem 60:252R–273R
6. Brown SD (1990) Chemometrics. Anal Chem 62:84R–101R
7. Brown SD, Bear RS, Blank TB (1992) Chemometrics. Anal Chem 64:22R–49R
8. Brereton RG (1987) Chemometrics in Analytical Chemistry: A Review, Analyst. 112:1635–1657
9. Kateman G, Pijpers FW (1981) Quality Control in Analytical Chemistry. Wiley, New York
10. Massart DL, Dijkstra A, Kaufman L (1978) Evaluation and Optimisation of Laboratory Methods and Analytical Procedures, Elsevier, Amsterdam

11. Miller JC, Miller JN (1993) Statistics for Analytical Chemistry – Third Edition, Ellis Horwood PTR Prentice Hall, New York
12. Youden WJ (1951) Statistical methods for Chemists, Chapman and Hall, London
13. Davies OL, Goldsmith PL (1972) Statistical Methods for Research and Production, Oliver and Boyd, Edinburgh
14. Geladi P, Kowalski BR (1986) Partial Least Squares Regression: A Tutorial, Anal Chim Acta 185:1–17
15. Wold S, Martens H, Wold H (1983) The Multivariate Calibration Problem in Chemistry solved by the PLS method in Proc Conf Matrix Pencils (eds. Ruhe A, Kågström B). Springer Verlag, Heidelberg 286–293
16. Brown PJ (1982) Multivariate Calibration (with discussion). J Roy Stat Soc Ser B 44:287–321
17. Malinowski E (1991) Factor Analysis in Chemistry – Second Edition. Wiley, New York
18. Hamilton JC, Gemperline PJ (1990) Mixture Analysis Using Factor Analysis Part II: Self Modelling Curve Resolution, J Chemometrics. 4:1–13
19. Lewi PJ (1987) Spectral Map Analysis: Factorial Analysis of Contrasts Especially for Log Ratios, Chem Int Lab Syst 5:105–116
20. Liang Y-Z, Kvalheim OM, Rahmani A, Brereton RG (1993) Resolution of Strongly Overlapping two-way Multicomponent Data by Means of Heuristic Evolving Latent Projections, J Chemometrics. 7:15–43
21. Brereton RG, Rahmani A, Liang Y-Z, Kvalheim OM (1994) Investigation of the Allomerization Reaction of Chlorophyll a: Use of Diode Array High Performance Liquid Chromatography, Mass Spectrometry and Chemometric Factor Analysis for the detection of early products, Photochem Photobiol 59:99–110
22. Liang Y-Z, Brereton RG, Kvalheim OM, Rahmani A (1993) Use of Chemometric Factor Analysis for Chromatographic Integration: Application to Diode array High Performance Liquid Chromatography of Mixtures of Chlorophyll a degradation products, Analyst 118:779–790
23. Box GEP, Hunter WG, Hunter JS (1978) Statistics for Experimentalists, Wiley, New York
24. Brereton RG, Gurden SP, Groves JA (1995) Use of Eigenvalues For Determining The Number Of Components In Window Factor Analysis Of Spectroscopic And Chromatographic Data, *Chemometrics Intell. Lab. Systems*, 27:73-87
25. Wold S (1976) Pattern Recognition by means of Disjoint Principal Components Models, Pattern Recognition 8:127–139
26. Musumarra G, Wold S, Gronowitz S (1981) Application of PCA to ^{13}C shifts of Chalcones and their Thiophene and Furan Analogues: A Useful Tool for Shift Assignment and for the Study of Substituent Effects, Org Magn Reson 17:118–123
27. Blomquist G, Johansson E, Söderström B, Wold S (1979) Classification of Fungi by means of Pyrolysis – Gas Chromatography, J Chromatogr 173:19–32
28. Wold S, Johansson E, Jellum E, Bjørnson I, Nesbakken R (1981) Application of SIMCA multivariate data analysis to the Classification of Gas Chromatographic Profiles of Human Brain Tissues, Anal Chim Acta 133m:251–259
29. Thomsen JU, Meyer M (1989) Pattern Recognition of ^{1}H Spectra of Sugar Alditols Using a Neural Network, J Magn Reson 84:212–217
30. Murray-Rust P, Bland R (1978) Computer Retrieval of Molecular Geometry II Variance and its Interpretation, Acta Cryst B34:2527–2533
31. Taylor R (1985) The Cambridge Data File in Crystallographic Computing 3. Data Collection, Structure Determination, Proteins and Databases (Sheldrick GM, Krüger C, Goddard R, eds.), Clarendon, Oxford pp. 96–105
32. Zupan J (editor) (1986) Computer Supported Spectroscopic databases. Ellis Horwood Chichester
33. Bush Bl, Nachbar RB (1993) Sample Distance Partial Least Squares: PLS optimized for many variable, J Comput Aided Mol Design 7:587–619
34. Hudson B, Livingstone DJ, Rahr E (1989) Pattern Recognition methods for the analysis of computed molecular properties. J Comput Aided Mol Design 3:55–65
35. Camilleri P, Livingstone DJ, Murphy JA, Manallack DT (1993) Chiral Chromatography and Multivariate Quantitative Structure Property Relationships of benzimidazole sulphoxides. J Comput Aided Mol Design 7:61–69
36. Tosata ML, Cesareo D, Passerini L, Clementi S (1988) PLS assessment of the performance of short-term tests for carcinogens, J Chemometrics. 2:171–187
37. Hammett LP (1940) Physical Organic Chemistry, McGraw-Hill, New York
38. Muller A, Steele D, (1990) On the Extraction of spectra of components from spectra of mixtures. A development in factor theory-I, Spectrochimica Acta 46A, 5:817–842
39. Sibisi S, Skilling J, Brereton RG, Laue ED, Staunton J (1984) Maximum Entropy Signal processing in practical NMR Spectroscopy, Nature 311:446–447

40. Laue ED, Skilling J, Staunton J, Sibisi S, Brereton RG (1985) Maximum Entropy Methods in Nuclear Magnetic Resonance Spectroscopy, J Magn Reson 62:437–452
41. Stephenson DS (1988) Linear Predictions and Maximum Entropy Methods in NMR Spectroscopy, Progr in NMR Spectrosc 20:515–626
42. Ferrige AG, Seddon MJ, Skilling J, Ordsmith N (1992) The Application of 'Max Ent' to High Resolution Mass Spectrometry, Res Commun In Mass Spectrom 6:765–770
43. Buck B, Macaulay VA (eds.) 1991 Maximum Entropy in Action. Clarendon Press, Oxford
44. Jansson PA (1984) Deconvolution with Applications to Spectroscopy, Academic Press, New York
45. Gans P (1992) Data Fitting in the Chemical Sciences, Wiley
46. Press WH, Teukolsky SA, Vetterling WT, Flannery BP (1992) Numerical Recipes in C-second edition, Cambridge University Press
47. Press WH, Flannery BP, Teukolsky SA, Vetterling WT (1989) Numerical Recipes (FORTRAN version), Cambridge University Press
48. Press WH, Flannery BP, Teukolsky SA, Vetterling WT (1989) Numerical Recipes (Pascal version), Cambridge University Press
49. Date CJ (1981) An Introduction to Database Systems – third edition, Addison Wesley, Reading Mass
50. Allus MA, Brereton RG, Nickless G (1988) Chemometric Studies on the effect of toxic metals on plants: the use of response surface methodology to investigate the influence of Tl, Cd and Ag on the growth of cabbage seedlings, Environmental Pollution 52:169–181
51. Allus MA, Brereton RG (1990) Chemometric methods for the study of the influence of toxic metals on the growth of plants: use of experimental design and response surface methodology, Int J Env Anal Chem 38:279–304
52. Chemometrics Int Lab Systems 2:15–243 (1987)
53. Davis JC (1986) Statistics and Data Analysis in Geology – second edition, Wiley New York
54. Brereton RG (1987) Spectral Analysis of Multivariate Geochemical Time Series. Chemometrics Int Lab Systems 2:177–185
55. Avila F, Myers DE (1991) Correspondence Analysis applied to Environmental Dataset: a study of the Chautauua lake, Chemometrics Int Lab Systems 11:229–249
56. Mellinger M (1987) Interpretation of Lithogeochemistry using Correspondence Analysis, Chemometrics Int Lab Systems 2:93–108
57. Hopke PK, Lamb RE, Nutusch DF (1980) Multielement Characterization of urban roadway dust, Env Sci Techn 14:164–172
58. Piepponen S, Linström R (1989) Data Analysis of heavy metal pollution in the sea by using Principal Components Analysis and Partial least Squares. Chemometrics Int Lab Systems, 7:163–170
59. Kvalheim OM, Øygard K, Grahl-Nielsen O (1983) SIMCA Multivariate Analysis of Blue Mussel Components in Environmental Pollution Studies, Anal Chim Acta 150:145–152
60. Brereton RG (1994) Chemometrics in Trace Analysis in Analysis of Contaminants in Edible Aquatic Organisms (eds. Kapenciuk J, Ray P), VCH, New York, Chapter 3
61. Devillers J, Chambon P, Zakarya D, Chastrette M, Chambon R (1987) A predictive structure-toxicity model with Daphnia Magna, Chemosphere 16:1149–1163
62. Könemann H (1981) Quantitative Structure Activity Relationships in fish toxicity studies Part 1. A relationship for 50 industrial pollutants, Toxicology 19:209–221
63. Vogt NB, Bye E, Thrane KE, Jacobsen T, Benestad C (1987) Composition Activity Relationships Part 1. Exploratory Multivariate Analysis of elements, polycyclic aromatic hydrocarbons and mutagenicity in air samples, Chemometrics Int Lab Systems 6:31–47
64. Vogt NB, Bye E, Thrane KE, Jacobsen T, Benestad C (1987) Composition Activity Relationships Part II. Indirect and Direct Mutagens. Multivariate Dose-Response Regression, Chemometrics Int Lab Systems 6:127–142
65. Krzanowski WJ (1987) Selection of Variables to preserve Multivariate Data Structure using Principal Components Analysis, Applied Statistician 36:22–33
66. Joliffe IT (1973) Discarding Variables in a Principal Components Analysis space: 2 Real Data, Applied Statistician 22:21–31
67. McCabe GP (1984) Principal Variables, Technometrics 26:137–144
68. Massart DL, Brereton RG, Dessy RE, Hopke PK, Spiegelman CH, Wegscheider W (eds.) (1990) Chemometrics tutorials. Elsevier, Amsterdam
69. Brereton RG, Scott DR, Massart DL, Dessy RE, Hopke PK, Spiegelman CH, Wegscheider W (eds.) (1992) Chemometrics tutorials II. Elsevier, Amsterdam

70. Sharaf MA, Illman DL, Kowalski BR (1986) Chemometrics. Wiley, New York
71. Massart DL, Vandeginste BGM, Deming SN, Michotte Y, Kaufman L (1988) Chemometrics: A Textbook, Elsevier, Amsterdam
72. Brereton RG (1993) Chemometrics: Application of mathematics and statistics to Laboratory Systems. Ellis Horwood, Chichester
73. Strouf O (1986) Chemical Pattern Recognition Research Studies Press, Letchwoth
74. Willett P (1987) Similarity and Clustering in Chemical Information Systems, Research Studies Press, Letchworth
75. Lewi PJ (1982) Multivariate Analysis in Industrial Practice, Research Studies Press, Letchworth
76. Commans D, Broeckaert I (1986) Potential Pattern Recognition, Research Studies Press, Letchworth
77. Bawden D, Ioffe IT (eds.) (1988) Application of Pattern Recognition to Catalytic Research, Research Studies Press, Letchworth
78. Buydens LMC, Melssen WJ (eds.) (1994) Chemometrics: Exploring and Exploiting Chemical Information, University of Nijmegen
79. Kowalski BR (ed.) (1984) Chemometrics: Mathematics and Statistics in Chemistry NATO ASI Series C, Mathematical and Physical Sciences, Vol. 138, Reidel, Dordrecht
80. Kowalski BR (ed.) (1977) Chemometrics: Theory and Application, ACS Symposium Series 52, American Chemical Society, Washington DC
81. Meloun M, Militky J, Forina M (1992) Chemometrics for Analytical Chemistry Vol. 1, Ellis Horwood, Chichester
82. Hopke PK (1991) Receptor Modelling for Air Quality Measurements, Elsevier, Amsterdam
83. Jurs PC, Isenhour T (1975) Chemical Applications of Pattern Recognition, Wiley, New York
84. Varmuza K (1980) Pattern Recognition in Chemistry, Springer-Verlag, Berlin
85. Wolff DD, Parsons MIL (1983) Pattern Recognition Approach to Data Interpretation, Plenum, New York
86. Massart DL, Kaufman L (1983) The Interpretation of Analytical Chemical Data by the Use of Cluster Analysis, Wiley, New York
87. Brereton RG (ed.) (1992) Multivariate Pattern Recognition in Chemometrics, illustrated by case studies, Elsevier, Amsterdam
88. Martens H, Næs T (1989) Multivariate Calibration, Wiley, Chichester
89. Deming SN, Morgan SL (1993) Experimental Design: A Chemometric Approach – second edition, Elsevier, Amsterdam
90. Morgan E (1991) Chemometrics: Experimental Design (Analytical Chemistry by Open Learning), Wiley, Chichester
91. Bayne CK, Rubin IB (1986) Practical Experimental Designs and Optimization Methods for Chemistry, VCH, Florida
92. Spiegelman CH, Waters CH, Sacks J (eds.) (1985) J Res Natl Bureau Standards Special Issue, 90, vol. 6
93. Chemometrics Int Lab Systems 7:11–194 (1989)
94. Chemometrics Int Lab Systems 14:1–427 (1992)
95. Wolff DD, Parsons MIL (1983) Pattern Recognition Approach to Data Interpretation. Plenum, New York pp.11–13
96. Dunn WJ (1987) SIMCA3B, Chemometrics Int Lab Systems 1:126–127
97. Malinowski E (1991) Factor Analysis in Chemistry – Second Edition, Wiley, New York pp. 326–328
98. Quantum Chemistry Program Exchange, Chemometrics Int Lab Systems 1:122 (1987)
99. Hopke PK, Dharmavaram S (1986) Recent Improvements to FANTASIA, a target transformation factor analysis program. Comput Chem 10:163–164
100. Brereton RG (1986) SAS (Statistical Analysis System), Chemometrics Int Lab Systems 1:9
101. SAS Introductory Guide Third Edition, SAS Institute, Cary, NC, (1985)
102. SAS Language and Procedures: Usage, Version 6, First Edition, SAS Institute, Cary NC (1991)
103. Dixon WJ (ed. in chief) (1983) BMDP Statistical Software, Univ California Press, Berkeley, Calif.
104. SPSSX BASICS, SPSS Inc, Chicago, (1984)
105. Bratchell N (1987) GENSTAT, Chemometrics Int Lab Systems 1:300–301
106. Wolff DD, Parsons MIL (1983) Pattern Recognition Approach to Data Interpretation, Plenum, New York p.13
107. Newton HJ (1991) S-Plus for Unix and DOS, Chemometrics Int Lab Systems 11:255–256
108. Denham MC (1993) S-Plus, J Chemomet 559–566

109. Spiegelman CH (1989) SYSTAT, Chemometrics Int Lab Systems 6:89
110. Spiegelman CH (1990) Statistical Software Packages for the Macintosh, Chemometrics Int Lab Systems 9:115–117
111. Harvill JL (1993) MINITAB Statistical Software, Chemometrics Int Lab Systems 18:111–112
112. Stone M (1974) Cross-validatory choice and assessment of Statistical prediction, J Roy Stat Soc B36:111–147
113. Wold S (1987) Cross-validatory estimation of the Number of Components in Principal Component Models, Technometrics 20:397–405
114. Krzanowski WJ (1987) Cross-validation in Principal Components Analysis. Biometrics 43:575–584
115. Wold H (1966) Estimation of Principal Components and Related Models by Iterative Least Squares, in Multivariate Analysis (Krishnaiah P ed.) Academic Press. Orlando
116. Erskine RL (1987) Ein*Sight for multivariate analysis, Chemometrics Int Lab Systems 1:302–303
117. Thielemans A, Massart DL (1987) Spectramap, Chemometrics Int Lab Systems 1:206–207
118. Iverson KE (1980) Notation as a tool of thought (1979 Turing Award lecture) Communications for the Association for Computing Machines 23:444–465
119. Kvalheim OM, Karstang TV (1987) A general program for Multivariate Data Analysis, Chemometrics Int Lab Systems 2:235–237
120. Vogt NB (1989) PARVUS-50 programs for pattern recognition, Chemometrics Int Lab Systems 6:173–175
121. Bratchell N, MacFie HJH (1987) CLEOPATRA – an extendable library of programs to assist the teaching of chemometrics, Chemometrics Int Lab Systems 1:124–126
122. Lunn AD, McNeil DR (1991) Computer-Interactive Data Analysis, Wiley, Chichester
123. Brown SD (1994) Unscrambler II, J Chemomet 8:175–176
124. Tyssø V, Esbensen K, Martens H (1987) UNSCRAMBLER – an interactive program for multivariate calibration and prediction, Chemometrics Int Lab Systems 2:239–243
125. Wegscheider W (1993) Unscrambler II. Chemometrics Int Lab Systems 19:269–270
126. Blackburn M, Nam ICS Newsletter 8 July 1994, p.2–414
127. O'Haver TC (1989) Teaching and Learning Chemometrics with MatLab, Chemometrics Int Lab Systems 6:95–103
128. Malinowski E (1991) Factor Analysis in Chemistry – Second Edition, Wiley, New York pp. 329–333
129. Deane JR in Brereton RG (ed.) (1992) Multivariate Pattern Recognition in Chemometrics, illustrated by case studies, Elsevier Amsterdam Chapter 5
130. Malinowski ER (1977) Determination of the Number of Factors and Experimental Error in a Data Matrix, Anal Chem 49:612–617
131. Wagener JL (1980) FORTRAN 77: Principles of Programming, Wiley, New York
132. Agelhoff R, Mojena R (1981) Applied FORTRAN 77 featuring Structured Programming, Wadsworth, Brlmont, CA
133. Ledgard HF, Chmura L (1978) FORTRAN with Style, Hayden, Rochelle Park, New York
134. NAG FORTRAN library: Introductory Guide Mark 13, NAG, Oxford (1988)
135. Kerrigan JF (1991) From Fortran to C, Windcrest/McGraw-Hill, Blue Ridge Summit, PA
136. Oualline S (1991) Practical C Programming, O'Reilly and Associates Sebastopol, CA
137. Hansen A (1988) Learn C Now, Microsoft Press, Redmond, WA
138. Huggins E (1983) Mastetring Pascal Programming, MacMillan, Houndmills
139. Graham I (1991) Object Oriented Methods, Adison Wesley, Wokingham
140. Meyer B (1988) Object-oriented Software Construction, Prentice Hall, Hemel Hempstead
141. Brereton RG Object Oriented programming on Personal Computers, Analyst, in press
142. Petzold C (1992) Programming Windows 3.1 Third Edition, Microsoft Press, Redmond, WA
143. Sedgewick R (1990) Algorithms in C++, Addison Wesley, Reading, MA
144. Schildt H (1991) C++ The complete reference, McGraw Hill, Berkely, CA
145. Mansfield R (1992) The Visual Guide to Visual Basic for Windows, Ventura Press, Chapel Hill, NC
146. Socha J (1992) Learning Programming and Visual Basic with John Socha, Sybex, San Francisco, CA
147. Microsoft Visual Basic Programmers' Guide Version 3, Microsoft (1993)
148. Using Toolbook, Asymetrix Corporation, Bellevue, Washington (1991)

Experimental Design and Optimization

Ramon A. Olivero[1], *John M. Nocerino*[2], *and Stanley N. Deming*[3]

[1] Lockheed Environmental Systems & Technologies Company, 980 Kelly Johnson Drive, Las Vegas, NV 89119, USA
[2] United States Environmental Protection Agency Environmental Monitoring Systems Laboratory Las Vegas, P.O. Box 93478, Las Vegas, NV 89193-3478, USA
[3] University of Houston, 4800 Calhoun, Houston, TX 77204-5641, USA

Summary

The application of statistical experimental design and optimization (SEDOP) to environmental chemistry research is presented. The use of SEDOP approaches for environmental research has the potential to increase the amount of information and the reliability of results, at a cost comparable to, or lower than, traditional approaches. We demonstrate how researchers can attain these benefits by adhering to a systematic program of design and execution of experiments, including the analysis and interpretation of results. The lack of general knowledge about experimental statistical techniques had hindered their widespread application in the environmental field. To benefit from the SEDOP advantages, the United States Environmental Protection Agency (USEPA) has an ongoing project to investigate applications of statistical design to environmental chemistry problems. There exist standard experimental arrangements (designs) to address all phases of a research program, from identifying important effects, to modeling the behavior of the experimental system of interest, to optimizing the operating conditions (e.g., minimizing waste or maximizing reproducibility). The most useful standard design arrangements (both for system characterization and optimization) are introduced, together with a discussion of their applicability to pollutant analysis as well as their strengths and weaknesses. Practical environmental applications from the literature are presented and discussed from the perspective of the approaches and techniques that they illustrate. Examples include optimization of analyte extraction, instrument calibration, method comparison, ruggedness testing, selection of indicator contaminants, and pollution prevention. The implementation of statistical experimental design today is greatly facilitated by the use of available software for the selection of designs, the planning of experiments, the analysis of data, and the graphical presentation of results.

List of Symbols and Abbreviations

ANOVA	analysis of variance
B	simplex vertex with the best response
$\boldsymbol{B}$	matrix of model parameters
$\hat{\boldsymbol{B}}$	matrix of estimated model parameters
Ba	barium
C_R	simplex contraction vertex (towards reflection vertex)
C_W	simplex contraction vertex (towards worse response vertex)
DOE	design of experiments
E	simplex expansion vertex
EVOP	evolutionary operation
f	number of distinct factor level combinations
$f(x)$	mathematical function on variable
F	random variable of F distribution
GC	gas chromatograph
GC/MS	gas chromatograph/mass spectrometer
GSAM	generalized standard addition method
HCl	hydrochloric acid
HNO_3	nitric acid
i	individual experiment identifier
k	number of factors
l	number of levels
LC/MS	liquid chromatograph/mass spectrometer
ml	milliliter
N	simplex vertex with the next-to-best response; noise
n	total number of observations
ORD	Office of Research and Development
p	number of parameters, number of fractional design generators
$\overline{P}$	centroid of the remaining hyperface for a simplex
PAH	polynuclear aromatic hydrocarbon
PCB	polychlorinated biphenyl
PLS	partial least squares
r	residual, error
R	simplex reflection vertex
s_{AB}	pooled standard deviation
s_y	sample standard deviation
S	signal
s_y^2	sample variance
s_r^2	variance of the residuals
Sb	antimony
SEDOP	statistical experimental design and optimization
SFE	supercritical fluid extraction
S/N	signal-to-noise ratio
SS_f	sum of squares of treatment factor effects

SS_r	sum of squares of residuals
SS_y	sum of squares of the responses
SPE	solid phase extraction
SRM	standard reference material
t	random variable of Student's t distribution
t_o	observed Student's t statistic
USEPA	United States Environmental Protection Agency
$\boldsymbol{V}$	variance-covariance matrix
$\boldsymbol{W}$	simplex vertex with the worst response
x	factor level
$\boldsymbol{X}$	design matrix
y	response value
$\hat{y}$	estimated response value
$\overline{y}$	mean response value
β	model parameter
ε	error, residual
μ	population mean
ν	number of degrees of freedom
$\sum$	summation
%R	percent recovery
°C	degrees Celsius

Introduction

An introduction to experimental design and optimization is presented for environmental scientists and engineers as tools for the development and evaluation of methods for monitoring chemicals in the environment. Knowing the quality of environmental measurements is crucial to understanding and protecting the environment. A key component of the quality of environmental data is the reliability of the methods used to produce them. Environmental researchers can obtain the desired information for method development in an efficient manner through the systematic planning of experiments and the analysis of results. Statistical experimental design and optimization techniques (collectively referred to here as SEDOP) provide a framework for a systematic approach to experimentation. The use of SEDOP techniques in chemistry, also referred to as the design of experiments (DOE), is a subject of chemometrics. Chemometrics is the study of the application of mathematical, statistical, and computational methods to chemical data for the extraction of useful information [1]. Researchers are growing accustomed to the use of chemometric methods for environmental data analysis [2, 3] (hence the term environmetrics [4]), but the use of chemometric techniques during the planning phase of experiments is not as common.

The most efficient and reliable approach to the acquisition of experimental data is to introduce statistical considerations at the planning and design stage. The application of seemingly sophisticated statistical methods to the analysis of data acquired from poorly-designed experiments may lead to erroneous conclusions about the results. Beebe and Pell recommend that chemometrics should be used at every phase of an experimental study [5]. This approach requires the clear establishment of the relationships among the desired information and the way that the experiments are to be conducted, the way that the data are to be analyzed, and the way that the results are to be interpreted.

The application of experimental design in the complex framework of environmental chemistry varies according to the research project's objectives and its stage of development. Even if the overall goal might be the optimization and understanding of a process, several sub-goals are frequently encountered in environmental chemistry research for analytical applications: increasing accuracy, precision, or specificity; maximizing sensitivity; minimizing the limit of reliable measurement (e.g., detection limits); assessing reproducibility (such as interlaboratory testing); optimizing particular performance characteristics; testing for method ruggedness (tolerance); and improving the understanding of a system's behavior. In general, experimentation has the goals of determining the effect of a factor (variable) upon an output, minimizing the variability for an experimental system, and optimizing an output.

We discuss herein some of the available SEDOP methodologies and software tools, review how these techniques have been applied in environmental chemistry research, and provide guidelines for the application of experimental design to an environmental experimental program.

System Theory and Response Surfaces

A conceptual model for the formulation of experimental design applications is based on system theory [6]. A system consists of inputs transformed by a process to produce outputs. The inputs are called "factors" and the outputs are called "responses." An analytical system is represented in Fig. 1. Conceptual modeling of the elements involved in the experiment as components of a system helps establish a framework for definitions, relationships, and methodologies. Choosing the "levels" (values) for the factors is a fundamental aspect of experimental design. A given set of levels for the factors in the system (called a treatment combination) constitutes an experiment, with the resulting values for the responses being a product of the system's transform.

The domain of values that factors can take is termed the "factor space." Factors may be controllable by the researcher (with an acceptable degree of precision) or not. From that perspective, factors are classified as controllable or non-controllable. Quantitative factors are those with a continuum of values, such as the acidity of a chemical solution, conveyor speed, or pumping rate; qualitative factors have discrete values, such as the type of acid used in a process, the physical configuration of a pilot plant, or a field sampling tool type. Most of the factors involved in chemical research are quantitative rather than qualitative. The conceptual modeling step for the physical system of interest identifies known factors and known responses.

The experimental system, as defined here, includes all elements related to the study, including equipment, materials, human intervention, procedures, and factor settings. In the field of experimental environmental chemistry, the physical system of interest could represent a measurement method for pollutant concentrations, an industrial chemical process in which waste should be minimized, or a pilot

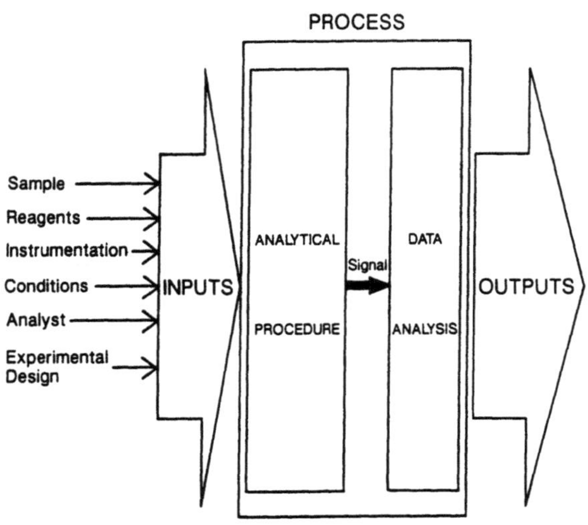

Fig. 1. Conceptual model of an experimental system as applied to environmental analytical chemistry with input (factors), process, and output (response) components

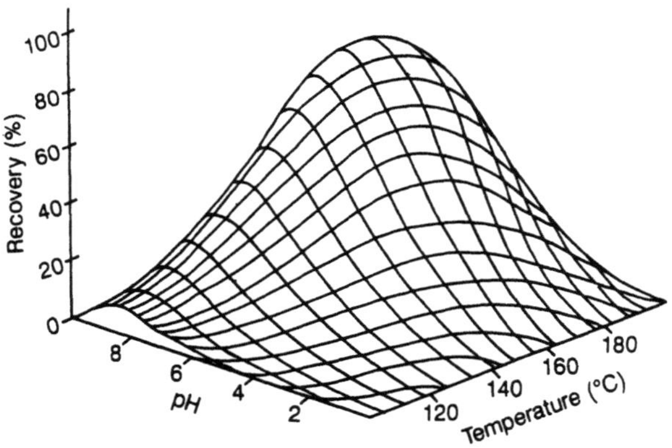

Fig. 2. Pseudo three-dimensional representation of a response surface for two factors (temperature and pH). The response (percent recovery) is represented by the vertical axis

plant for remediation of contaminated material. The environmental chemist or engineer might typically study parameters related to analytical conditions, material processing procedures, and sampling methodologies.

Response values can be related to factor levels by a mathematical function (model). The graphical representation of such a model is the "response surface." Figure 2 shows an example of a response surface for a response (percent recovery) as a function of two factors (temperature and pH). This response surface (an elongated dome with a maximum, tilted with respect to the main axes) is being used as an example here, but experimental systems can have a number of shapes, including minima, saddles, ridges, planes, and others. The common calibration curve used in analytical chemistry is a response surface with one factor (usually concentration). Obviously, the shape of the response surface is not known to the researcher at the outset of a study. Since experimental measurements are usually relatively few and are affected by noise, the estimate of a system's response surface is subject to uncertainty. One of the main purposes of statistical experimental design is to model at least some portion of a response surface to distinguish the effects of important factors over noise.

Experimental Design

Researchers want answers to specific questions concerning the reliability of results, such as what is the most appropriate experimental arrangement to obtain the desired information, what is the probability of obtaining the same results if an experiment is repeated, what is the probability that the observed results are due to the attributed causes, what is the probability that the observed results are not a purely random occurrence, what are the optimal operating conditions for a process, and how well can we predict the results for experiments that have not been conducted? Statistical approaches to experimental design can provide a quantitative measure of the reliability of results and a degree of confidence in the conclusions.

The term experimental design has been used in a general way to include intuitive decisions taken by the researcher to fix the levels of the factors based on experience and system constraints. Some factors will not be varied in a study. Fixed factors then define the framework of the system (e.g., equipment, technique, and materials).

Statistical experimental design refers in particular to the selection of appropriate treatment combinations for the factors that will be varied (treatment factors). These factors are the subject of statistical analysis and their values are explicitly set for a given experiment. At a later time during a project some of the fixed factors might be added as treatment factors in the study. For example, the objective of a project could be to study what the best flow rate is for a solid-phase extraction (SPE) for sample preparation, if it has been pre-determined that SPE is the method to be used. In this case, flow rate might be a treatment factor and extraction type a fixed factor. Alternatively, the project could study if SPE is a good choice at all, as compared to liquid-liquid extraction, in which case extraction type is a treatment factor.

Traditionally, the only resource available to the researchers at the planning stage of the work has been heuristic knowledge (based on chemical principles, experience, and common sense), as a substitute for a knowledge of experimental statistics. If the factor levels were varied non-systematically (intuitively or randomly) to gain an understanding of an experimental system, it might take a large number of experiments to obtain any useful information. Figure 3 depicts

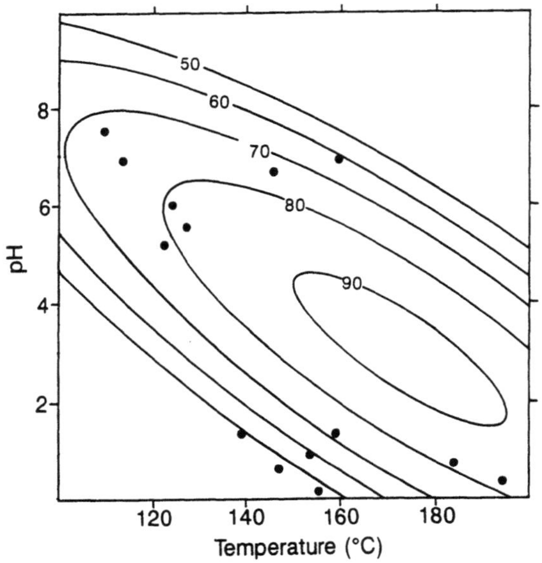

Fig. 3. Example of random experimentation by intuitively varying more than one factor at a time (shotgun approach). *Dots* represent the conditions for an individual experiment (i.e., the settings of temperature and pH). This is a two-dimensional representation of a response surface (see Fig. 2) by iso-response contours (isopleths) for the response of interest, percent recovery in this case. Treatment combinations in an isopleth yield the same percent recovery

an example of this "shotgun approach" (using a two-dimensional representation of the response surface in Fig. 2).

The most common systematic approach is to vary one of the factors in a series of experiments while all of the other factors are held constant (one-factor-at-a-time approach) [7]. An example is shown in Fig. 4. This approach is widely taught as a systematic alternative to intuitively changing several factors at a time (shotgun approach), but it could also produce misleading results in many actual situations. A major flaw of the one-factor-at-a-time approach is that it is unable to reveal any information about interactions between factors. Interactions are interdependencies among the factors, present when the effect of a factor depends on the level of one or more of the other factors. In our example this condition is indicated by the fact that the main axis of the response surface is tilted with respect to the factor level axes. In the case depicted in Fig. 4, an apparent maximal percent recovery was found for a temperature of 115 °C and then the pH was varied to find its maximal effect upon recovery at that temperature. This approach fails to locate the true optimum. The apparent optimal conditions are on a ridge, again due to the interaction between factors. A better approach might be to vary more than one factor at a time, based on a statistical design, and analyze the results accordingly.

The application of statistical concepts to the design of experiments can be traced back to the pioneering work of R.A. Fisher, started in England in the 1920s [8]. Experimental design methods have been used extensively in agricultural experimentation, social sciences, and manufacturing [9, 10]. In fact, many of today's concepts and nomenclature for statistical experimental design can be traced back to the agricultural heritage. Chemists have also originated experimental design techniques to serve specific chemical applications [11].

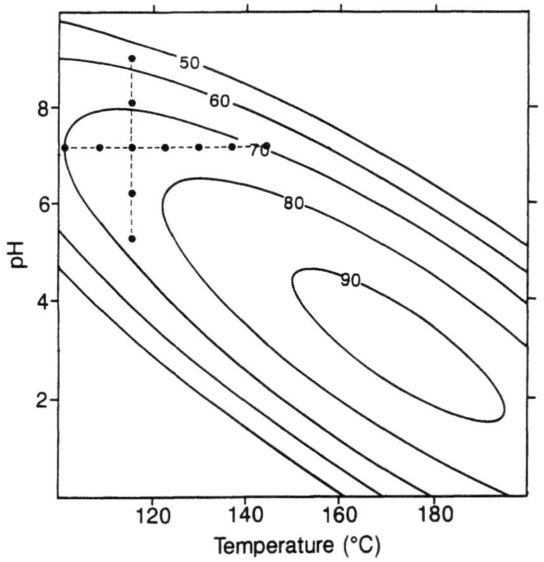

Fig. 4. Example of one-factor-at-a-time experiment. The temperature is varied while the pH is kept constant, and then the temperature is fixed for the highest recovery found while the pH is varied

Experimental Statistics Concepts

Statistically based experimental design provides tools to address common problems that are inherent to experimentation and hard to address by a purely intuitive design of experiments. These include failing to quantify experimental error (i.e., uncertainty or noise), wrongly assigning causal relationships between variables or responses that appear to be correlated, and "confounding" of individual factor effects because of the complexity of the system being studied [10].

In the framework of statistical experimental design, mathematical modeling provides a tool for simulating the physical behavior of the system, while statistics provides a methodology for quantifying how well the physical process is known.

A well-designed experiment should provide the information sought at a reasonable cost. The desired (or needed) information and a reasonable cost are to be defined according to the researcher's view of the problem. For example, the number of levels included in the design (i.e., number of different factor values tried) must be enough to cover the portion of the factor space that is of interest, but not so many that the design would not be cost-effective.

For statistical considerations, experiments can be described mathematically by an equation of the form

$$y = f(x) + r \tag{1}$$

where y is an observed response value, $f(x)$ is a function model that describes the effects of a set of independent variables (factors), x, on the response variable, and r is the error term (uncertainty) that represents departures of the actual data from the model. The r term is also called the "residual" or "error" (sometimes represented by ε in the literature). For statistical considerations, the magnitude of the observed responses in a set of experiments ($i = 1, \ldots, n$) can be represented by its total sum of squares (related to the variance of the data set): $SS_y = \sum_{i=1}^{n} y_i^2$. Sums of squares have the property of being additive; thus, from Eq. (1):

$$SS_y = SS_f + SS_r \tag{2}$$

where SS_f is the sum of squares due to the treatment factor effects, $f(x)$, and SS_r is the sum of squares due to the residuals (error). The formulas for the calculation of SS_f and SS_r are omitted in the interest of simplicity, but they can be found in the general experimental statistics literature [10]. SS_f and SS_r can be broken up into additive components depending on the experimental design and model used. The sums of squares will be discussed later in the context of their relevance to the practical aspects of statistical experimental design. For the purpose of distinguishing signal over noise as well as possible, a good experimental design will make SS_f as large as experimentally feasible ($SS_f \gg SS_r$).

To increase the magnitude of SS_f, researchers can, among other things, use a wide range for factor values (e.g., a wide range of temperature and pH) and include more replicated experiments. It is usually possible in chemical research to obtain an a priori estimation of the experimental error; this permits the placement of the experimental points at a distance large enough to distinguish between the effects and the error [12]. To decrease the error contribution, improvements to the experimental system and procedures should be implemented. From an experimental design viewpoint, as many of the potentially influential factors (and their interactions) as possible should be taken into account in the design and analysis by including them in $f(x)$. Potentially influential factors that are not of interest (e.g., laboratory technician skill) become noise factors. They can be excluded from the noise term by fixing their values for the entire study if possible. When this arrangement is not possible, experiments can be run in blocks respective to the noise factors, thus separating their effect (a technique called "blocking").

It should be noted that, when trying to model an experimental system, r could contain contributions not only from random experimental error, but also from any systematic factor effects not taken into account in the experimental design or the proposed model's $f(x)$. The deviations of the experimental results from the mathematical model not due to purely experimental error are called the "lack-of-fit" contributions.

Experimental Uncertainty and Replication

Experimental measurements are always affected by error. A prerequisite for statistical analysis of experimental results is that the system under study be in "statistical control", meaning that if any given treatment combination is repeated many times, the different responses obtained will be affected only by a small random error (i.e., follow a random pattern of variability about their mean). A set of values that meet this criterion often have a "normal" statistical distribution. This occurs when the system responses are due to the underlying transform modified only by relatively minor disturbances that happen randomly (noise) and are not attributable to a systematic cause. If the data follow a normal distribution, then a simple parametric statistical comparison of the results to standard statistical distributions can be done to determine if the observed effects may have occurred by mere chance or are real (at a preset level of confidence). For statistical purposes, it is further assumed that all of the experiments have the same error variance. Another aspect of statistical control is that the factor effects be distinguishable over the noise. Researchers should make sure that the system is in statistical control before inferences are made from the statistical analyses of the results or any possible conclusions will have a low confidence. This might happen in a case where the magnitude of the noise variability is comparable or exceeds the magnitude of the factor effects ($SS_f \approx SS_r$ or $SS_f < SS_r$).

The only way to assess the experimental uncertainty is by running "replicates," which are repetitions of the same treatment combinations. The more replicates included in a design, the better the estimate of noise variability, $SS_r/(n - f)$, where n is the total number of experiments and f the number of distinct treatment combinations ($n - f$ is the total number of replicates). For laboratory replicates, if collected under proper care, the sample variance will be due to purely experimental error and the sample distribution should approximate a normal distribution. For pollutant concentrations at hazardous waste sites, data commonly follow a log-normal distribution.

Blocking and Randomization

Because there are unknown and uncontrollable factors (that have not been blocked), their effects should be minimized by randomization. One of the major assumptions of the normal distribution of errors is that the determinations are random; that is, any one of the outcomes has the same probability of occurring. The test order should be randomized. This can be accomplished by designing the experiments in such a way that the order in which they are carried out is determined by a random drawing (i.e., a coin toss, random number table, or computer random number generator). Restrictions to randomization will occur when the levels of one or more factors are experimentally difficult to change [13].

Statistical Testing

A set of values of the same kind (i.e., those that belong to the same statistical population) represent a statistical sample of the population if the values were obtained under statistical control. Characteristics of a statistical sample are its mean value and its spread (i.e., variance or standard deviation).

The data set is assumed to be a sample from a general population, and values of the statistics from the sample are compared with the expected values of parameters from the population. If each datum of the set is statistically independent and its collection follows a random pattern, then descriptive statistics of the sample, such as the mean (average), $\bar{y}$, the variance, s_y^2, and the standard deviation, s_y, can be used to compare that set with other sets meeting similar requirements. The comparison is made in the light of an appropriate reference distribution (found in tables of basic statistics texts) which represents what the data set would be like if it met the test criteria.

Tests of significance can determine (with a pre-established risk of being wrong) if a set of values belongs to a given population. This procedure is based on the formulation of a hypothesis (H_0, the "null hypothesis"), and its "alternative hypothesis" (H_1). A statistical test of significance will disprove the null hypothesis or not (thus accepting the alternate hypothesis or not). Statistical analysis allows researchers to determine confidence intervals (as a range of values) with the condition that this range will contain the true value of a calculated parameter (such as a mean) with a selected probability.

There are two key elements for statistical tests of significance and confidence interval determination: the level of risk that the researcher is willing to accept and the number of values in the set(s) being tested. Flatman and Mullins [14] discuss how to set the levels of confidence for environmental studies, both for the risk of wrongly accepting a false null hypothesis and the risk of wrongly rejecting a true null hypothesis. The number of observations determines the degrees of freedom for statistical parameters, which are a measure of the representativeness for the parameter value. Degrees of freedom are the number of independent comparisons that can be made in a data set. Mathematically, the degrees of freedom, v, are given by the number of independent observations, n, minus the number of population parameters, p, which must be estimated from the sample, that is $v = n - p$. For example, it is completely unreliable to try to determine which of two analytical methods gives the most accurate results from only one analysis of a reference standard by each analytical method. The one analysis ($n = 1$) for each of the two methods would be used to represent the mean ($p = 1$) for each method, leaving no other degrees of freedom ($v = n - p = 0$) for other comparisons (such as variance). Properly obtained sets of results for each analytical method might be statistically tested with a comfortable level of confidence. When testing for the statistical significance of a factor effect over noise, the ratio of the variances $(\frac{SS_f/(p-1)}{SS_r/(n-p)})$ is compared against standard F-test values. If the calculated ratio is larger than the appropriate distribution table value, the effect is considered statistically significant. Hence, the need for $\frac{SS_f}{p-1} \gg \frac{SS_r}{n-p}$.

Statistical results should be interpreted with caution. For example, a statistically significant correlation between two variables does not necessarily prove a cause and effect. The variables could merely be concomitant because they both depend on another common underlying factor.

The procedure described is the common parametric test procedure. Non-parametric statistics could be used when there is an appropriate reference set of historical data and the normal distribution assumptions are known to be false in the particular situation [9].

Experimental Design Strategies

After the experimental system is defined, the next step is to focus on which particular factors and responses are to be studied and why. Depending on what is known and what needs to be known about the system, researchers may be interested in screening for important factors, quantifying factors effects, comparing two or more methods (systems), optimizing a system's response, or gaining a formal understanding of the system's transform (e.g., through mathematical modeling).

A sensible strategy is to divide the study of a complex system into stages, with limited questions being answered at each stage. For the sake of efficiency, a practical sub-set of the known factors should be chosen as the object of study. If this cannot be done with the available information, a preliminary set of experiments can help identify what factors are important. Once that is known, other designs can be used to answer more fundamental questions, such as what is the best setting for the main factors or how does each factor influence the system's response.

The various types of research strategies described below often require or benefit from the use of statistical designs. A typical approach to a sequential SEDOP study is to screen the factors, looking for important effects (statistically significant), and then search for the optimum combination of the important factors [15].

Screening for important factors is often one of the first steps in understanding the conditions that determine the performance of a system. It is good practice to try to identify which factors are important and which factors are not. This will allow informed decisions to be made about which variables should be controlled or studied further and which can be neglected without affecting the system's performance. This approach saves time and cost. Identifying which factors contribute to imprecision is crucial to method improvement.

Modeling the system is necessary when a fundamental understanding of the underlying principles and relationships among the relevant factors is desired. Calibration relates specific instrument responses to the properties of interest (e.g., concentration). It is probably the most common type of modeling in the environmental laboratory. Calibration curves are usually linear relationships. While method development techniques are applied only at certain specific stages, calibration is used throughout the life of the method to assure the day-to-day accuracy of the assay.

Optimization is often involved in analytical method development (finding the factor setting that produce an optimal response of interest). Typically, optimization is attempted after the system has been well characterized by previous studies. An alternative, and possibly more efficient, approach is to try to optimize the conditions first and then determine which factors are of importance in the region of the optimum. If necessary, additional experiments can be carried out after the optimization to obtain a better understanding of how the system behaves in the region of the optimum [16].

Comparison of methods is utilized to decide between alternative analytical techniques or to establish their equivalence. Comparison of methods is common in method improvement or validation, and in interlaboratory studies. This type of comparative study is present at the first stages of method development to evaluate the reproducibility and accuracy of the candidate method. It is also useful to test the ruggedness and interlaboratory performance of the procedure in the final validation process, after the method has been developed and optimized.

Statistical Experimental Design Methodologies

There are well-documented experimental designs and optimization techniques that can enable the environmental researcher to achieve the goal of the study with efficiency and reliability. Some statistical designs provide quantitative measures of confidence in the results. Other techniques help place the experiments in the factor space in a way that will maximize the chances of obtaining meaningful results, facilitating the statistical, mathematical, or intuitive analysis of the results after the experiments are conducted.

Setting up a group of experiments (an experimental plan) involves selecting the design type according to the project's current goals and then choosing the specific factor levels (treatment combinations) for the experiments. Researchers could use standard design arrangements or develop custom designs based on the statistical theory of optimal design [17].

Design characteristics, applicability, and strategies for standard designs and optimal designs are presented here to guide researchers in the planning process. The omitted details for the application of, and numeric calculations for, each design can be found elsewhere in the literature and appropriate references are included with each design. Mastering the procedural details is less important today, since powerful computer software that takes care of the record keeping and calculations is becoming increasingly available. A summary of the strengths and weaknesses of each design is included to help researchers in the design selection process.

Most of these designs are orthogonal and balanced [18], simplifying the analysis through mathematical relationships and permitting the assessment of main factor effects (since there is an equal number of test runs under each level of each factor). Orthogonality simplifies the statistical analysis and assures that the factors are uncorrelated, rendering independent variances. Even though standard designs specify settings for the factors at each treatment combination, sometimes system limitations prevent the execution of experiments exactly as planned. Unfortunately, deviations from the prescribed setting for the factor levels or any missing experiments will destroy the orthogonality properties, precluding the standard analysis of the results and possibly reducing the value of the information obtained. If the deviations from the standard factor values are known, they can be taken into account at the time of data analysis to obtain results that are more representative of the system's behavior [19].

The standard experimental design types discussed here are calibration designs, comparison of means, randomized paired comparisons, Latin squares, full factorial designs, fractional factorial designs, saturated fractional factorial designs, Plackett-Burman designs, Box-Behnken designs, central composite designs, and simplex mixture designs. Although these designs have been proven to be very effective in chemical research, many of them have not been explored to their full potential in environmental chemical applications.

Comparison of Means

During method development, improvement, and modification, researchers are frequently faced with the task of deciding between two or more alternatives. For example, the decision might be between two different methods, or the decision might involve two or more different variations of the same procedure. Comparison of means (a completely randomized design) and block designs (see next section) are useful for statistical inference about the significance of presumed effects due to changes in experimental conditions.

Statistical inference can be used to determine the difference, or lack of difference, between two or more sets of data at a given level of confidence. These significance tests can be used as effective and reliable tools for comparing treatments.

For the comparison of two treatments (corresponding to sample data sets A and B), with means, $\overline{y}_A$ and $\overline{y}_B$, respectively, it is possible to establish if the difference between these means is statistically significant or might be attributed to random variations. The standard deviations of the data sets are assumed to be equal.

The null hypothesis states that there is no difference between the treatments, $H_0 : \overline{y}_A - \overline{y}_B = 0$. An acceptable level of confidence (usually 95%) is specified for the test. Disproving this null hypothesis strongly suggests that the alternative hypothesis, $H_1 : \overline{y}_A - \overline{y}_B \neq 0$, is true at the given level of confidence, and that there is strong evidence to deny that the treatments are equivalent.

To try to disprove the null hypothesis, it is necessary to determine the frequency that the observed difference between the averages would occur by chance alone. This is done by calculating the statistics for the samples and comparing their value to a reference distribution, usually by reading the appropriate entry in statistical distribution tables (e.g., Student's t-table). These tables also take into account the number of degrees of freedom in the sample (which depends on the number of observations) and the desired percent level of confidence in the result. The procedure described is the common parametric test procedure. Nonparametric statistics could be used when there is an appropriate reference set of historical data and the normal distribution assumptions are known to be false in the particular situation [9].

The Student's t distribution approximates the normal distribution very closely for a number of degrees of freedom larger than 15 [10]. For the comparison of two sample means, with the null hypothesis, $H_0 : \mu_A - \mu_B = 0$, and alternative hypothesis, $H_1 : \mu_A - \mu_B \neq 0$, the Student's t statistic can be calculated from the sample

$$t_o = \frac{\overline{y}_B - \overline{y}_A}{S_{AB}\sqrt{1/n_A + 1/n_B}} \tag{3}$$

where S_{AB} is the pooled standard deviation, estimate

$$S_{AB} = \sqrt{\frac{(n_A - 1)\ S_A^2 + (n_B - 1)\ S_B^2}{n_A + n_B - 2}}\ . \tag{4}$$

A Student's t table should be consulted to find at what level of confidence the test statistic, t_o, becomes larger than the tabular t value for the corresponding number of degrees of freedom $(n_A + n_B - 2)$. If the probability of the difference between the sample averages occurring by chance is large, then the alternative hypothesis cannot be accepted that there is no statistically significant difference between the treatments.

The chemical meaning of this finding (that there is no highly probable, statistically significant difference between the compared treatments) will depend on the experimental objectives. If a new method is being validated in comparison with one known to be good, then, under such findings, the new method cannot be assumed to be different from the reference method at the given level of confidence. The same holds true for the interlaboratory testing of a method, resulting in the acceptance of the procedure as rugged. In contrast, if the purpose of the experiments is to determine if a new detector is more sensitive than an existing one (i.e., the detector has a larger response for the same concentration of analyte), then it would be desirable to disprove the null hypothesis, thus demonstrating statistically that there is a difference between the two detectors. A larger number of experiments may be necessary or the noise (s_y) may need to be reduced to prove the hypothesis with greater confidence.

The experimental design consideration for the application of this methodology is that the statistical independence of the determination should be assured. The conditions for two treatments should be such that they differ only in the characteristics being compared, holding everything else constant. If the conditions of statistical independence and freedom from bias are not met, then we can try to overcome this difficulty by blocking and randomization.

Block Designs

Randomized Block

Sometimes in comparative experimentation there are factors that cannot be held constant for all of the experiments and are not included in the factors being studied. These factors are called background factors. If the sources of variation are known and controllable, then blocking can be used to assure that all of the factor levels are applied to all of the different experimental conditions. For example, in comparing two sample cleanup eluent mixtures for analysis of extractables in soil matrices, it might not be possible to assure that every batch of sorbant material will be the same. An experiment could be designed in such a way that the same number of runs (experiments) with each eluent is performed on each batch of sorbant. All of the runs made on a batch of sorbant represents a block of experiments.

Blocking can be a naturally occurring phenomenon. Time is a common blocking factor. Because runs performed close together in time are likely to be more similar when time trends are present, it is advisable to design experiments blocked in time (i.e., blocks performed consecutively) whenever possible.

The use of more than one instrument, batch, or technician, indicates a need for blocking, and the treatments within each block (experimental units) should be randomized.

Randomized Paired Comparisons

A commonly occurring blocking pattern is the use of blocks of pairs of runs performed in random order. In this case, for the comparison of two treatments, each block of two runs will contain one of each treatment. This design is known as a randomized paired comparison and is the smallest blocking plan. The analysis of the data is not performed in terms of comparing the statistics of the two treatments, but by comparing the statistics of the differences between paired runs to the reference distribution.

Randomized paired comparisons are applicable only to the situation of one factor with two levels. For more complex situations other blocking patterns might need to be employed.

Latin Squares

Latin square designs can estimate the effect of one factor while minimizing the effects of two interfering background factors. The Graeco-Latin square design is used to block against three background factors, and hyper-Graeco-Latin squares are used for blocking against more than three background factors.

Latin square designs are employed when there are two background variables (e.g., technician and sample type) which each have the same l number of levels as the number of treatments under study resulting in a $l \times l$ design. As is common with blocking designs, this is particularly applicable to categorical or qualitative factors (those with values that cannot be ordered, as opposed to quantitative factors, with numerically ordered values).

Latin square designs have the characteristic that the levels of the primary factor (treatments) are completely randomized with respect to the background variables, and each combination occurs once along each row and column. Figure 5 shows an example of a Latin square design. In this example, technician and sample type are background factors that are expected to be uncorrelated with each other. The effect of the method, though, is expected to depend heavily on

TECHNICIAN \ SAMPLE TYPE	1	2	3	4
1	*A*	*D*	*C*	*B*
2	*B*	*A*	*D*	*C*
3	*C*	*B*	*A*	*D*
4	*D*	*C*	*B*	*A*

METHODS: *A*, *B*, *C*, and *D*.

Fig. 5. Latin squares design for the comparison of four methods (A, B, C, and D). The unwanted effects of different technicians and sample types are separated by using them as blocks (four technicians and four sample types as background variables). Each method is run once with each technician and each sample type

sample type. The Latin square design and its analysis make possible the resolution of the method effect by itself. The analysis of Latin square experiments allows the calculation of a pure variance component for the primary factor or treatments, as well as for each of the background variables.

The randomized block design can only block a single background factor. Some designs are generalizations of the Latin square – Graeco-Latin squares and hyper-Graeco-Latin squares allow blocking of several background factors simultaneously. Other block designs, such as balanced incomplete block designs and Youden squares, accommodate for a number of treatments larger than the number of experimental units in each block [8].

Full Factorial Design

Block designs are useful in experiments in which many levels of a single factor are being used. For experiments involving several or many factors, usually at only two or three levels, factorial type designs are often preferred.

Full, or complete, factorial designs "are experiments which include all combinations of several different sets of treatments of factors" [20]. Factorial designs are commonly represented for ease of analysis by coding the levels of the factors (such as -1 for the low level and $+1$ for the high level in two-level designs). By far the most commonly used design is the two-level (2^k), followed by three-level (3^k) factorial. Figure 6 shows a two level, two factor (2^2), full factorial design overlaid on response contours. Compare this figure to the one-factor-at-a-time approach (Fig. 4). The 2^2 factorial covers more of the factor space, whereas the one-at-a-time experiment is stuck on a ridge. The one-factor-at-a-time procedure is not able to detect interactions because each factor is

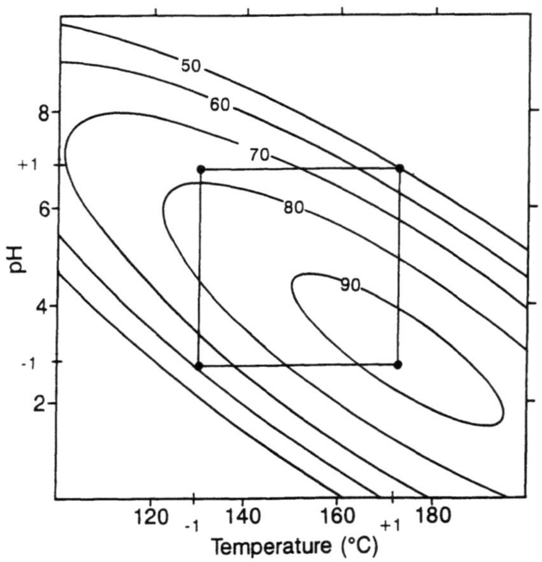

Fig. 6. Full factorial design for two factors at two levels. Each factor level appears in combination with all levels of the other factor

changed while holding everything else constant [21]. Even in the case of no interaction, there is much better ground for making conclusions about the main effect of a factor, since the effect has been observed in a variety of experimental conditions. Note also that the main characteristics of the response surface are more effectively represented with only four experiments for the 2^2 factorial design, whereas the one-factor-at-a-time approach may require many more experiments.

The experimental data can be represented by the equation

$$y_i = \boldsymbol{f}(x_i)\boldsymbol{B} + r_i \tag{5}$$

where y_i is the measured response from the i-th experimental run, x_i is a vector of predictor variables for the i-th run, $\boldsymbol{f}$ is a vector of p functions that model how the response depends on x_i, $\boldsymbol{B}$ is a vector of p unknown parameters, and r_i is the experimental error for the i-th run [8].

A linear model for the effects of two factors and their interaction would be

$$y_{1i} = \beta_1 x_{1i} + \beta_2 x_{2i} + \beta_{12} x_{1i} x_{2i} + r_{1i} \,. \tag{6}$$

Full factorials are arranged in such a way that all of the measurements contribute information about the effect of changing the factor levels. This is a major advantage of the full factorial design.

For quantitative variables, the parameters, in matrix form, $\hat{\boldsymbol{B}}$, may be estimated by the least squares method. The estimates of the parameters can thus be obtained through matrix operations on the design, $\boldsymbol{X}$, and response, $\boldsymbol{Y}$, matrices:

$$\hat{\boldsymbol{B}} = (\boldsymbol{X}'\boldsymbol{X})^{-1}(\boldsymbol{X}'\boldsymbol{Y}) \,. \tag{7}$$

This permits the estimation of the responses according to the model. An analysis of variance (ANOVA) will give information about the appropriateness of the model employed and the experimental precision.

If the residuals are uncorrelated and their variance is constant, then the variance-covariance matrix, $\boldsymbol{V}$, of the least square estimators, $\hat{\boldsymbol{B}}$, is

$$\boldsymbol{V} = var(\hat{\boldsymbol{B}}) = s_{\mathrm{r}}^2(\boldsymbol{X}'\boldsymbol{X})^{-1} \tag{8}$$

where $\boldsymbol{X}$ is the $n \times p$ matrix whose i-th row is $f(x_{\mathrm{i}})$ and s_{r}^2 is the variance of the residuals, or $SS_{\mathrm{r}}/(n-p)$, where SS_{r} is the sum-of-squares of the residuals and $(n-p)$ are the degrees of freedom.

The variance of the estimated response at x is

$$var(\hat{y}) = s^2 f(x)(\boldsymbol{X}'\boldsymbol{X})^{-1} f(x)\,. \tag{9}$$

Since both Eqs. (8) and (9) show that the experimental design depends on the $p \times p$ dispersion matrix, $(\boldsymbol{X}'\boldsymbol{X})^{-1}$, a good experimental design will make this matrix small, [8] usually by covering a broad domain of experimental conditions and by carrying out a reasonably large number of experiments [6]. To minimize the variance of residuals (make s_{r}^2 small), it is recommended to use appropriate mathematical models, and to carry out precise and repeatable experiments. The amount of variation due to lack of experimental reproducibility is called the purely experimental uncertainty (or pure error).

The statistical significance of the effects can be determined in several ways. A simple method is to use a Student's t test to evaluate the difference between the mean responses at the two levels of a given variable, even for qualitative (categorical) variables.

Statistical tests for significance must be interpreted with caution and within the context of the known practical aspects of the experimental system being studied. For example, sometimes a model seems to have both highly significant factor effects and a highly significant lack-of-fit [6]. This situation may arise when a model fits the data well and at the same time the experimental results are very reproducible. The latter condition might exaggerate the relative importance of any lack-of-fit (the test is based on a ratio between the variances for lack-of-fit and purely experimental uncertainty). In this case, the statistically significant lack-of-fit is of no practical significance and can be ignored.

Full factorials have the disadvantage that as the number of factors under study, k, at a number of levels, l, increases, the number of experimental runs, n, increases exponentially ($n = l^{k}$). For a higher number of factors, or for the preliminary screening of factor effects, fractional factorial designs are used.

The factor combinations in a full factorial are designed in such a way that information is obtained about the main effects, and all possible factor combination interactions. For a system with k factors, there are $k(k-1)/2$ possible two-factor (second order) interactions. However, in many applications, only the main effects are of interest, and, in chemical systems, interactions involving more than two factors are uncommon. As the number of factors or levels increases, the full factorial becomes inefficient in terms of the amount of useful information added per run [7].

Screening Designs

Fractional factorials, saturated fractional factorials, Plackett-Burman, and Taguchi are fractional designs, i.e., not every combination of the factor levels is included. They are used for screening significant factors. These partial designs do not have the degrees of freedom that would be required to assess the separate effects of higher-order interactions. When interactions are assumed to be negligible or are of no interest, the few degrees of freedom available are used to obtain information about the useful lower order effects. The idea is to design the experiment with fewer level combinations (experiments) such that some of the studied lower order and main effects will be confounded or confused with higher order effects. Because of this confounding, one cannot tell if the higher- or lower-ordered term contributes more to the given effect (thus, the higher- and lower-ordered terms are called aliases). However, since it is unlikely that higher-ordered terms contribute very much, the effect is usually attributed to the lower-ordered term of the aliased factors and their interactions. This is possible in chemical experimentation where the experimental error is usually relatively small. For disciplines like biology or agriculture, a large number of experiments may be required [22].

It should be noted that if the confounded interactions had an effect then there will be ambiguities and errors in interpreting the results (viz., discerning which alias caused the effect), which is the drawback of this otherwise more efficient design [20]. The ability to gather factor effect information in less runs is not gratis.

Fractional Factorial Designs

Fractional or "partial" factorials [8] constitute a general category that includes designs with different levels of fractionality. For example (see Fig. 7), a half-fractional factorial is equivalent to half of a complete factorial (two different half-fractional designs can be obtained from one complete factorial, each with half the number of experiments). A drawback for this more efficient design is that it prevents the possibility of discerning the main effects confounded with the factor interaction effects. To observe the interaction effects we would have to run the other half of the complete factorial.

Fractional factorial designs are generally designated by l^{k-p}, where l is the number of levels, k is the number of factors, and p denotes the degree of fractionation (p is the number of "generators") [10]. A two level fractional factorial, be represented as 2^{k-p}. Thus, 2^{-p} is the fraction of the full factorial, 2^k. For

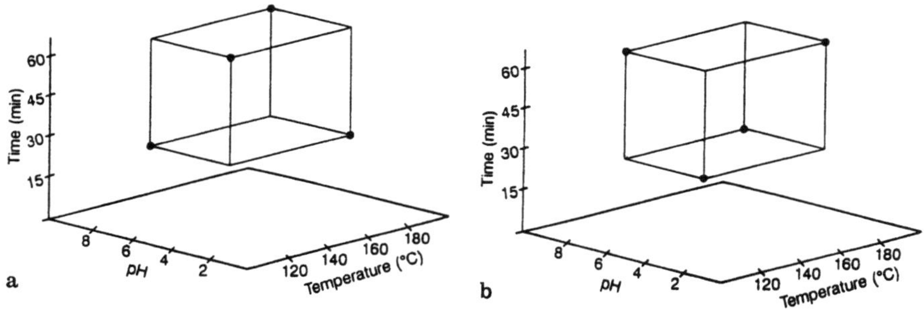

Fig. 7a, b. Possible fractional factorial designs for three factors at two levels (temperature, pH, and extraction time). Each design (**a** or **b**) is a half-fractional factorial (only half of the possible eight factor level combinations in a 2^3 full factorial are used). The treatment combinations included in the design are indicated by *dots*

example, a 2^{7-4} fractional factorial design is a $1/16$ (2^{-4}) fraction of the 2^7 full factorial design.

The statistical analysis of fractional factorial experiments is similar to the analysis of full factorial designs. Fractional factorials are recommended for quick and economical screening determinations like collaborative tests and method revision [21].

Saturated Fractional Factorial Designs

A fractional factorial design in which each degree of freedom is used to estimate an effect is said to be saturated. The saturated fractional factorial design is obtained by associating (confounding) every interaction with a factor [23]. This class of fractional factorials can be made to be very efficient (i.e., low proportion of experiments to factors), allowing for the exploration of the main effects of the k factors in $k + 1$ runs. This is useful when there are many variables to screen for effects (e.g., 15 variables can be screened in 16 experiments with a 2^{15-11} design).

A potential disadvantage of saturated designs is that they can be constructed only for cases where the number of factors, k, is equal to a power of two minus one (i.e., $k = 3, 7, 15, \ldots$). This type of design should be used only in situations where any interactions are known to be negligible, since the calculated effect for each of the added factors will be averaged with that of its confounded interactions and there is no way to separate the two effects without more experimentation. Screening studies are a good application for fractional designs, and the missing experiments can be run later to complete a full factorial, if needed.

Plackett-Burman Designs

Plackett-Burman designs were introduced in 1946 by R. Plackett and J. Burman [24]. They constitute a variation on saturated fractional designs. Again, k factors can be studied in $k + 1$ runs (only the main effects are estimated). These designs can be used only when $k + 1$ is a multiple of 4 (i.e., $k = 3, 7, 11, \ldots$). This enlarges the opportunities offered by the saturated factorial design, which it complements. The orthogonality property is a characteristic of Plackett-Burman designs.

Taguchi Designs

This is a group of designs, originally developed by Plackett and Burman [24] and widely applied by Taguchi [25] for process quality improvement, aimed at finding the combination of levels for controllable factors that will minimize the effects of noise factors and thus achieve a more rugged performance of the system. These designs are based on the concept of parameter design. Parameter design focuses on the noise portion of system behavior (i.e., minimization of the error variance). Noise factors, which are uncontrollable during system operation under typical conditions, can be included in the experimental design at controlled levels to find ways of minimizing their effects through the control of the other system factors. A goal of Taguchi designs is to maximize the signal-to-noise ratio (S/N), where the signal (S) represents the desired response and the noise (N) represents the uncontrollable variability. Taguchi designs are highly efficient and are as easy to analyze as saturated fractional factorial experiments.

Calibration Designs

Calibration designs relate an instrument response to a chemical property (or another relevant input) in a quantitative way, along with an estimated measurement of precision [26].

Fundamentally, calibration is a special application of curve fitting (modeling, when the fundamental function is known). It allows the determination of the parameters of a mathematical equation that will allow the prediction of the system's response for a given setting of the factor level. The fitted model is then used in an inverse way to predict the chemical property of the analyte from the responses [27].

Calibrations can be classified as direct, indirect, internal standard, and standard addition. The calibration design mode chosen for a given analysis will depend on the available knowledge about the system. Direct calibration is the common calibration method in which the model that relates responses to concentration is known approximately. When the model is unknown, it might be estimated statistically by least squares fitting (indirect calibration). When a drift of the system with time is known to exist, internal reference methods can be used to correct for the deviation. Finally, if matrix effects are present, the generalized standard addition method (GSAM) can serve as a tool to compensate for these matrix effects [28].

The final objective of calibration designs can be stated as making inferences about an unknown concentration vector, $\boldsymbol{X}$, from an observed response vector, $\boldsymbol{Y}$ [28]. The relationship between $\boldsymbol{Y}$ and $\boldsymbol{X}$ is calibrated with the experimental data, y_i, $x_i (i = 1, 2, \ldots, n)$, where i is one of a total of n experiments. This is the inverse of the more common situation of trying to predict $\boldsymbol{Y}$ from $\boldsymbol{X}$.

By far the most common model used in calibration is the straight line (first order linear model):

$$y_{1i} = \beta_0 + \beta_1 x_{1i} + r_{1i} \tag{10}$$

where β_0 and β_1 are model parameters (in this case, the straight-line intercept and slope, respectively).

Models are commonly fitted by the method of least squares. To obtain useful results, enough levels and replicates for degrees of freedom need to be included in the design. The degrees of freedom needed to determine the variance of residuals is $n - p$, the degrees of freedom needed to determine the variance due to lack of fit is $f - p$, and the degrees of freedom needed to determine the variance due to purely experimental uncertainty is $n - f$. Therefore, it is desirable to have $n > f > p$. A rule of thumb is to have at least three degrees of freedom for determining each of the variances, requiring that $f \geq p + 3$ and $n \geq f + 3$. For the model described by Eq. (10), which contains two parameters ($p = 2$), it is recommended that at least five distinctly different factor combinations (levels) be used ($f = p + 3$), with at least three of those combinations being replicated, for a total of no less than eight ($n = f + 3$) experiments.

If a model is proven inadequate (i.e., non-statistically significant estimated parameters), the data can be fitted to other models in search of a proper fit. Again, as more parameters are added to the model, more degrees of freedom will be required.

Response Surface Modeling Designs

The quantitation of the effects of factors and their interactions that can be obtained from factorial designs is very useful information. Many times, however, the analyst is interested in getting a more fundamental idea of how the system under study works. Building an appropriate mathematical model to describe the system's behavior and determining its parameters will provide this deeper understanding. Such a model predicts responses for untried treatment combinations within the explored region. Another dimension, and probably the most important for the chemist, is the ability to drive the system to its optimal conditions by calculating the treatment combination that would maximize (or minimize) the model response. The goal is to estimate the form of the system's response by a function. When plotted (on two factors) that function may look something like Fig. 2.

The set of techniques that allows us to explore and predict the system's response through a mathematical model is called "response surface methodology" [29, 30]. Response surface designs are used to generate the necessary data to estimate the parameters in a model similar to Eq. (11) or a sub-set of it [13], which is an example of a full "second order" model for two factors:

$$y_{1i} = \beta_{11}x_{1i}^2 + \beta_{22}x_{2i}^2 + \beta_1 x_{1i} + \beta_2 x_{2i} + \beta_{12}x_{1i}x_{2i} + r_{1i} \,. \tag{11}$$

The linear model is fitted to the data by a least squares method in most cases. The model should be linear for the parameters, but factors can have higher-order terms to model curvature and interaction. The variables involved have to be quantitative and continuous.

Full factorial, fractional factorial, Box-Behnken, and central composite designs fall into the general category of response-surface designs. Rubin et al. [20] recommend the use of response surface designs after the number of factors to be tested has been reduced and there is some idea of the area where the optimum values for those factors is located.

Central Composite Designs

Central composite designs are built out of the corresponding full factorial design (levels low and high) with a larger superimposed star design, giving it a spherical shape (Fig. 8). The star design is made out of two axial treatment combinations at the lowest and highest levels for each factor, keeping all of the others fixed at the mid-point, plus a center point experimental unit at that middle level for all factors.

The levels of the axial experimental points are set at the same distance from the center point as the inner points. For three factors, the inner points and the axial points are on the surface of a sphere. This can give central composite designs the properties of rotatability [18].

It is advisable to use a central composite design after running a factorial design when no major effects were indicated or a better understanding is desired [20]. A factorial design could have indicated a lack of factor effects if it was positioned in an area of the response surface with a relatively high curvature. The central composite design could be easily completed from the already performed factorial design by carrying out the missing experiments. One use of central composite designs is to determine relationships in the region of the optimum, usually after screening out the unimportant variables.

The statistical analysis of central composite designs is more elaborate than the analysis of factorials. On the other hand, the central composite is overall a

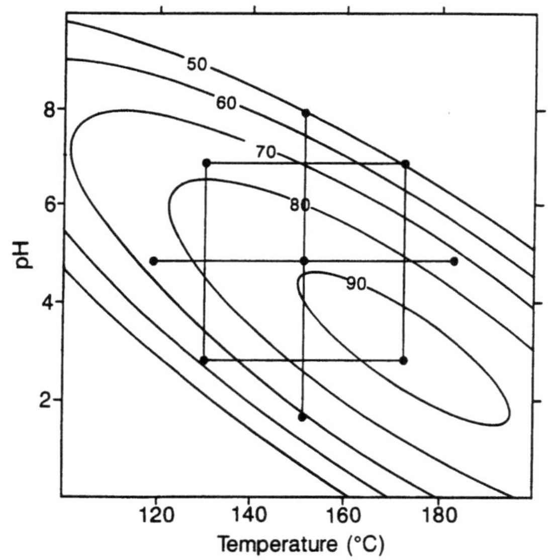

Fig. 8. Central composite design on two factors. Each factor appears at five levels, which provides for the accurate determination of curvature. The full factorial pattern (*center four points*) allows for the determination of interaction effects

very efficient design and its analysis allows one to detect any inappropriateness in the model being tested (lack-of-fit). The central composite design gives a better estimation of the model parameters than do the factorial designs and provides a check for variance constancy [13]. The full second order model (including quadratic and interaction terms) can be fitted with the number of degrees of freedom provided by the central composite treatment combinations [7].

Box-Behnken Designs

Full factorials become inefficient not only with an increase in the number of factors, but also as the number of levels employed increases. Box-Behnken is a type of fractional design that allows the use of three levels for designs with more than two factors [31].

Box-Behnken designs are constituted by an assembly of sub-sets of 2^2 factorials for each possible pair of factors with all of the other factors held constant at the midpoint level, plus a center point which is usually replicated (see Fig. 9).

The need for three, and higher, level designs arises when the analyst wants to determine the effect of quadratic, and higher-ordered, terms in a linear model similar to the one given below for two factors (extendable to any number of factors):

$$y_{1i} = \beta_{11}x_{2i}^2 + \beta_1 x_{1i} + \beta_2 x_{2i} + r_{1i} \ . \qquad (12)$$

These quadratic terms account for the curvature of the response, i.e., when the effect of a factor causes an increase and then a decrease (or vice versa) in the

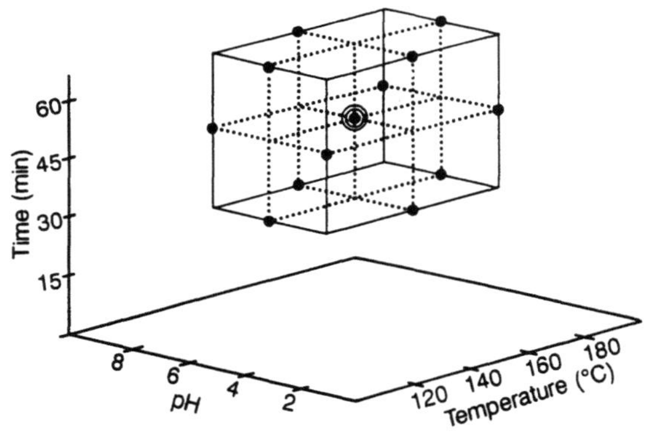

Fig. 9. Box-Behnken design on three factors (temperature, pH, and extraction time). Each factor is at three levels (a center point experiment with three replicates is included)

response. This type of effect cannot be picked up from a two-point (implying a straight line) determination, as in two-level designs. Box-Behnken designs provide information on the quadratic terms, as well as on the main effects, and on the interaction terms.

Simplex Mixture Designs

Mixture designs apply to experimental situations in which the response depends on the relative amounts of the variables, but not on the absolute amount of each [8]. They are used to evaluate formulations and to determine the effects of the different components in them. This is particularly true of many chemical systems whose behavior depends on the mole fractions, or proportions, of the components.

If $X_i, \ldots, X_k$ denotes the k factors, measured as proportions (ratios), then for each experimental run we have $0 \leq x_j \leq 1$ and $\sum_{j=1}^{k} x_j = 1$ for all $j = 1, \ldots, k$.

Given this constraint, for three factors, the feasible region for experimentation is a triangle (see Fig. 10a). In general, such a region is called a $(k-1)$-dimensional simplex, where k is the number of factors [13]. Figure 10b shows an example of a mixture experimental design.

Scheffé [32] introduced a family of models for mixture designs and proposed the class of lattice designs. In this arrangement the runs are placed in a uniform lattice of points around the mixture simplex, allowing one to explore the response throughout the whole simplex design.

In many mixture experiments, some or all of the components are subject to additional constraints, either independently (e.g., minimum or maximum concentrations), or in combination (e.g., mixing ratios). These added constraints reduce the feasible factor space within which it may be not possible to set up a regular lattice. A design approach to this type of system is to make runs at the extreme points and centroids of the constrained region (extreme vertices designs).

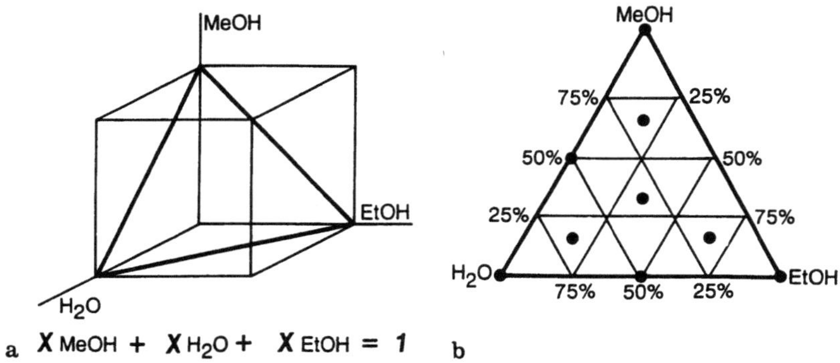

Fig. 10. **a** Factor space plane for a three-component mixture. **b** Simplex mixture design for the formulation of a mixture of water, methanol, and ethanol. The proportions of the mixture components add up to unity

Optimal Designs

Optimal designs are statistical experimental designs tailored to offer the best performance with respect to a given criterion [17]. The design could be optimal with respect to the minimization of variance (and therefore the uncertainty of the results), the setting of factor levels, the minimization of experimental variance, or the minimization of the number of experimental runs to achieve a given confidence.

The experimental design affects the variance-covariance matrix and the precision of the parameter estimates. For example, for a straight-line model, $y_{1i} = \beta_0 + \beta_1 x_{1i} + r_{1i}$, the most precise estimate of β_0 in a two-level design is obtained when the two experiments are centered around $x_1 = 0$. The factor level setting that will maximize the precision of the β_1 parameter estimates is that with the two experiments placed as far apart as possible. In most real situations, practical boundaries impose a limit in the spread of factor levels [6].

Experimental designs generally seek to cover as much factor space as possible (e.g., maximizing $|X'X|$) while still being in a workable and modelable range of factor space. These two aspects are complements and must be kept in balance; e.g., if $|X'X|^{-1}$ is made small, another level may have to be added to the treatment combinations to get the higher order effects in the model.

Response Optimization Designs

Optimization brings a process to its peak of performance within the standards set as acceptable. The property taken as the optimization indicator depends on the objectives of the study. For example, in gas chromatography we could try to optimize the overall resolution of the chromatogram by using a desirability function [33], or we could optimize the separation of only two analytes of interest, or we could attempt to minimize the time required for an analysis without failing to meet some minimum performance requirement. An optimum is the combination of factor settings that yields a response (maximal or minimal) exceeding response values in its vicinity. Systems can have more than one optimum, with a global optimum and local optima. A graphical representation of the optimization process is that of searching for the "mound" of a response surface, such as the one depicted in Fig. 2. In actuality, each system will have its characteristic response surface, which could include minima, ridges, and other topographical features.

In this section we discuss only response surface methods for optimization and not other techniques such as linear programming.

Response surface modeling can be applied to system optimization. If a second-order (or higher) model has been successfully fitted to the experimental responses, the location of the optimum can be determined mathematically. Canonical analysis [10] can be applied to simplify the higher-order equations that describe the response surface, eliminating first-order and interaction terms. This procedure involves setting the derivatives with respect to each factor to zero. A simpler interpretation of the response surface shape and the effect of individual factors can then be made. The resulting simplified equations indicate the location of the stationary point.

Evolutionary Operations

Evolutionary operation (EVOP) [34] is an optimization method based on carrying out a series of factorial designs moving in the direction of the calculated improvement (steepest ascent up the mound). Figure 11 shows the progress of a box-type EVOP on a response surface. The size of the factorial design can be systematically reduced to achieve better precision; however, doing this could reduce the chances of ever getting to the optimum. For a two-factor situation, a typical experimental pattern for each iteration is a two-level factorial (2^2) with a center point (five treatment combinations). Additional runs as replicates of the center point could be added for the assessment of precision.

Steepest Ascent

The technique of steepest ascent (or descent) moves a pattern of experiments in the direction of the optimum response (largest slope) in sequential, iterative sets of experiments. The direction on the response surface in which the experimental pattern will be moved for each iteration can be determined by finding the local portion of a response isopleth (contour line). The direction perpendicular to

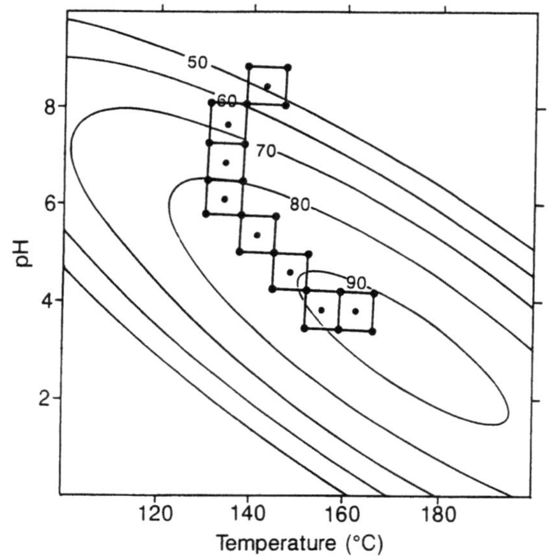

Fig. 11. Progress of a box-type evolutionary operations optimization study. After the first five-point factorial (*with center point*), each subsequent iteration is placed in the direction of the largest effect

the isopleth will indicate the general direction of maximal (or minimal) gain in response. Experimental runs are continued along the path of steepest ascent (or descent) until progress stops (no change in response) or factor interactions and curvature start having a pronounced effect. This approach has a better chance of success when previous studies have revealed that the response fits a linear first-order model reasonably well. Interaction and curvature (second-order effects) may prevent the experiment iterations from following the most efficient path in the response surface or produce erratic results (e.g., large residuals) [35].

The steepest ascent or modeling approaches to optimization are not very appropriate for applications where the cost of the potentially large number of experiments could be prohibitive.

Sequential Simplex

The sequential simplex was originally introduced as a technique to optimize industrial manufacturing processes [36]. Since then, this technique has been extended to general system optimization. Simplex optimization has had wide utilization in analytical chemistry experimentation [37, 38].

A simplex is a geometric figure in hyperspace with one more vertex than the number of dimensions in the factor space. Thus, each vertex represents a different treatment (different factor levels). This geometric figure has the property of being transformed into a new simplex by the replacement of only one vertex. Use of the sequential simplex does not require building and fitting a mathematical model of the response surface.

The initial simplex for k factors takes only $k + 1$ experiments, and only one more experiment at a time is necessary to move across the response surface (see Fig. 12). This makes it a very efficient technique in terms of the number of experiments required.

The basic progress mechanism to find the optimal response for the fixed size simplex (see Fig. 13a) is to "reflect" itself into a potentially better response area. The first step is to evaluate the response at all of the initial vertexes and identify the best (B), next-to-the-best (N), and the worst (W) responses. For a fixed-size simplex, the next step is to eliminate the worst response vertex and reflect it into the new vertex (R) across the remaining face's centroid ($\overline{P}$) to create the new simplex. The reflection step is repeated to advance along the response surface.

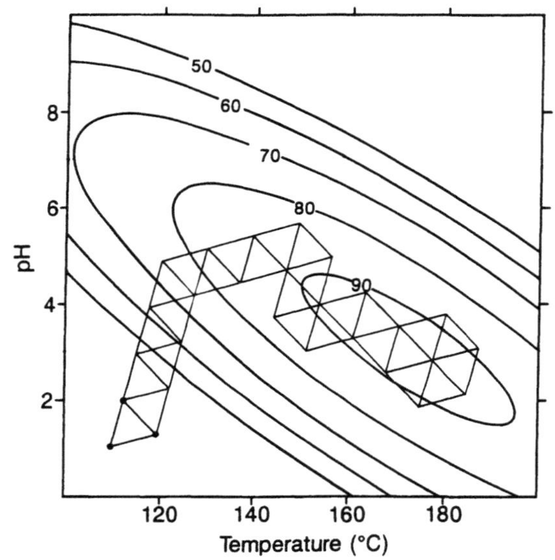

Fig. 12. Example of a two-dimensional fixed-size simplex progressing on a response surface toward the optimum. After the first three-point simplex, each subsequent iteration is accomplished with one additional experiment

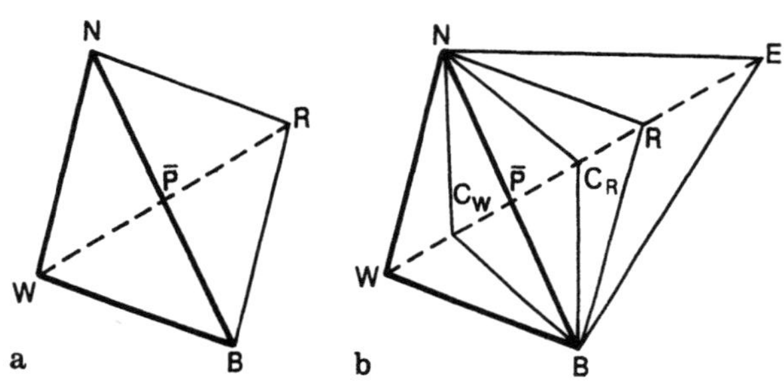

Fig. 13. a Fixed-size reflection (R) through the centroid of the opposing face to the worst-response vertex (W), $\overline{P} \cdot B$ is the best response vertex and N is the next-to-the-worst vertex response. **b** Variable-size simplex movements: reflection (R), contraction toward worse-response vertex (C_W), contraction toward reflection vertex (C_R), and expansion (E)

The responses have to be at least rankable, but not necessarily quantitative, in nature. For the simplex to progress well the process must be in statistical control (to be able to see factor effects over the noise).

The variable-size simplex, introduced as a modification of the original algorithm [39], can make faster progress towards an optimum than the fixed-size simplex. In this version, the simplex has an option to move in one of four different ways (see Fig. 13b), and each move is still done by replacing the worst vertex by a new one through its opposed face's centroid ($\overline{P}$). The response corresponding to the reflection is evaluated before every move and if it falls between the current best (B) and the next-to-the-best (N) vertexes it is taken as the new vertex. When the new (R) response is better than the prior best, an expansion across a double distance is evaluated. If the new response, R, is still better than B, the expansion (E) would become the new vertex, enlarging the size of the simplex and speeding the progress.

For new R vertexes with a response less optimal than the current N response the simplex will contract by half the distance, toward W if the R response was worst than the response at W (C_W contraction), or toward R if it falls between the W and N responses (C_R contraction); this is particularly useful in the region of the optimum, where the simplex collapses onto the local maximum (or minimum) and pinpoints the optimum with precision (see Fig. 14). In many cases it is profitable to start with a variable-size simplex as big as possible and let it *collapse* into the optimum.

Fitting non-linear models is another application of the sequential simplex method, where the parameter values that optimize the fit are sought. Thus, the variables for the simplex are the parameters of the model and the response is the

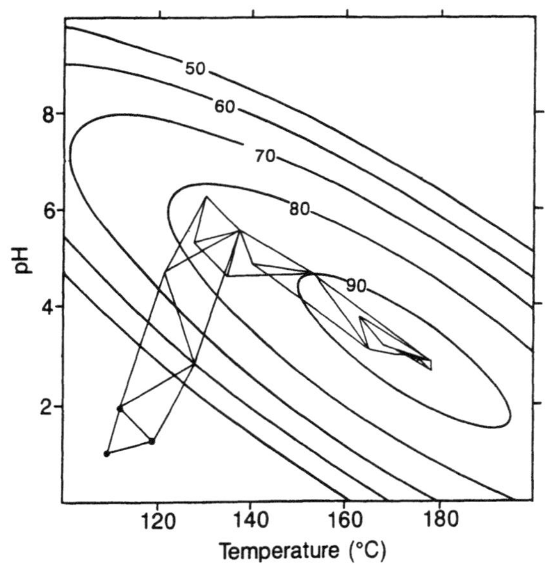

Fig. 14. Example of a two-dimensional variable-size simplex progressing on a response surface toward the optimum. The size of the simplex can adjust with each iteration for more efficient progress and a more accurate determination of the optimum

sums of squares of the residuals (SS_r). The goal of the simplex is to locate a set of parameter estimates that give minimum residuals (i.e., the best fit) [40].

Environmental Applications of SEDOP

At present, few applications of SEDOP to the field of environmental science can be found in the literature, as was the case for the field of chemistry a few years ago, before the methods and benefits became better known [18].

The United States Environmental Protection Agency (USEPA), international entities, and private industry develop, evaluate, and implement analytical procedures for environmental pollutant analysis. Some procedures are officially approved by government regulating entities for the analysis of specific hazardous substances. Since method development, validation, and approval is a long and costly process (and official method changes require considerable effort), it is important that method development and evaluation be done in an efficient and reliable manner.

Because of space constraints, the details for the following applications of SEDOP are not given; however, those details may be found in the literature references provided. Such references may serve as useful templates for similar studies under consideration.

Optimization of Analyte Extraction Procedures

Optimization of a Supercritical Fluid Extraction Procedure

A multi-phase experimental design approach was implemented by Ho and Tang [41] to optimize the supercritical fluid extraction (SFE) of 29 environmental pollutants from a liquid-solid extraction cartridge. Compounds included polynuclear aromatic hydrocarbons (PAHs) and organochlorine pesticides. These compounds are representative of larger classes of compounds of environmental concern. A 2^3 factorial experimental design was initially performed to study the relative importance of three SFE variables: pressure, temperature, and extraction time. Pressure was found to have a statistically significant effect on the recovery of all of the compounds studied. The next most influential variable was found to be extraction time (with a statistically significant effect on the recovery of only some of the PAHs). Temperature was the least influential factor. Following the factorial design, a variable-size sequential simplex was used to optimize the SFE conditions to obtain the best overall recoveries of the compounds studied. For those analytes still demonstrating low recoveries in pure supercritical carbon dioxide after the optimization, the addition of methanol as a modifier afforded quantitative recoveries.

Assessment of the Effects of Premeasured and Mixed Acid Composition in Microwave Digestion Recoveries

Experimental designs to assess the effects of pressure and mixed acids composition on antimony (Sb) and barium (Ba) recoveries during microwave digestion were performed [42]. A 2^3 full factorial and a mixture design were used. For two acids (HCl and HNO_3) the experiments were performed in a constrained mixture space: the total acid solution volume (including the two acids and water) could not be greater than 10 ml. Experimental data based on those designs revealed that, in all cases, high acid concentrations (either HCl and/or HNO_3) resulted in enhanced recoveries. Pressure did not seem to be a factor for Sb recovery but did seem to be important for Ba recovery.

A Paired-Comparison Design for Solvent Extraction Glassware

A randomized paired-comparison design was used to assess and compare the accuracy and precision of two types of solvent extraction glassware for semi-volatile pollutants (including pesticides) [40]. The objective was to determine if a new extraction apparatus performed better than the one in current use. Splits for six samples a day (spiked with eleven indicator analytes) were extracted using each apparatus and analyzed during six different days. Analyses were performed as close as possible in time by gas chromatography/mass spectrometry (GC/MS) for each extract pair. For accuracy, the differences in experimental percent recovery for each analyte were computed and a t-test at the 95% confidence level was performed to ascertain if the observed differences were simply due to sampling fluctuations. For precision, percent relative standard deviations were computed for the data obtained from each type of glassware. There was a risk that the analyses performed during a day were correlated with one another and did not represent truly independent samples in the way expected when computing the standard deviation. The magnitude of analytical variability, which could mask any extraction differences, was assessed by calculating intra-analysis-day and total variance components. It was found that the intra- and inter-analysis-day variances were comparable. According to the study, the new extraction glassware produced statistically significant lower recoveries and higher recovery variability than the current glassware for all but one of the analytes tested. There was also some indication of lower day-to-day reproducibility with the new apparatus for some analytes, judging by the higher degree of differences between the intra-day and the total variability for this type of glassware.

Instrument Optimization

Sequential simplex optimization was applied to a particle beam liquid chromatograph/mass spectrometer (LC/MS). This instrument has unique advantages for the analysis of non-volatile pollutants, such as chlorinated phenoxy acid herbicides and PAHs with high molecular weights. The optimization study focused on the factors that affect the performance of the particle beam interface. In this section of the instrument the mobile phase is desolvated and a particle beam of the analyte is directed into the ion source of the mass spectrometer. The response to be maximized was total ion current. The factors studied were capillary distance relative to the nebulizer orifice, desolvation chamber temperature, probe distance to the source, and helium flow rate into the nebulizer. A variable-size simplex was used for the optimization. The study was concluded after the response ceased to increase significantly for several iterations. Best response was considered to be the optimum. Plots of the response vs the individual factor settings showed that only one of the four factors (capillary distance) had a defined optimal level, apparently without interaction with the other factors. This condition caused the simplex to wander about a local optimum once the capillary distance factor had been optimized. The effects of the other three factors in the response (relative to noise) were not significant enough to allow the simplex to find and follow slopes of response improvement on the response surface based on those factors.

Calibration

Because temperature is one of the most important factors in microwave digestion for the analysis of inorganic analytes, the functional relationship between microwave power setting (x_i) and actual watts delivered (y_i) must be known for calibration purposes. As a result of improved control over digestion conditions, the intralaboratory and interlaboratory variability of inorganic analysis should be reduced, which is particularly important for environmental trace metal analysis. Given a known relationship, the desired watts can be delivered to a sample being digested by the proper adjustment of the microwave setting either manually or by computer (robotic) control. Several studies have applied experimental design techniques to the characterization of microwave oven power calibration. It has been found that each individual microwave digestion instrument may present different power calibration characteristics [43].

An experimental design [40] included two different calibration experiments at thirteen levels, each with a several days interval between each calibration to determine long-term repeatability and trends. Completely randomized triplicates were used to estimate purely experimental uncertainty (short-term repeatability) for a total of 39 experiments for each calibration. This many data points provided more than enough degrees of freedom for the statistical testing of model lack-of-fit. A third calibration set at eight new levels (in triplicates) was carried out for confirmation of the results.

Several calibration models were fitted to the microwave data, including the straight-line model not constrained to go through the origin, $y_{1i} = \beta_0 + \beta_1 x_{1i} + r_{1i}$, the constrained straight-line model, $y_{1i} = \beta_1 x_{1i} + r_{1i}$, and the second-order model, $y_{1i} = \beta_0 + \beta_1 x_{1i} + \beta_{11} x_{1i}^2 + r_{1i}$, using the method of least squares. A straight line calibration curve relating watts delivered to microwave power setting was found to have a good fit for the three data sets. In statistical terms, the F-ratio for the lack-of-fit was relatively small and not very significant, failing to show that the lack of fit was real. Even though a straight line calibration curve constrained to go through the origin fitted as well as the unconstrained straight line for the resulting data set (the β_1 term was significant, but β_0 was not and the goodness of fit was comparable for both models), it was recommended to use the constrained line form. This was because the uncertainties of the constrained straight line were assumed to be proportional to the response; the data did not seem to support this assumption. The uncertainties of the unconstrained straight line were more constant across the domain of the factor space, and probably better represented the situation investigated in these studies. It was found that for some microwave ovens the region of higher power setting (90–100%) followed a straight line with a different slope than the lower power range, and therefore two calibration curves were required. This phenomenon, identified through the calibration design, was determined to be due to two separate electronic circuits controlling the microwave power.

Figure 15 shows the characteristics of the data, including the fitted line and confidence intervals. The repeatability of the actual watts delivered (to assess the

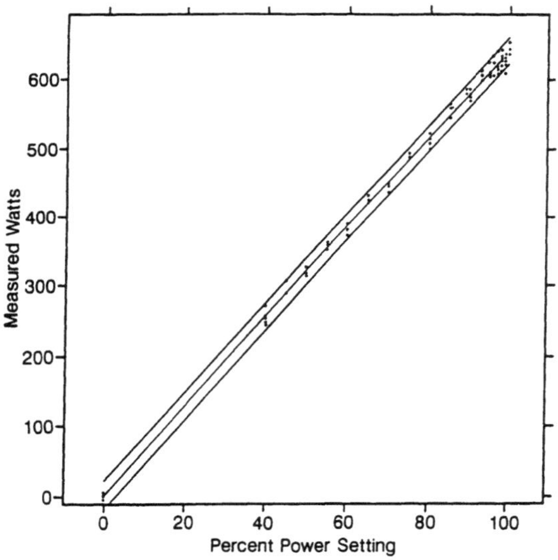

Fig. 15. Calibration curve (straight-line model unconstrained through the origin) with 95% confidence (prediction) intervals for microwave power setting experiments [40]

ruggedness of associated analytical methods) was determined as the maximum confidence (prediction) interval in the estimated response over the domain of the experiment for one new measurement of power at a single factor level (it was found to be ±20 W over a range of 0 to 600 W).

Ruggedness Testing

An approach to ruggedness (tolerance) testing for a method is to run experiments over a range of levels expected to be encountered during actual method use for factors suspected of having an effect on response variability. The response for a rugged (robust) method does not vary that much with changes in the factors of interest. A ruggedness study for microwave digestion power settings was conducted to determine the effect of variations in the two power settings required for sample digestion using EPA Method 3015 metal analysis in a water matrix by inductively-coupled plasma/atomic emission spectroscopy [40]. Three-level, two-factor (3^2) full factorial experimental designs were employed, with the two power settings as factors. Separate sets of experiments were run for unlined microwave digestion vessels and lined vessels [42]. One of the samples in the study was split into three additional aliquots to give an estimate of the reproducibility of the system. Since the elements analyzed for were at the parts-per-billion level (environmental trace elements), the relationship between the reported concentration and the digestion power settings was not modeled. Variability at this low concentration level is primarily due to noise [44, 45]. Instead, various second order models were fitted to nine relative cumulative response values for each of the treatment combinations. The fact that none of the models tried had a good fit indicated that any existing variation was not controlled by the tested factors for the unlined-vessel experiments. Therefore, the method was rugged with respect to the power settings over the range investigated. For lined-vessels, second-order models showed a good fit, revealing that there was an effect of power setting on recovery. Thus, the method was found to be sensitive to the type of digestion vessel used in the microwave portion of the analytical procedure. A similar evaluation of EPA Method 3051 (soil matrix) showed that the method was not rugged over the range of microwave digestion power settings investigated for both the lined and unlined vessels [42].

Single-factor ruggedness testing (one microwave digestion power setting) was conducted for EPA Method 3050 [40]. In this case a one-factor calibration design had to be used. Model fitting revealed that there was no statistically significant correlation between the measured element concentration and the power setting, but plots of the data revealed that there was up to a 20% change in element concentration as the microwave digestion power was varied by ±10%. The method was deemed to be not rugged over the microwave power settings investigated.

Method Comparison

The high cost of laboratory analysis of environmental samples and the need for faster decision-making concerning pollution remediation has raised interest in field analytical methods. Since laboratory methods have been established, characterized, and accepted for most pollutants of interest, new field methods and instruments are frequently evaluated through a comparison with existing laboratory methods.

Gas Chromatograph/Mass Spectrometer Comparison

A mobile analyzer was evaluated as an alternative to laboratory-based technologies currently employed in the characterization and cleanup of hazardous waste sites [46,47]. The mobile instrument consisted of a portable mass spectrometer optionally coupled to a portable gas chromatograph (GC) for the analysis of pollutants in soil, water, and air. Natural soil samples contaminated with PAHs and polychlorinated biphenyls (PCBs) and surface water samples spiked with selected volatile organic compounds were collected and analyzed on-site. Sample replicates were analyzed by the mobile GC/MS instrument and by confirmatory laboratories using standard USEPA methodologies. This allowed the comparison of data acquired using both methodologies.

Two experimental designs were carried out to investigate the comparability of the methods. In experiment 1, duplicate standard reference materials (SRMs) for PAHs and PCBs were analyzed each day for 15 consecutive days by both the mobile instrument and the confirmatory laboratory. In experiment 2, seven replicates of each sample were analyzed by both the mobile instrument and the confirmatory laboratory each day with a total of five samples for each sample type. This experiment was designed to provide information on the inter-method variability for each sample. A series of t-tests and F tests were used to analyze the experimental results.

For experiment 1, a pooled t-test for the equality of the means between the two laboratories and an F test for equality of between-batch precision for the two laboratories were used. For experiment 2, separate t-tests were performed for each analyte. The type of t-test used depended on the result of the F test. When the variances between the laboratories were equivalent to a statistically significant confidence level, a standard t-test which assumes equal variance was used. When the variances were statistically not equivalent, a t-test accommodating unequal variance was used [48]. The null hypothesis of equal means was not rejected for ten cases (batch-analyte combinations). Comparisons were not possible for all of the analytes due to a lack of data points above the detection limit, as often happens with natural environmental samples.

The results were mixed, with each method having significantly greater mean concentration values for some analytes (26 cases for the mobile instrument and

33 cases for the confirmatory laboratory). In general, it was concluded that the mobile instrument data was comparable to the confirmatory method data, although the mobile environmental monitor data quality was poorer (lower between-batch precision).

Field Portable Gas Chromatograph Comparison

The performance of a field portable, multi-port GC for the analysis of volatile organic compounds was compared to EPA Method 502.2 – "Volatile Organic Compounds in Water by Purge and Trap Capillary Column Gas Chromatography with Photoionization and Electrolytic Conductivity Detectors in Series" [49]. A multiphase experimental design was conducted. Test samples were set up for paired analysis to study various aspects of the field analysis. A set of samples was measured by portable GC and also sent out for laboratory analysis by Method 502.2, while another set was prepared in the field and splits were measured at both locations. A third set of samples was used to determine differences between two field analysis modes (on-line and off-line) at the site's pump and treatment plant. A final set of samples was measured before and after transportation to determine the combined effects of time and transportation. A total of 411 paired analyses were performed. Measurements were converted to percent recoveries (%R) so that the results could be expressed on the same scale. The differences between the sample pairs were assessed by paired *t*-tests with a hypothesis of no difference in mean %R. The major advantages of the paired comparison *t*-test are simplicity and ease of interpretation. The paired-sample *t*-test amounts to blocking all other effects, and focusing only on the one effect of interest. Robustness is attained at the cost of degrees of freedom. An analysis of variance (ANOVA) was used to evaluate sample data with respect to the time of sample collection, the day-to-day variability, the mode of analysis (on-line/off-line, field/laboratory), and the GC sample port.

The results were also analyzed by fitting a straight-line model to each set of results and calculating the ratios between pairs of results for each sample (by analyte). The cost of replicates was prohibitive, which precluded more sophisticated statistical analysis. Complete equivalency between analytical results would be evidenced by a straight line with the slope equal to unity or the ratios equal to unity. To discern the effect of time, the ratios were plotted against the time between the corresponding analyses, which was expected to reveal any time trends in the differences. It was found that the field and laboratory results were consistent (although a proportional interlaboratory bias existed) for the analytes in the study. Analysis, both at the laboratory and at the field, under different conditions showed excellent intralaboratory agreement. Time and/or transportation effects appeared to be minimal for the samples investigated.

Selection of Indicator Compounds

Tosato et al. [50] predicted that the aquatic toxicities of 100 mono-substituted benzene compounds could be represented by the experimental values of eight compounds. The critical compounds were identified using a factorial design. This is a strategy for assessing and ranking the hazards of chemicals for which little or no data on their toxic effects are available. Missing data for chemicals were predicted on the basis of test data generated for a minimum number of specific compounds that were adequate representatives of the relevant series. Modeling of the data for the prediction was done by partial lest squares (PLS) [51], a multivariate modeling algorithm. The reliability of the predictions was verified by comparing predicted and experimental toxicities of additional compounds.

Pollution Prevention

Statistical designs have also been applied to the reduction of environmental pollution from chemical processes. An industrial process was optimized for increased yield and reduced by-products using a three-level factorial-type design [52]. Five factors were investigated, including the relative proportion of two reagents, the relative proportion of solvent, the temperature of the reactor, and the reaction time. Regression models [10] and PLS were used to analyze the data. The total waste reduction exceeded 90%, while at the same time the economics improved due to an increased throughput, less severe reaction conditions, and an increased yield of reaction product. Previous attempts to improve the process with a one-factor-at-a-time approach had failed to produce sufficient improvements.

Guidelines for the Application of Statistical Experimental Design

The following gives some guidelines for a general approach to the SEDOP process, which is summarized as a flow chart in Fig. 16. Experimental designs for environmental chemical applications should take into account all factors and sources of variability of interest. Major relevant effects could include matrix effects, sub-sampling effects, sample preparation effects, sample analysis effects, and data treatment effects. Field studies have the additional effects of sample collection, preservation, transport, and holding time. Spiked samples and SRMs are affected by spiking procedures. The stability of the latter should be always taken into account when they are used in a study.

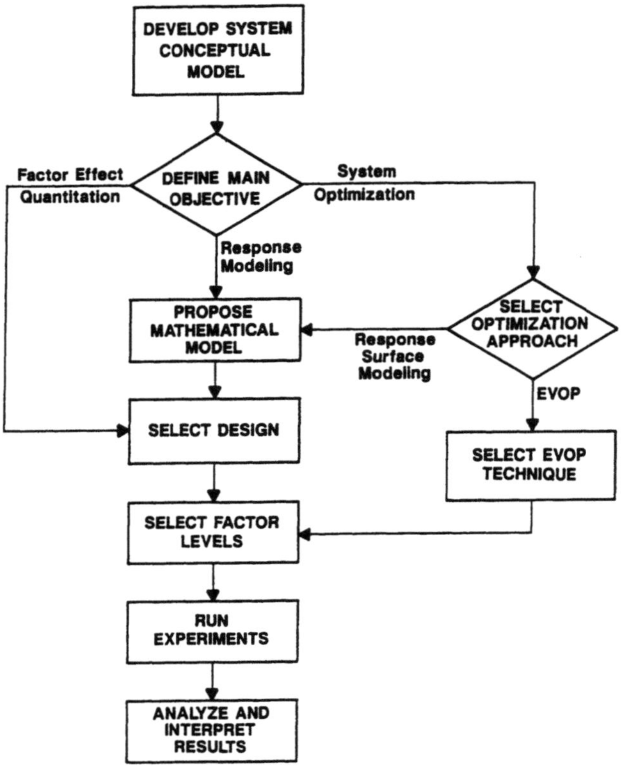

Fig. 16. A general step-wise approach to statistical experimental design and optimization. The process might be repeated iteratively during a study as sub-goals are attained and new questions emerge based on the information being gathered

General Approach and Implementation Considerations of SEDOP

The time to formulate an experimental design is well before the study begins. Experimental design should not become a remediation device to be used midway into an experiment, attempting to save a study that is heading out of control. After all, experimental design is an up-front planning tool and is integral to good quality assurance. As such, the project startup is the time to ask questions (formulate hypotheses) and scrutinize the study.

Any experimental design can be constructed if the environmental chemist expresses three key elements: common sense, good chemical intuition, and a concentrated effort to decide exactly what the study should accomplish. It is this last element that usually needs the most work. Remember, meaning is given to the data depending on the questions (experimental design) asked [53]. Questioning is the first step in the scientific process. Sound logic and statistical reasoning will help provide appropriate answers through experimentation.

Researchers are encouraged to learn the basic principles of experimental statistics and their applicability to their research problems. Box et al. state that "it is possible for scientists to conduct an investigation without statistics, [but] a good scientist becomes a much better one if he uses statistical methods" [10].

It has been our experience to approach environmental studies as a team of scientists and chemometricians. However, these guidelines are appropriate whether working together as a team or as an individual scientist.

An obstacle to the widespread use of designed experiments is that the systematic design of experiments often appears to be more complex than a purely intuitive approach. Researchers using a specific experimental design for the first time will usually not know all of the proper questions to ask or the appropriate answers to those questions [13]. Problems arise both from the ignorance of the existence of the methodologies and from their misapplication: researchers know what they want to accomplish, but have no clear knowledge of how to do it; this is the commonly encountered "what-to-how" gap.

The first step is to have a general information-gathering meeting with all the people involved in the study (managers, scientists, chemometricians, and technicians). Ask the members what they think the goals and objectives (including data quality objectives) of the study should be. Ask them what the study is to accomplish. It may be surprising to all of the participants as to the disparity of the answers. A consensus must be reached before further progress can be made.

How specific the route is to the answers depends upon the detail given in those questions. It is actually somewhat difficult (or at least frustrating) to design an experiment for an objective as vague as "we want to develop a method for lead." Such ambiguity, although it is a start, could result in a wide array of designs which may not gather enough information (a waste of effort) or too much information (a waste of resources). Further discussion of the problem may show the way to proceed.

A more specific objective may be "we want to develop a rapid semi-quantitative screening environmental field method for lead in water covering the ppb to ppm range (the design may not have to be that rugged); the method is to be developed for an urban area (a specific client is indicated) because inner-city children have recently developed high blood lead levels (there may be resource implications)."

Much more complex designs may be required if the detailed objective is "we want to develop a quantitative environmental laboratory method for lead in water, 'good' to four significant figures with a detection limit down to 50 ppt, and with a linear dynamic range of five orders of magnitude; the method will be used to compare the performance (must be rugged) of several contracted laboratories (could be legal implications)." However, even the best experimental design may not be able to assist with over-ambitious objectives.

All of these details may specify the type of design. It all depends upon how the questions are asked. Some pertinent questions are:

- what are the resources available (personnel, time, budget, equipment)?;
- what are the constraints (e.g., safety boundaries)?;

– what is an acceptable level of quality (precision, accuracy, ruggedness)?;
– what is the importance of false positive and false negative (power) rates?;
– who is the end user (a visit to the client may be warranted)?; and
– are there regulatory or organizational implications?

Be sure that all parties involved have some time to re-evaluate the objectives of the study before proceeding. It may be useful to give out a questionnaire after this meeting.

After the general meeting, gather specific historical technical information (from the scientists and technicians doing the actual experiments) regarding what preliminary experiments were done, what went *wrong* in their view, what went *right*, whether or not the system is in statistical control, and whether or not there is an historical system response variance that can be used to develop a design. Observe the system under actual operating conditions. Ask questions and get diagrams.

A specific meeting with the same participants from the general meeting should now be conducted. (If a questionnaire was handed out, go over it with the group.) Determine what are the specific questions to be answered (the hypotheses to be tested). Identify the components of the system:

– define the system;
– identify all input variables that may be considered as factors (chemical intuition is important here);
– specify factor ranges (intuition, preliminary data, safety boundaries);
– speculate on possible input contributions for noise and uncontrollable variables;
– identify all output variables, specifying the response(s) of interest to the study; and
– specify how the study will be conducted (by whom, over what period of time).

Several designs are usually possible and the choice is a compromise between information gain and cost. For example, if a design that gathers data about the effect of four factors of interest is much more costly than a design that gathers data about only two factors, then a researcher might seriously consider if the research objectives could be accomplished to a satisfactory degree with the two-factor design. Sometimes other cost tradeoffs occur, such as using a less concentrated (or cheaper) reagent and compensating by running the process at a higher temperature.

The next step may be to do a preliminary study (for familiarization) if the investigators are not used to operating the system under the conditions outlined at the specific meeting. The chemometrician is encouraged to participate at this point, at least as a casual observer. This study should demonstrate if the system is in statistical control.

An optimization study might then be performed only if there are relatively few factors (less than four or five) to be investigated and there still is not a general sense of where the factor space should be for further experimentation. Otherwise, a screening design such as a Plackett-Burman or a fractional factorial might be considered to determine which factors have the largest effects upon the response of interest. This will cut down on the number of factors to be investigated, resulting in fewer experiments. Recall that only the main effects will be revealed by these designs. It is not possible at this point to resolve any factor interaction effects because of the confounding between aliases in these designs. However, these interaction effects can be resolved later with other designs. If detailed information about the system is not required beyond finding the region of the optimum or what the main effects are, then this is a permissible place to stop.

When a more in-depth understanding of the system is desired (modeling for interaction and higher-ordered factor effects over noise), then more complex designs may be tried, although it is probably best to run a simplex optimization design before using Box-Behnken or central composite designs when modeling higher-order effects.

It is recommended that factorial designs with three-to-seven center points be used. The center points can be used to estimate the variance at the mid-point of the conditions of the design. This estimate can be used to represent an historical variance, in the ANOVA for model testing, or to see if the system is in statistical control; it can also be used to look at curvature in modeling factor effects.

Optimization experiments are sometimes run at this stage when an objective of the study is to search a specified factor space for the global maximum (e.g., maximize extraction percent recovery) or the global minimum (e.g., minimize model residuals) on a response surface.

Box-Behnken or central composite designs might now be used to model higher-order effects upon the response in the area of the optimum to gain a better understanding of the system and to be able to make predictions under somewhat different conditions in that region of factor space.

If the analysis of the experimental results meets the design objectives, then the conditions of the system may be specified. Those conditions would then be used to determine other performance measures, such as a method detection limit, a method linear dynamic range, precision, and accuracy.

An intralaboratory ruggedness study could be done to substantiate the previous findings. This should be done to examine the tolerance of factors, i.e., to see if perturbations in the factor settings have a pronounced effect upon the response – this should be fairly predictable based upon the results. A rugged method should lie on a flat portion of the response surface, such as at an optimum. Ruggedness studies may also be performed to see if the variance remains constant over the factor domain.

An interlaboratory validation study [54] is finally done to confirm that the study holds true and is reliable for the general scientific community. For environmental chemical interlaboratory validation studies, there are three variance components to consider: sample variance, intralaboratory variance, and interlaboratory variance. Sample variance is often overlooked. It is imperative that all of the laboratories receive identical samples. Therefore, care must be taken in sample preparation and transportation and holding-time effects must be estimated (a separate study in itself). Intralaboratory variance should already be quantified and in control (be sure that each laboratory follows exactly the same conditions for every set of replicate analyses). Interlaboratory variance should be small if the method is rugged (robust) and is what should be measured if the other two variances (sample and intralaboratory) are controlled. It is a measure of variation in operators, environment, reagents, equipment, time and possible unknown or uncontrollable factors. If the interlaboratory variation is large and the method is still regarded as being rugged, then experimental designs can be used to explore the effects of these factors.

Some Common Pitfalls

The intuitive phase of designing experiments can narrow down the elements of the problem left to be studied by statistical designs, but enough margin should be allowed for the SEDOP approach to operate. If the design has adequately covered the factor space to explore the system, valuable information that the researcher's experience and intuition may have discounted, a priori, could be unveiled.

The system should be proven to be in statistical control. Error variability should be demonstrated to be relatively small with respect to factor effects before any quantitative assessment of factor effects, modeling, or optimization are attempted. For example, an instrument evaluation study based on a Box-Behnken design had to be aborted after the initial data was gathered due to the high variability observed in the results for the center point replicates [42]. The subject of the study was a newly-designed thermal desorption GC/MS instrument for the analysis of PAHs in soil. If the data analysis had been completed, only the effect of noise would have been modeled. The recommended course of action was to perform mechanical improvements and conduct other studies that would help bring the instrument under control (e.g., determination of sources of variability). Common sources of variability, such as different operator's skills, should be identified and neutralized or accounted for in the design (e.g., using blocking).

Common problems encountered in method comparison studies (and other environmental applications of statistical design) are missing data, concentrations below the detection limit for one or more laboratories, or no spike added for particular compounds in an experimental batch [47]. These anomalies reduce both the number of data points available for statistical analysis and the degrees of freedom for the associated statistical tests. It is important to determine, use, and report the degrees of freedom available for each individual test performed in the data analysis.

Sometimes a shortage of enough degrees of freedom to lend confidence to the results is created by a poor design. A common case is calibration studies. Researchers should realize that even for a straight-line fit ($y_{1i} = \beta_0 + \beta_1 x_{1i} + r_{1i}$)

enough degrees of freedom are necessary to resolve the lack-of-fit of the model from the purely experimental error in the residuals through replicated experiments [55].

The experimental design (and the corresponding data analysis method) should be well established before the experiments are carried out. If the collected data reveal that the original experimental design was inadequate, then the study might have to be repeated or augmented with additional experiments. Experimental design planning may even reveal, before the experiments are started, that the researcher's questions cannot be answered satisfactorily (with statistical significance) or that more resources than those available will be necessary to reach reliable conclusions. A common outcome from the application of the SEDOP approach is a saving in the number of experiments and, consequently, cost. The use of statistical experimental design requires researchers to state the questions to be answered, which is also a benefit in itself.

Researchers should be committed to conduct the prescribed experiments fully and according to the specifications of the design. Sometimes, in practice, researchers take the liberty of making changes in the experiments without being aware of the impact for the statistical analysis of the results. Communication and consulting between the designers and the researchers at every stage of the project can avert many problems along this line. Advisors should make an effort to understand all of the relevant aspects of the system being studied as thoroughly as possible to avoid hidden flaws and impractical designs. For example, in a design tailored to determine the effect of holding time in the analysis of certain pesticide standards, it was discovered (after completion of the experiments) that the laboratory staff made the holding time control solutions from the same calibration standards used throughout the study. Thus, the week-to-week variability of these solutions were correlated and very limited conclusions could be drawn. When the researchers receive the experimental design from an advisor, and there is no interaction with the advisor until the results are sent back, there is a greater risk of major failures.

Armed with this knowledge, using existing software, and consulting with chemometricians and statisticians, environmental researchers should be able to start discovering and accessing the benefits of the SEDOP techniques on their own.

Experimental Design and Optimization Software

The availability of software that assists in the experimental design and optimization process gives researchers added flexibility and freedom. In fact, for iterative techniques, such as sequential simplex, it is almost impractical to depend on an external advisor for the analysis of each result before the researchers are able to plan for the next iteration of the study.

Microcomputer-based software for the set up, tracking, and data analysis of statistically-designed studies is widely available. Morgan [35] presented a critical summary of ten different software packages applicable to statistical design. Researchers are encouraged to use the available computer programs instead of relying on manual calculations or self-programmed software. A number of statistical computer packages include experimental design capabilities. Advantages of the use of these integrated packages are that, in general, they tend to have better graphics display capabilities (e.g., three-dimensional response surface plots) and that common statistical procedures are included. Some computer programs are designed exclusively for experimental design and have the capabilities for the analysis of one or several [56–58] design types. These programs can provide more flexibility and features in the experimental design area than general statistical packages.

Computer programs that help with the planning of statistically-designed studies have become available in the last few years. For example, programs are available to design a factorial-type experiment, optimize the treatment combinations for D-optimality, and then select a non-random order for carrying out the experiments to give the greatest rate of accumulation of information. Some programs with a multi-stage approach to planned experimentation include experiment definition, design and work sheet generation, analysis of collected data, and interpretation of results. Such software helps in the set up of an experimental design for the objectives of screening or response surface modeling. Also, there are expert systems for the selection of the appropriate experimental design types for a given research situation [58, 59]. It is relatively easy to analyze collected data, but no data analysis technique can extract non-existent information from a poorly-designed study. Even though much chemometric research focuses on data treatment techniques, the usefulness of those techniques is often related to the characteristics of the experimental design that was used to collect the data in the first place. Many different techniques for data analysis can easily be tried on existing data sets with computer time as the primary cost [8]. Nonetheless, Currie et al. [60] attribute some of the misuse of statistical techniques to the widespread use of microcomputers, which has made those techniques readily available to inexperienced users.

Notice. The U.S. Environmental Protection Agency (USEPA), through its Office of Research and Development (ORD), partially funded and collaborated in the research described here. It has been subjected to the Agency's peer review and has been approved as an EPA publication.

References

1. Kowalski B, Brown S, Vandeginste B (eds) (1987). Journal of Chemometrics 1(1)
2. Behar JV (ed) (1988) Proceedings of the workshop: Progress in Chemometrics. Chemometrics and Intelligent Laboratory Systems 3:15

3. Proceedings of the symposium: Chemometrics with Environmental Applications (1991). Journal of Chemometrics 5:3
4. Beer T (1991) Applied environmetrics hydrological tables. Applied Environmetrics, Victoria, Australia
5. Beebe KR, Pell RJ (1994) Today's Chemist at Work 21:24
6. Deming SN, Morgan SL (1990) Experimental design: a chemometric approach. Elsevier, Amsterdam
7. Barker TB (1985) Quality by experimental design. Marcel Dekker, New York
8. Steinberg DM, Hunter WG (1984). Technometrics 26(2):71
9. Natrella MG (1963) Experimental statistics. National Bureau of Standards, Washington, DC (Handbook 91)
10. Box GEP, Hunter WG, Hunter JS (1978) Statistics for experimenters: An introduction to design, data analysis, and model building. John Wiley & Sons, Inc., New York
11. Youden WJ, Steiner EH (1975) Statistical manual of the Association of Official Analytical Chemists. Association of Official Analytical Chemists, Washington, DC
12. Read DR (1954) Biometrics 10:1
13. Snee RD (1985) Journal of Quality Technology 17(4):222
14. Flatman GT, Mullins JW (1984) The Alpha and beta of Chemometrics in Breen JJ, Robinson PE (eds) Environmental applications of chemometrics (ACS Symposium series 292), American Chemical Society, Washington, DC
15. Driver RM (1970) Chem Brit 6:154
16. Deming SN (1985) Journal of Research of the National Bureau of Standards 90(6):479
17. Carlson R (1992) Design and optimization in organic synthesis. Elsevier, Amsterdam
18. Bayne CK, Rubin IB (1986) Practical experimental designs and optimization methods for chemists. VCH Publishers, Deerfield Beach
19. Milliken GA, Johnson DE (1984) Analysis of messy data, volumes 1 and 2. Van Nostrand Reinhold, New York
20. Rubin IB, Mitchell TJ, Goldstein G (1971) Anal Chem 43(6):717
21. Wernimont G (1968) Materials Research & Standards 9(9):8
22. Box GEP (1952) Analyst 77:879
23. John PWM (1971) Statistical design and analysis of experiments. MacMillan, New York
24. Plackett RL, Burman JP (1946) The Design of Optimum Multifactorial Experiments Biometrika 33:305
25. Ross PJ (1988) Taguchi techniques for quality engineering. McGraw-Hill, New York
26. Youden WJ (1951) Statistical methods for chemists. John Wiley & Sons, New York, pp 40–49
27. Draper NR, Smith H (1981) Applied regression analysis. John Wiley & Sons, New York
28. Brown PJ (1982) R. Statist. Soc. B44(3):287
29. Box GEP, Wilson KB (1951) Journal of the Royal Statistical Society, Series B 13:1
30. Hill WJ, Hunter WG (1966) Technometrics 8:571
31. Box GEP, Behnken DW (1960) Technometrics 2(4):455
32. Scheffé H (1958) Journal of the Royal Statistical Society, Series B 20:344
33. Harrington EC (1965) Industrial Quality Control (4):494
34. Box GEP, Draper NR (1969) Evolutionary operation. John Wiley & Sons, New York
35. Morgan E (1991) Chemometrics: Experimental design. John Wiley & Sons, Chichester
36. Spendley W, Hext GR, Himsworth FR (1962) Technometrics 4:441
37. Deming SN, Morgan SL (1983) Anal Chim Acta 150:183
38. Walters FH, Parker CR, Morgan SL, Deming SN (1991) Sequential simplex optimization. CRC Press, Inc., Boca Raton
39. Nelder JA, Mead R (1965) Computer Journal 7:308
40. Deming SN, Garner FC, Nocerino JM (1991) Annual report: Chemometric applications in quality assurance research (report number EPA/600//X-91/160). US Environmental Protection Agency, Las Vegas
41. Ho JS, Tang PH (1992) J Chromatogr Sci 30(9):344
42. Nocerino JM (1993) Annual report: Chemometric applications in quality assurance research (report number EPA/540/X-93/501) US Environmental Protection Agency, Las Vegas
43. Hillman DC, Nowinski P, Butler LC, Nocerino JM (1993) American Environmental Laboratory 5(3):28
44. Horwitz W (1982) Anal Chem 54:67A
45. Horwitz W (1985) Anal Chem 57:454
46. Silverstein ME, Klainer SMH, Ecker VA, Satterwhite G, Munslow WD (1991) Superfund Innovative Technology Evaluation Program project and quality assurance plan for demonstration of the Bruker mobile mass spectrometer. US Environmental Protection Agency, Las Vegas

47. Chaloud D, Silverstein M, Rosenfeld J, Hulse S (1991) Demonstration of the Bruker Mobile Environmental Monitor (report number EPA/600/X91/079). US Environmental Protection Agency, Las Vegas
48. Sokal RR, Rohlf FJ (1981) Biometry, 2nd edn. W.H. Freeman and Company, New York
49. Mayer C, Amick N, Davis C, Ecker V, Deming SN, Palasota J (1993) Automated on-site measurement of volatile organic compounds in water: A demonstration of the A+RT, Inc. Volatile Organic Analysis System (report number EPA/600/R-93/109). US Environmental Protection Agency, Las Vegas
50. Tosato ML, Vigano L, Skagerberg B, Clementi S (1991) "A New Strategy for Ranking Chemical Hazards: Framework and Application" Environ Sci Technol 25(4):695
51. Geladi P, Kowalski BR (1986) Anal Chim Acta 185:1
52. Wold S, Carlson R, Skägerberg B (1989) The Environmental Professional 11:127–131
53. Hendrix CD (1983) Chemtech 598
54. Wernimont GT, Spendley W (eds) (1985) Use of statistics to develop and evaluate analytical methods. Association of Official Analytical Chemists, Washington, DC
55. Deming SN (1986) Clin Chem 32:1702
56. Nachtsheim J (1987) Journal of Quality Technology 19(3):132–160
57. CAMO (1994) Unscrambler. CAMO USA, Redwood Shores, California
58. Olivero RA, Seshadri S, Deming SN (1993) Anal Chim Acta 277:441
59. Quinnell R (1993) EDN
60. Currie LA, Filliben JJ, DeVoe JR (1972) Anal Chem 44(5):497

Signal Processing and Correlation Techniques

H.C. Smit

University of Amsterdam, Laboratory for Analytical Chemistry, Nieuwe Achtergracht 166, 1018 WV Amsterdam, The Netherlands

Summary

The applicability of chemometric methods like correlation techniques and optimum signal processing is treated. Different approaches, mainly directed to improve the precision and to lowering the detection limits in (sub)trace analysis, are discussed and illustrated with examples.

List of Symbols and Abbreviations

H	entropy
$H(j\omega)$	complex frequency response
$h(i)$	chromatogram
I	number of injections
i	discrete time
j	complex parameter
M	PRBS length
$N(\omega)$	noise power spectral density
n	integer determining the number of clockperiods in a PRBS
$n(i)$	detector noise
p	probability of an event
$p(i)$	PRBS pattern
$R_{xy}(i)$	cross-correlogram of signals x and y
$S^*(j\omega)$	complex conjugate of the signal model in the frequency domain
T	system operator on input stimuli
t	time
$x'(t)\ldots x^n(t)$	set of input stimuli of a system
$y(i)$	detector output resulting from PRBS injection
$y'(t)\ldots y^m(t)$	set of output stimuli of a system
$y(t_i)$	discrete fitting function
$\Delta(i)$	Kronecker delta function
Δt	clockperiod of a PRBS
σ_i^2	variance in the data points
μ	time shift
χ^2	sum of weighted squared deviations of the signal amplitude values from the fitting function
ω	angular frequency
CC	correlation chromatography
CMCC	chemical modulation correlation chromatography
CCZE	correlation capillary zone electrophoresis
DCC	differential correlation chromatography
MEM	maximum entropy method
MF	matched filter
MLS	matched linear system
PRBS	pseudo random binary sequence
PSD	power spectral density
SD	standard deviation
S/N ratio	signal-to-noise ratio
SSCC	single-sequence correlation chromatography

Introduction

Analytical chemical information is crucial in environmental chemistry. Precise measurements of a variety of pollutants, both in air and water, in the presence of complex matrices is an important task of the analyst. Often low detection limits are required because of the low concentration of many species of interest. Selective low noise analytical methods are essential in this field. Usually a direct measurement of a component within the matrix is not possible; a preceding selectivity enhancement step is necessary, for instance a suitable specific chemical reaction preceding a spectroscopic measurement.

The application of chemometrical methods like multivariate calibration allows the simultaneous (spectroscopic) determination of more than one component. However, only the extension to a few components is possible and in general separation from the matrix is still necessary.

Chromatography is an important tool in environmental analysis. That means that the desired analytical information has to be extracted from (dynamic) signals, peaks in the case of chromatography. The application of multi-channel (e.g. diode arrays) detection implies information extraction from multidimensional signals. Other analytical techniques (FAA, electro-analytical methods) also produce dynamic signals.

The concept of a signal is directly related to system theory and it plays an important role in several scientific disciplines. In a simplified approach a system can be considered as a unit transforming a set of input stimuli, $x^1(t), x^2(t) \ldots x^n(t)$, to a set of output stimuli, $y^1(t), y^2(t) \ldots y^m(t)$, by some operation T (Fig. 1). In vector notation:

$$\underline{y}(t) = F\{\underline{x}(t)\} \,. \tag{1}$$

These stimuli are called signals. Signals include a vast variety of stimuli such as chemical concentrations, acoustic waves, electric voltages, etc. Mostly, a signal is considered as a function of time (dynamic signal), although this is not a necessary condition.

Problems in the processing of analytical chemical signals are caused by low signal-to-noise (S/N) ratios, because of the low concentrations to be measured, particularly in environmental chemistry, and by selectivity problems caused by a non-ideal separation. Noise can be defined as everything contributing to the uncertainty in the determination of the measurement value.

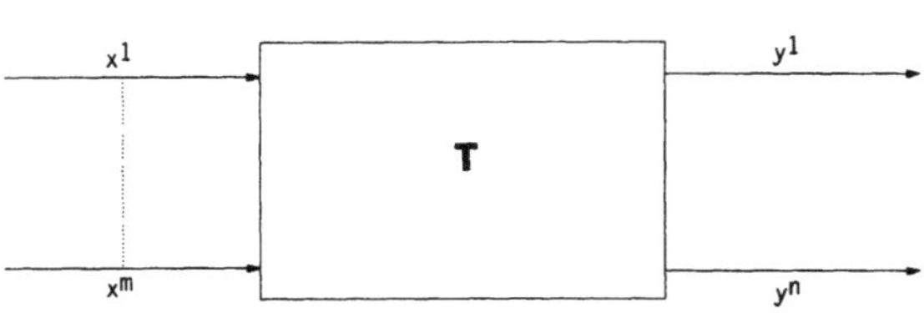

Fig. 1. System input and output signal vectors. T is the system operator

Permissible signals and noise can be divided into two partitions: deterministic, being known functions of time, and stochastic, being unknown or random functions of time. While in case of stochastic signals, no functional relation exists between the value of the signal at different times, statistical parameters are usually assumed to be known. Signal processing has apparently a mathematical/statistical nature.

Often, some signal parameters (height, area, etc.) being characteristic for the desired quantity (e.g. concentration) have to be determined from the total signal. Examples of simple signal processing are peak height determination and peak area determination. In case of more complex signal processing, more parameters of the signal have to be determined. Mostly this implies the formulation of mathematical models and an approximation of the measured (deterministic) signal. Real-world deterministic signal functions are seldom exactly known and one is often forced to accept an approximation that may be close to the true function, but not exactly equivalent. A measure of closeness has to be formulated.
If considerable noise is present, signal approximation is replaced by signal estimation. Optimum estimation can be formulated as: find the best estimate of a noisy signal value at each constant in time; or: find the best estimate of the relevant signal parameter. The latter is usual in analytical chemistry. In chromatography, for instance, quantitative evaluation implies the estimation of the "intensity" of peaks. Usually peak height or peak area is used as intensity parameter. However, neither is optimal with respect to the remaining uncertainty in the results.

Each signal processing procedure assumes some prior information, but in optimal estimation procedures all available and obtainable prior knowledge is used to maximize the precision. One may conclude that the final precision in the results depends on the S/N ratio, on the choice of the processing procedure (optimal/non-optimal) and, in case of advanced signal processing, on model errors, both with respect to signal models and statistical noise models.

One may suggest that the influence of the factors mentioned is increasing if the selectivity of the measurement method is decreasing. It is possible to calculate quantitatively this influence on the precision, for instance in case of unresolved chromatographic peaks. Maximizing the precision is not the only reason to optimize signal processing. Analytical chemical results are used, e.g., for setting or monitoring regulation pertaining to environmental guidelines.

Signal processing is an important part of the analytical procedure, having great influence on the final precision. Standardization and comparison of results from different laboratories are hardly possible if this part is not well defined and if the final analytical results are to a great extent dependent on unknown, uncontrollable and unpredictable software. "Integration" software in chromatography is notorious in that respect.

One has to strive for well defined standardized optimum signal processing, including uncertainty calculations. The latter is not yet common, not even in simple processing like integration, where the fundamental theory of uncertainty calculations has been derived quite a long time ago. In more advanced signal processing it is even more difficult, not to mention multi-variate data (signal) analysis, where uncertainty calculation is almost an unexplored area until now.

Analytical signal processing can be considered to belong to the class of chemometrical techniques because of the use of advanced mathematical and statistical methods. There is another way of using chemometrics in the development of analytical methods. An analytical system can be considered as an information channel. In some cases the usual input stimuli (input signals) are not optimal with respect to utilizing the information capacity of the analytical information channel; the information rate can be increased drastically using the proper input signals. Simple signal processing is out of the question then; the calculation and data handling capacities of computers have to be used. Examples of this approach are the different modes of *Correlation Chromatography* (CC), also known as *multiplex chromatography* and *multiple input chromatography* [1].

The most important advantage of CC over conventional chromatography is the rapid increase in the precision (S/N ratio) at the cost of sample in a relatively short time. Compared to preconcentration techniques, possible changes in the sample composition, a known disadvantage of preconcentration [2], are avoided, giving much more precise results. In addition, CC can be used to monitor changes in concentration [3], being a kind of continuous chromatography.

Correlation chromatography is essentially a *differential* technique, which can be used to eliminate the influence of matrix peaks or to determine very sensitively minor changes in the composition of samples as a function of place or time (Differrential CC (DCC)) [4]. In *Simultaneous CC* (SCC) more than one sample can be "separated" (no real separation takes place) in the same chromatographic column under exactly the same conditions [5]. CC requires a stationary system, and temperature programming or gradient elution is not possible. However, *Single-Sequence CC* (SSCC) is an intermediate between conventional single-injection chromatography and CC, allowing for instance the application of gradient elution [6]. *Chemical Modulation CC* (CMCC) adds extra selectivity to the method, apart from the selectivity from the column or a special detector [7]. Not only CC is possible, but the same principles can be used in *Correlation Capillary Zone Electrophoresis* (CCZE), with the same advantage. The considerable decrease of the detection limit is particularly important here. Work in this field is in progress [8].

Signal Processing: Intensity Determination, Estimation

Generally, quantitative evaluation of analytical signals implies intensity determination. Often the signal shape is not very important. The magnitude of some signal parameters, like the level of a constant signal, peak height, peak area, etc., are a direct measure for the desired quantitative analytical information, e.g.

concentration. Without noise the determination of amplitudes and peak areas is simple and straightforward. However, in the presence of noise, determination becomes an estimation. An (optimum) estimation procedure can be formulated as follows: find the best estimate of the noisy signal value at each instant of time, or find the best estimate of the relevant signal parameter. The "best estimate" means optimal with respect to some criterion, generally minimization of the "least squares".

If one signal parameter has to be estimated, the ensemble concept has to be introduced. Each estimation can be considered as one sample of parameter estimations of a set (ensemble) of similar signals plus noise. The square of the deviations from the true value has to be a minimum. Often the estimation can be improved by signal preprocessing, in general filtering, particularly smoothing. A filter is a device that is transparent for the (utility) signal but rejecting parts of the noise. It selects a particular frequency or a frequency range, called a passband, and suppresses ideally all the other frequencies. One has to keep in mind, however, that sometimes the result only serves cosmetic purposes: there is no reduction in the final uncertainty or the precision is even worsened [9].

Simple intensity parameter estimation like peak height or peak area is not optimum with respect to uncertainty in the results. An optimum estimation procedure uses all available and obtainable prior information to maximize the precision. This prior information consists of preknowledge of the signal and the noise using parameterized models. Signal models are mathematical expressions with adjustable parameters, describing the shape of the signal as a function of time. The parameters have to be determined in such a way that the functions fit the real peak shape satisfactorily.

The sum χ^2 of the squared deviations of the (discrete) signal amplitude values from the fitting function, weighted with the uncertainty in the datapoints, is used as goodness of fit criterion:

$$\chi^2 = \sum_{i=1}^{m} \left[\frac{1}{\sigma_i^2} [y_i - y(t_i)]^2 \right] \tag{2}$$

where m = number of data, y_i = datapoints, $y(t_i)$ = fitting function, and σ^2 = uncertainty (variance) in the datapoints. The fitting function can be linear or non-linear in the parameters.

Noise is modelled with the Probability Density Function (PDF), the Autocorrelation Function (ACF) and the Power Spectral Density (PSD), which can be derived directly from the ACF by Fourier transformation. The final uncertainty in the analytical result, defining the quality of the measurement, depends on a number of factors, e.g. the very important sampling errors, preprocessing of the sample, etc. Uncertainty due to signal processing is caused by measurement noise and, particularly in the case of advanced signal processing, by signal and noise model errors.

Some work on uncertainty in signal processing has been published. An example is the calculation of the influence of noise on signal integrating methods

[9]. For instance the uncertainty in peak area determination due to baseline noise can be expressed in the variance σ_I^2 of the integrated noise, integrated in an interval with time duration T. The variance σ_I^2 can be derived as a function of the properties (ACF, PSD) of the noise and the integration time T. An important result is that the uncertainty increases with increasing integration time. For instance in case of (bandlimited) white noise, σ_I^2 is proportional to the integration time. Low frequency noise is more dangerous in that respect, e.g. in case of flicker noise ($1/f$ noise), where the noise power is reversed proportional to the frequency, and the variance σ_I^2 is proportional to the square of the integration time.

Signal Estimation Methods

Regression Techniques

In signal processing the general name "linear and non-linear regression analysis" is adopted for signal and curve fitting techniques where the parameters (coefficients) of a model (function) are optimized with respect to a minimum value of χ^2. A linear function can be expressed as a sum of separate terms, each multiplied by one and only one coefficient. The most simple example is a straight line, $f(t) = a + bt$, where t = time, and a and b are constants. However, (orthogonal) polynomials [10] are also applied as linear fitting functions.

The procedure may act as a smoothing filter if the number of polynomial terms is limited. If the model is non-linear in the parameter, non-linear regression is required, which is a much more complicated procedure [11–13]. It implies simultaneously minimizing χ^2 with respect to each of the parameters k_j in the model:

$$\frac{\partial}{\partial k_j} \chi^2 = \frac{\partial}{\partial k_j} \sum \left[\frac{1}{\sigma_i^2} [y_i - y(t_i)]^2 \right] = 0 . \tag{3}$$

Assuming a model with n parameters a_j, χ must be taken as a continuous function of the parameters, describing a hypersurface in an n-dimensional space. The minimum value of χ^2 has to be found.

Several procedures are known, although the Levenberg-Marquardt algorithm is used most, combining the gradient-search method of least squares and the method of linearizing the fitting function. One problem can be that not the real optimum (minimum χ^2), but a local optimum, is found; therefore it is advisable to repeat the procedure with different initial values of the parameters. Another possible weak point is that non-linear regression assumes "white" noise and dominating low-frequency noise causes difficulties.

Figure 2a shows a fitted HPLC-chromatogram with phenols and nitro-phenols, using a Fraser-Suzuki peak model. Figure 2b shows the residuals, the difference

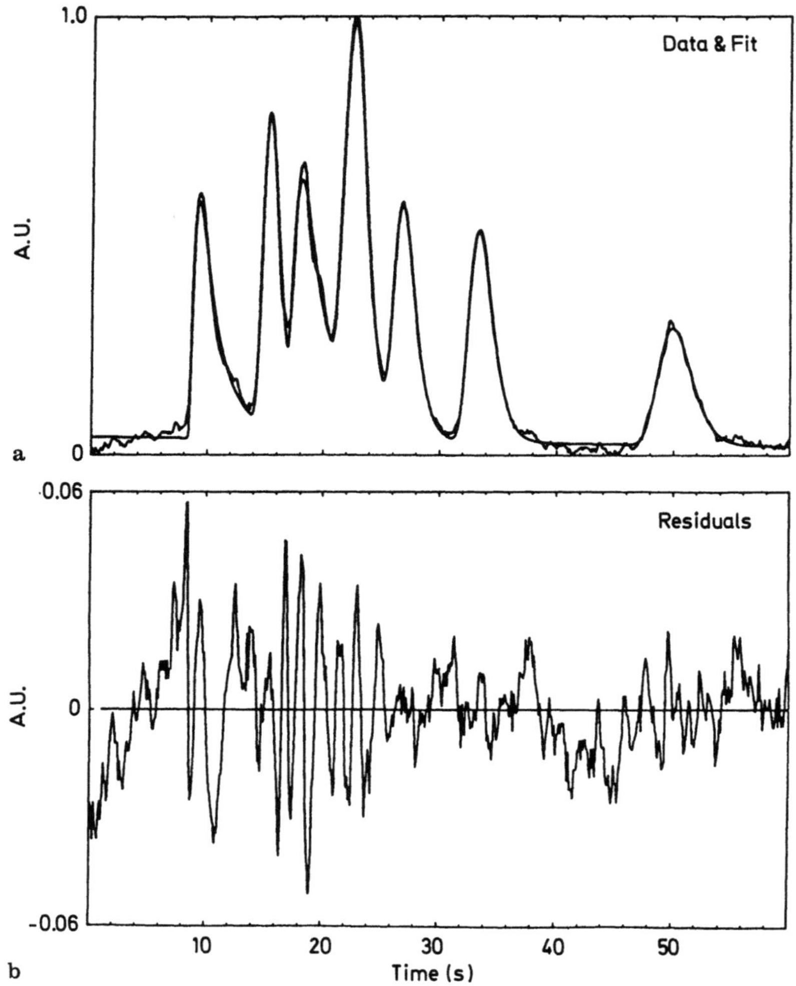

Fig. 2. **a** Non-linear regression fit of High Performance Liquid Chromatogram. Phenols and nitrophenols. **b** Residuals of the fit

between the fit, and the chromatogram after fitting. The χ^2 can be considered as a quantitative measure of the goodness of fit.

Correlation Detection

If, in contrast with (non-linear) regression, the signal shape is completely known except for the intensity, then correlation detection can be applied. In correlation detection the noise is also assumed to be white, i.e. the power of the noise is equally distributed along the frequency axis in frequency range of the utility signal; the PSD is flat. It is a relatively simple optimum signal processing procedure. The procedure involves shifting a completely known model peak shape along the time axis and calculating the integral of the product of the model shape and the real signal for each time shift. The amplitude (intensity) of the real signal is the only signal parameter not known in advance. Figure 3 shows the procedure applied to a noisy peak, resulting in a less noisy peak. The original peak shape is not maintained; the original peak is skewed, the resulting peak is symmetric. However, each point has a known relation to the other points of the peak, determined by the (known) original peak shape, and each point is also directly related to the desired intensity of the real peak.

Of course, the maximum (top) is optimum with respect to the minimum uncertainty, and is used as an estimate for the unknown intensity. The increase in the S/N ratio is not drastic, about 1.5–2, depending on the peak shape, and if compared with integration using optimum integration limits. However, in practice, the determination of optimum integration limits is difficult, as is known from several integration software packages in chromatography. The peak top determination in correlation detection is simpler and in comparison the final increase in precision is higher.

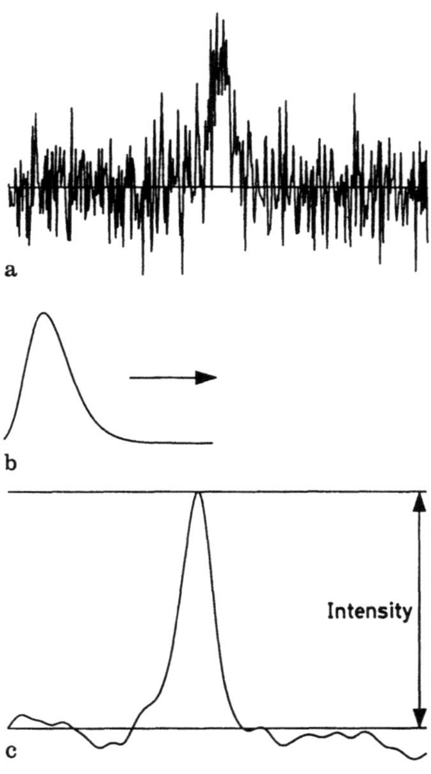

Fig. 3a–c. Correlation detection: **a** signal $s(t) =$ peak $f(t)+$ noise $n(t)$; **b** model function $f_1(t-\tau)$, where $\tau =$ time shift; **c** correlation detector output $=$ integral of the product $s(t).f_1(t-\tau)$ as a function of τ

Matched Filtering (Matched Linear Systems)

Matched filtering (MF), i.e. the application of matched linear systems (MLS), is a signal-processing method directly related to correlation detection. In the latter white noise is assumed. If, however, the noise is non-white and if not only the signal shape but also the power spectrum (power spectral density, PSD), is known, this extra pre-knowledge can be used for further reducing the uncertainty in the intensity determination by MF [14, 15]. The description of a signal can be approached via either the time domain or the frequency domain.

As already mentioned, if the shape of a signal is known exactly, then it is sufficient to know the amplitude of each frequency in the frequency spectrum of the signal obtained by Fourier transforming in order to determine the intensity of the signal, because a functional relationship exists between the spectral components given by the known signal model in the frequency domain. However, noise

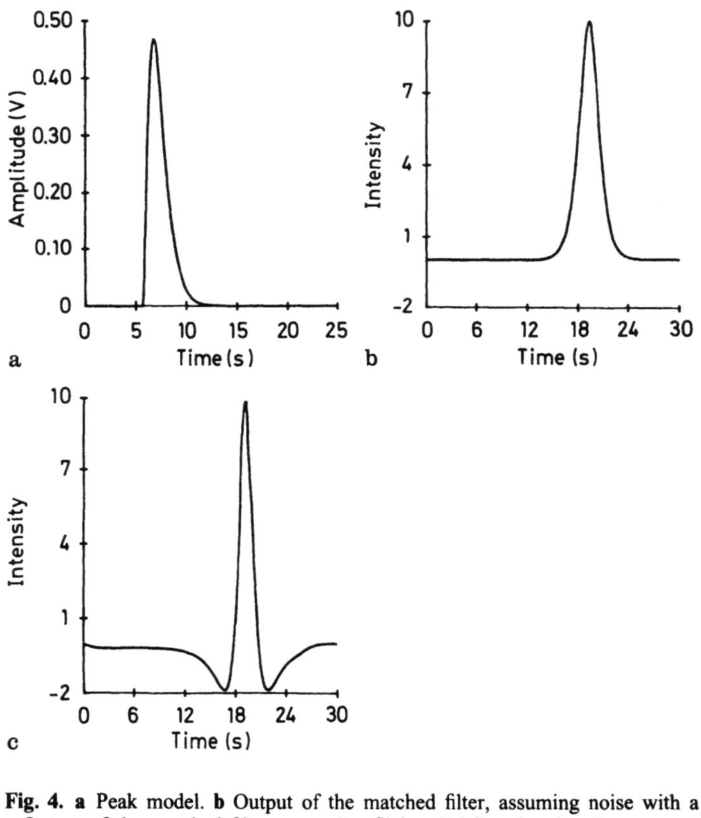

Fig. 4. a Peak model. **b** Output of the matched filter, assuming noise with a flat (white) spectrum. **c** Output of the matched filter, assuming flicker ($1/f$) noise, i.e. the power of the noise is inversely proportional to the frequency. The input peaks in **b** and **c** are taken to be almost noise-free in this instance to demonstrate the effect of matched filtering on the output peak shape

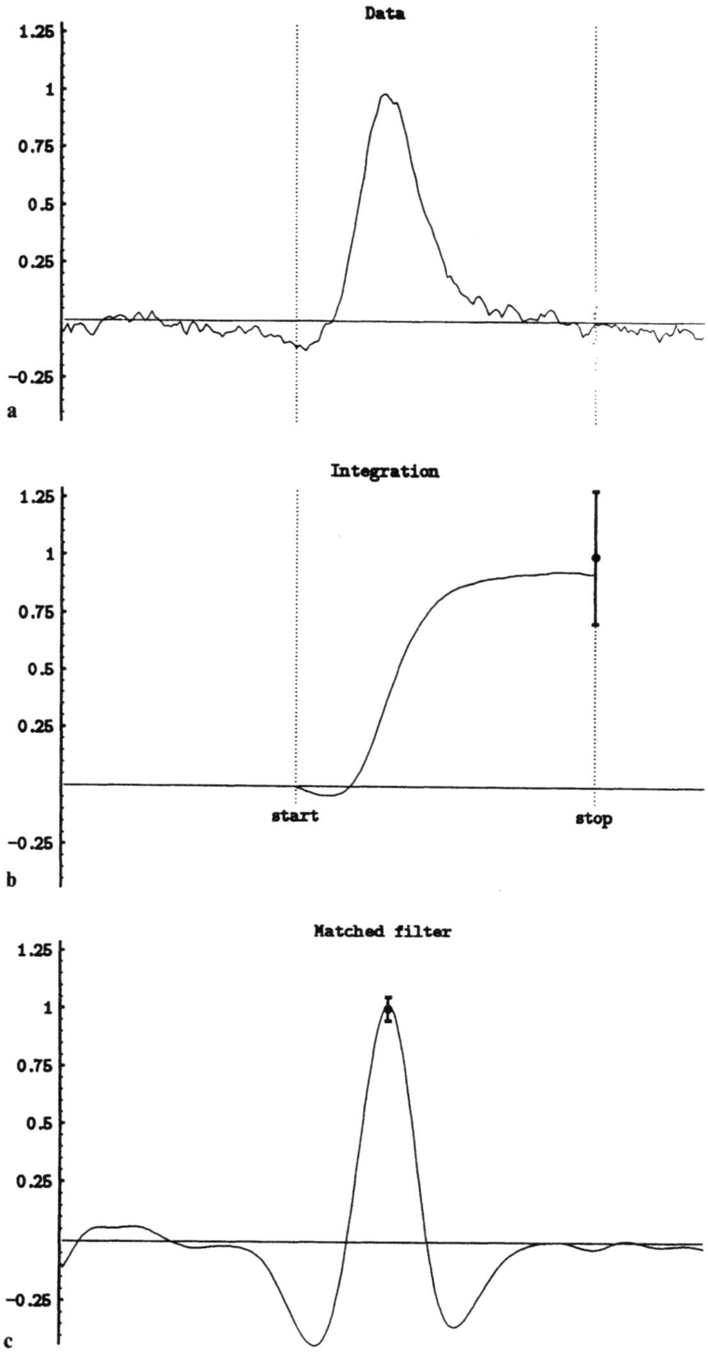

Fig. 5a–c. Signal intensity estimation by integration and matched filtering. The bars are indicating the confidence intervals: **a** noisy peak; **b** integral of the noisy peak; **c** matched filter output resulting from the noisy peak input

contributes to the total (signal plus noise) power in each part of the frequency spectrum; in case of non-white (coloured) noise it is different for each frequency.

A matched filter acts by selectively enhancing or suppressing certain frequencies in the total signal; the frequencies in the signal are weighted according to the ratio of signal power to noise power frequency. Not one single frequency but all relevant frequencies are used to determine the signal intensity.

The complex frequency response of an MF is:

$$H(j\omega)_{\mathrm{m}} = \frac{S^*(j\omega)}{N(\omega)} e^{-j\omega\mu} \quad (4)$$

where $\omega =$ (angular) frequency; $j =$ complex parameter $(j^2 = -1)$; $S^*(j\omega) =$ complex conjugate of the signal model in the frequency domain; $N(\omega) =$ power spectral density of the noise; and $\mu =$ time shift introduced by the MF.

In Fig. 4 a realistic signal model (peak) and the resulting (noise-free) output of a "matched" filter adapted for white noise and for $1/f$ or flicker noise with dominating low frequencies are shown. In this example no noise is present in the input signal to show the influence of matched filtering on the resulting signal shape. Note that the output signal is broadened and symmetric, even in case of a skewed input signal. However, only the output peak top with the highest S/N ratio is required for the desired intensity information.

Figure 5 shows the intensity estimation of a noisy peak (Fig. 5a), by integration (area determination, Fig. 5b) and by matched filtering (Fig. 5c). The confidence intervals for both methods are depicted by the bars. Obviously, matched filtering is superior to integration, as could be expected.

Maximum Entropy Method

A signal-recovering method of growing importance in analytical chemistry is the Maximum Entropy Method (MEM), derived from Shannons information theory [16] and reexamined by Jaynes [17]. MEM has been successfully applied to a number of problems, e.g. the reduction of noise, image restoration, and the determination of unknown parameters from incomplete data. It is rather easy to explain *how* MEM is done, but it is difficult to understand *why* it works. Maybe a good starting point for making MEM understandable in a simple way is emphasizing the relation to the old principle of insufficient reason in statistics, described in terms of the probability of events. This principle states that without any prior knowledge it *must* be assumed that all events have equal probabilities. The base of this statement is that probability is interpreted here as a measure of the state of knowledge about the events. If the events would not have the same probability, then any change in (the indices of) the events would give different probabilities without a change in the state of knowledge.

Entropy is directly related to probability. *Probability* is a measure of the uncertainty about occurrence or non-occurrence of an event in a single performance of an experiment. One can assign a measure of uncertainty not to the occurrence or non-occurrence of not a single event, but of any event of a partition (collection of mutually exclusive events) of the underlying experiment. This measure is called *entropy* of the probability distribution and is denoted by H. The relationship between H and the probabilities p_j is

$$H = -\sum_{j=1}^{n} p_j^2 \log p_j \,. \tag{5}$$

An important property of entropy [Brillouin] is that it can be proved that it assumes a maximum value when all probabilities of the events are equal. In this given example maximizing the entropy is conceptually equivalent to the insufficient reason principle and results in the most probable solution. More generally, MEM is defined as the determination of the probabilities of the events of a partition, subject to given constraints, by maximizing the entropy. These constraints may be phrased as expected values, but MEM has applications in non-probabilistic problems as well. MEM is justified by the same reasoning as in the given example of equal probabilities; the resulting solution is the most probable one.

A typical advantage of MEM is the drastic simplification of the analysis in practice, and therefore MEM is a valuable tool in the solution of applied problems. The arising problems can generally be solved numerically. The solution involves the determination of a function several parameters. Well known variational techniques can be used, involving Lagrange multipliers and Euler's equations.

Summarizing, MEM is a robust signal-recovering method, improving the precision of the detection. It is particularly useful in analytical chemical applications, when limited a priori information concerning the expected signal is present.

Correlation Techniques in Separation Methods

Correlation Chromatography

A separation system like a chromatograph or a capillary zone electrophoresis system can be considered as an information channel. However, in these techniques the information capacity of this channel is very little utilized. The input stimulus, an approximately pulse-shaped injection with limited amplitude (concentration) and time duration, results in an impulse response, e.g. a chromatogram, but during the separation only a small part of the column is occupied. The components in the input sample are diluted considerably.

The application of correlation techniques allows the use of multiple injections as an input, resulting in a much more efficient use of the separation power of the column, finally resulting in a drastic reduction of the noise and the detection limit. This is particularly important in environmental analysis, where low concentrations of components of interest occur in the presence of a complex matrix. Correlation Chromatography is a typical example of this chemometric technique, with impressive results in (ultra-) trace analysis [18, 19].

A schematic set-up of a CC-system, with mechanical valves controlling the multiple injection, is shown in Fig. 6a. This mechanical modulation can be replaced by a chemical modulation system (Fig. 6b). The response is a massive noise-like group of fused peaks with a greatly raised baseline; separate peaks cannot be visualised. Using the known input function, the resulting output and correlation techniques, the computer can calculate a "correlogram" almost similar to a normal chromatogram, although with a much higher S/N ratio. The longer the system is run the higher the S/N ratio will be.

In trace analysis, trace compounds, otherwise not detectable by conventional single injection techniques, can be measured at the cost of a larger amount of sample and a longer analysis time. In general, in environmental analysis sufficient sample is present. To get an impression of the possibilities of the method, one can say that an increase by a factor of 2 in the analysis time compared to conventional chromatography decreases the detection limit by about one order of magnitude. A further reduction is about proportional to the square root of the time. The multiple injection pattern at the input has to fulfil certain demands.

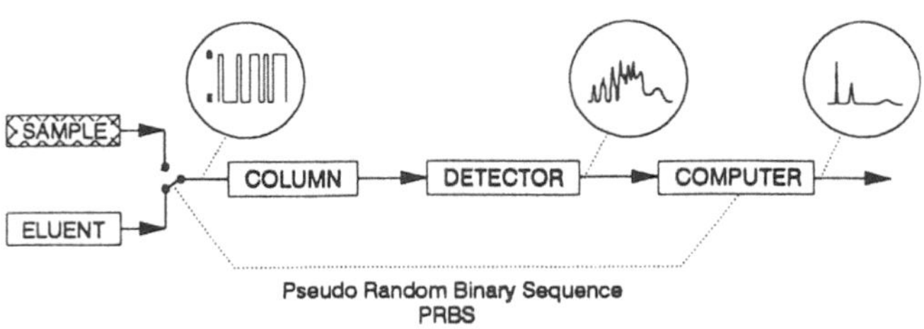

a MECHANICAL MODULATION SYSTEM

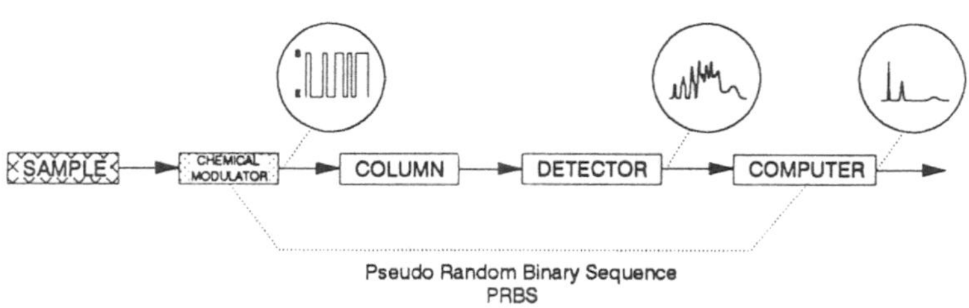

b CHEMICAL MODULATION SYSTEM

Fig. 6. a Set-up of a correlation chromatograph, mechanical modulation system. **b** Chemical modulation system

Usually a so-called Pseudo Random Binary Sequence (PRBS) pattern $p(i)$ is used, where i is the discrete time. A PRBS is a binary "noise" with only two levels, $+1$ and -1, or 1 and 0, corresponding to injection of sample or mobile phase, respectively.

A PRBS has a specific length M, the sequence length of $2^n - 1$ periods (n is a positive integer) controlled by a clock determining the minimum time Δt of the 1 or 0 state. M clockperiods correspond to $I = 2^n - 1$ injections. Essentially, a PRBS is a logic function combining the properties of a true (binary) random pattern and a reproducible deterministic (periodic) pattern. It allows low estimate variance of statistical quantities such as correlation functions if taken over an integral number of sequences. Besides, the signal power of a PRBS, determining the power of the detector response, is much higher than that of the conventional impulse-like injection function with similar amplitude. The power is equally spread over the frequency range of the chromatographic system if the clockperiod is chosen sufficiently short. This white noise property is an essential condition for CC application. The binary levels can be used to control simple on/off valves.

The detector signal resulting from the multiple injections is built up of chromatograms $h(i)$ shifted in time, plus detector noise $n(i)$:

$$y(i) = \sum_{j=0}^{M-1} [h(j), p(i-j)] + n(i) . \tag{6}$$

The PRBS length has to be equal to or longer than the time duration of a comparable single injection chromatogram. After one pseudo-random sequence, the pre-sequence, the detector signal becomes circular.

A correlogram, almost similar to the conventional chromatogram, can be calculated using the inverse of the PRBS, defined as the function $p^{-1}(i)$, producing a Kronecker delta function $\Delta(i)$ after circularly cross-correlating with $p(i)$:

$$R_{p^{-1}p}(i)\frac{1}{M} = -p^{-1}(j+i) \cdot p(j) = \Delta(i) \tag{7}$$

with $\Delta(i) = 1$ for $i = 0$ and $\Delta(i) = 0$ for $i \neq 0$.

The cross-correlation of the detector signal $y(i)$ with the inverse $p^{-1}(i)$ results in a correlogram:

$$R_{p^{-1}y}(k) = \frac{1}{M} \sum_{i=0}^{M-1} [p^{-1}(i+k) \cdot y(i)] . \tag{8}$$

Combining Eqs. (6) and (8) gives

$$R_{p^{-1}y}(k) = \sum_{j=0}^{M-1} \left\{ h(j) \frac{1}{M} \sum_{i=0}^{M-1} [p^{-1}(i+k) \cdot p(i-j)] \right\}$$

$$+ \frac{1}{M} \sum_{i=0}^{M-1} p^{-1}(i+k) \cdot n(i) . \tag{9}$$

One of the special properties of a PRBS is that the inverse calculated with one point per period and levels 1 and 0 is the same PRBS, but with levels $+M/I$ and $-M/I$. Considering the levels for $p^{-1}(i+k)$, $p^{-1}(1+k) \cdot n(i)$ can be replaced by $(M/I) \cdot n(i, k)$.
Inserting Eq. (7) into Eq. (9) gives

$$R_{p^{-1}y}(k) = \sum_{i=0}^{M-1} [h(j)\Delta(k-j) + \frac{1}{M} \sum_{i=0}^{M-1} \frac{M}{I} \cdot n(i,k) . \qquad (10)$$

The first term is a convolution of the impulse response (chromatogram) and a Kronecker delta function resulting in the same impulse response. Concerning the second term, adding M non-correlated points for every k results in noise with a standard deviation (SD) of $M^{1/2}$ times the original SD of the noise:

$$R_{p^{-1}y}(k) = h(k) + \frac{M^{1/2}}{I} n(k) \approx h(k) + \sqrt{2/I}\, n(k) . \qquad (11)$$

The autocorrelogram of the input sequence is sometimes referred to as the "virtual injection" profile because it can be proved that the cross-correlogram is identical to a chromatogram obtained from an injection with a profile equal to that autocorrelogram. The procedure can be continued with an arbitrary number of sequences. The longer the system is run the higher the S/N ratio will be, assuming the chromatographic system is stationary. Ultratrace compounds can be detected precisely without the necessity of cumbersome and often irreproducible preconcentration step.

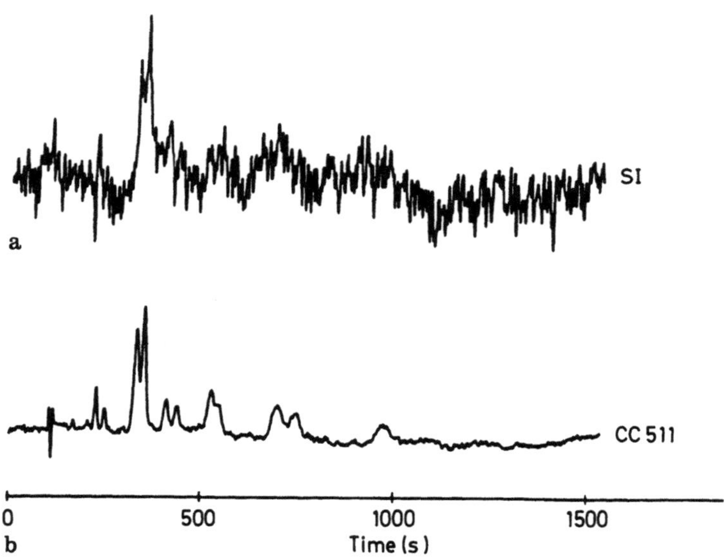

Fig. 7. **a** Single injection (SI) chromatogram of polynuclear aromatic hydrocarbons (PAHs). **b** Corresponding correlogram of the PAHs

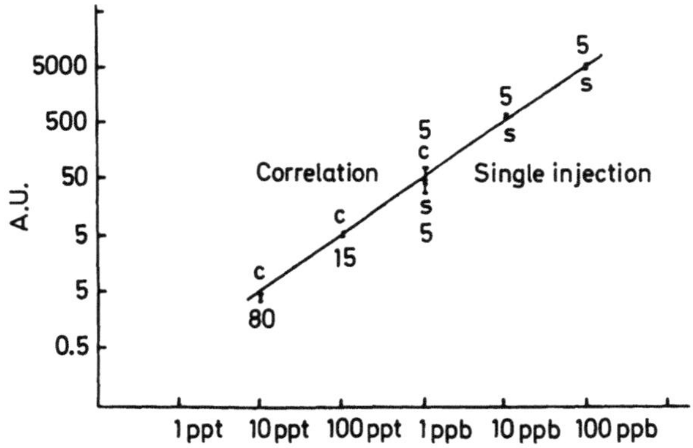

Fig. 8. HPLC calibration graph of phenol with fluorimetric detection

Figure 7 shows a typical example of CC applied in environmental analysis [20]. An HPLC-trace and the corresponding correlogram of a mixture of polynuclear aromatic hydrocarbons (PAHs) is shown.

In Fig. 8, a calibration graph, determined with CC as well as with single injection chromatography, demonstrates the analytical performance of CC. Phenol was measured over five decades of concentrations: 0.01–100 μg l^{-1}, with a conventional HPLC-system with fluorimetric detection. The two higher concentrations (10–100 μg l^{-1}) were determined by conventional reversed phase HPLC, the two lower concentrations (0.01–0.1 μg l^{-1}) by correlation HPLC with 16 and 3 sequences of correlation time, respectively. The bars indicated on the calibration graph represent the peak area $\pm 3\sigma_I$, when σ_I is the standard deviation of the integrated noise. The inner bars at the 1 μg l^{-1} level represent the correlation results and the outer bars the single injection results.

The detection limit for the single injection experiments, defined as $3\sigma_I$, was about 0.5 μg l^{-1}, the detection with the 10 ng l^{-1} concentration was estimated to be 3 ng l^{-1} (= 3 ppt = 3 parts per trillion) [19].

Instrumental and Computer Requirements

The injection device is the most noteworthy modification of the separation system to be used with multiple injections. Such a device has to satisfy high demands concerning reproducibility, switching speeds, no wear and tear problems, absence of memory effects, and computer controllability. Incorrect injection causes disturbances (ghost peaks, correlation noise), proportional to the amplitudes of real peaks, at specific relative positions on the time axis. Correlation noise limits the determination of traces in the presence of main components.

At the moment a reliable, accurate and simple injection system, meeting all the demands of correlation HPLC, is obtainable. Concerning the computer hardware requirements, a standard PC with 640-k byte memory plus a hardware card for controlling the switching valves and an AD-converter (12 bits, minimum sampling frequency 10 Hz) including an anti-aliasing filter is sufficient.

Appropriate software is essential, it has to include the generation of an arbitrary number of PRBSs with selectable duration of the clockperiod and $2^n - 1$ clockperiods in a sequence; n is an integer usually between 5 and 12. Furthermore, a straightforward cross-correlation algorithm has to be included, which can eventually be replaced by an off-line Hadamard transform procedure, speeding up the calculation.

Several features like the display of the selected parameters (clock-period, sequence length, number of sequences, detector signal, the on-line calculated chromatogram, and the injection PRBS) make such a program much more user-friendly.

Comparison with Single Injection Chromatography

The much lower detection limits without preconcentration compared to single injection chromatography makes CC very suitable for trace analysis. Besides, the accuracy and reproducibility of the determination satisfies high demands compared to preconcentration techniques, generally showing poor performance in that respect. A mostly unknown property is that in CC non-linear affected separations (due to e.g. a non-linear distribution isotherm), improve drastically, although some correlation noise may arise [25].

Some disadvantages of CC are the required extra sample, the extra time, the high demands on the injection system, and the possible correlation noise. An important disadvantage may be that in CC stability (stationarity) of the system is required; gradient elution or programmed temperature techniques are out of the question.

Derived Correlation Techniques

As already mentioned, quite a number of modifications of CC, each with special properties and each with potential applicability in environmental analysis, are developed: SSCC, DCC, SCC, CMCC.

Single Sequence CC can be considered as an intermediate between single injection chromatography and CC. It overcomes the difficulty that in CC stationarity of the system (no varying condition influencing the retention time, the peak shape, etc.) is required. In SSCC a large volume sample is injected into the column. The width of this rectangular input pulse is acceptable with respect to peak broadening and resolution of the broader peaks with longer retention times, but is too large for a satisfactory resolution of narrow peaks.

The method implies that rectangular injection shape is modulated with a fine structure according to a PRBS. The first part of the chromatogram is calculated by a deconvolution procedure without a loss of resolution, and the last part can be processed in the conventional way. This opens the possibility of applying gradient elution or programmed temperature after elution of the narrow peaks. For the whole chromatogram the S/N increases considerably, but not as much as in normal CC.

In Fig. 9 the results are shown of injections of different concentrations of a mixture of *m*-dihydroxybenzene, *o*-dihydroxybenzene, *p*-cresol, *o*-cresol, 2,3-dimethylphenol, 2,4-dihydroxyphenol, and toluene, both in SSCC and in single injection HPLC. The eluent was methanol-water (50:50, v/v) [6]. Concentration

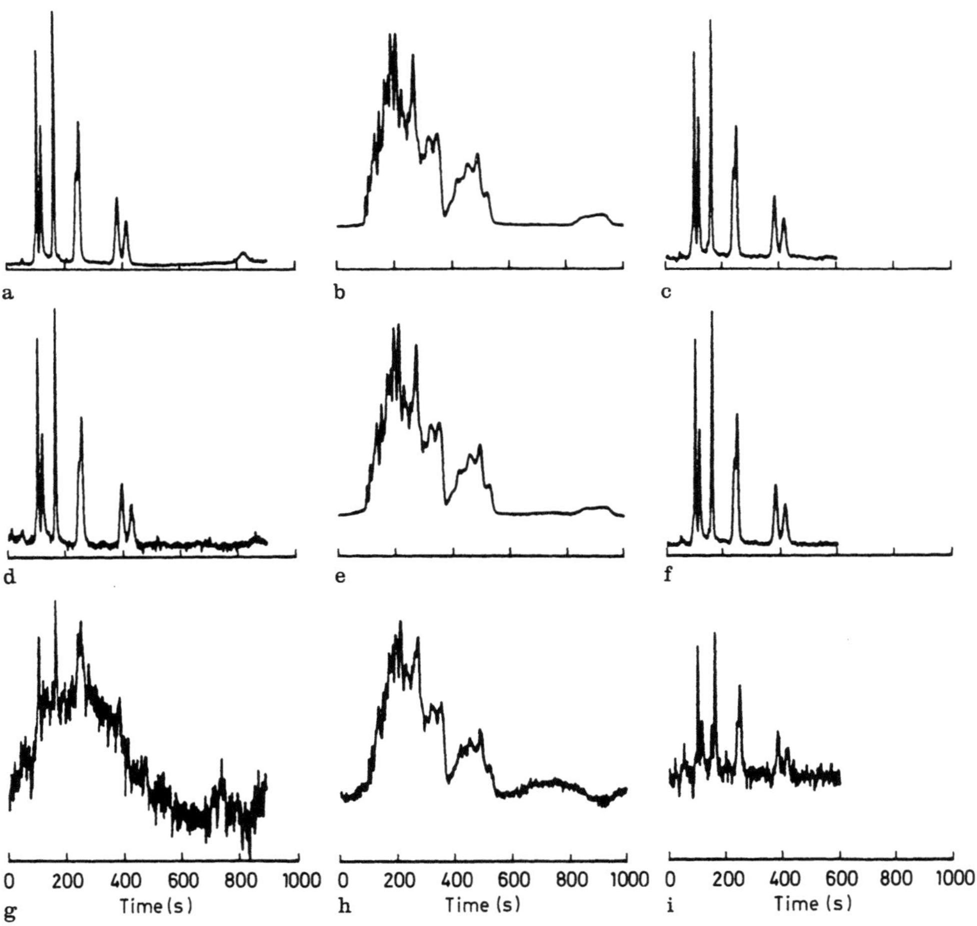

Fig. 9a, d, g. Chromatograms. **b, e, h** SSCC detector signals. **c, f, i** SS correlograms. These are for different concentrations – see text

from top to bottom: 3–5 μg ml^{-1}, 0.3–0.5 μg ml^{-1}, and 30–50 ng ml^{-1} for each component.

In CC only differences between sample and adapted eluent or possibly another sample are measured, resulting in a correlogram with positive and even negative peaks. This property can be used to suppress to a great extent the correlation noise directly due to (unimportant) main components in the correlogram, interfering in the determination of trace components. Often these main components are well known and the concentrations of these components in the eluent can easily be made almost equivalent to the concentrations in the sample. The result is a drastic reduction of the correlation noise, either caused by imperfect injections or by a non-linear distribution isotherm.

Differential Correlation Chromatography can be useful in environmental analysis, particularly in trace analysis of samples with a relatively complex matrix. A possible application of DCC is the monitoring of potential sources of pollution at the subtrace level by determining the difference in concentration before and after the source of pollution [4]. Another application is monitoring variations in concentrations with time, of course after taking some precautions concerning conservation of the samples.

A peculiar modification of CC is Simultaneous Correlation Chromatography. The principle of this technique is injecting not one but a number of different samples, each according to a pseudo-random pattern, mutually completely uncorrelated. If required, the same components, of course with generally different concentrations, may be present in the samples. The different correlograms are calculated by cross-correlating the very complex output with the corresponding input pattern [5]. In practice a long PRBS with a length equal to the sum of the duration of the n different corresponding chromatograms has to be used. SCC does not reduce the analysis time, because the duration of the correlogram is n times the duration of one chromatogram. All different samples are injected according to the same long PRBS, however, each with a different time shift equal to an integral number of chromatogram durations. A possible application is high precision chromatography, where calibration and measurements are affected simultaneously in the same column under identical conditions. The noise reduction property of CC is maintained and, a very accurate calibration and determination can be achieved.

A relatively novel application of correlation techniques in separation methods is Correlation Capillary Zone Electrophoresis. In Capillary Zone Electrophoresis (CZE) the high detection limit is a serious problem due to the required small amount of sample and the small detector volume. The application of correlation techniques is obvious and work in this field is in progress. Some preliminary results, showing the possibilities of CCZE, are presented [8]. Figure 10 shows a set-up of a CCZE-system. Most attention has to be paid to the development of a suitable injection system, where electro kinetic injection is probably most suitable. Typical problems to be overcome are the influence of the high-voltage switching in the sensitive detector and the required stationary of the system.

In Chemical Modulation Correlation Chromatography the mechanical modulation by switching valves is replaced by a chemical conversion of components

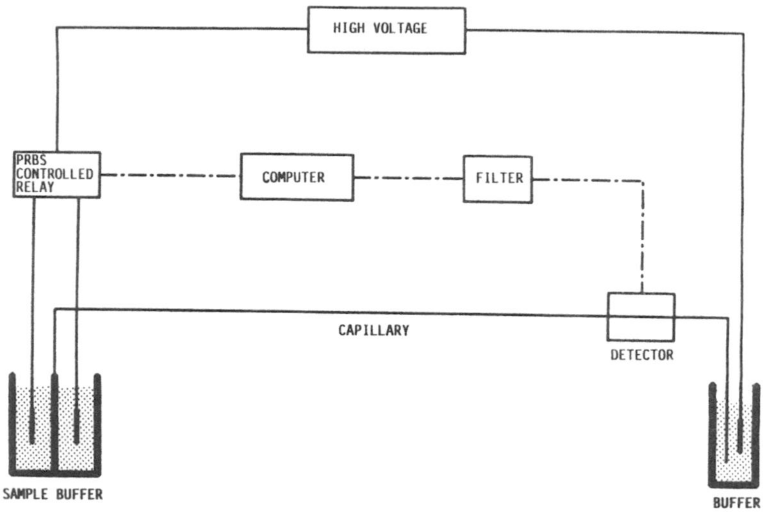

Fig. 10. Set-up of a correlation CZE system

in the sample, also controlled by a PRBS (Fig. 6b). Extra selectivity is added to the system because the chemical modulation can be specific for one or more components. There is no separate carrier gas or eluent: the analyte, e.g. ambient air or water, and possibly modified for optimum separation, is used as the mobile phase, which may be an advantage but which may also reduce the flexibility. No moving parts are present, but the chemical modulator is generally more sensitive to pollution. Much pioneering work in this field has been done by Phillips et al. [3, 7]; several destructive and non-destructive modulators were developed.

In the literature many principles of chemical modulation, both in GC and in HPLC, are described, e.g. thermal decomposition modulation [21, 22], thermal-catalytic modulation, spark modulation [23], and electrochemical concentration modulation [24]. The (potential) possibilities are promising for applications in environmental analysis, as can be concluded from the results obtained so far. For instance, traces of methane in ambient air were monitored with a thermal-catalytic modulator, hydrocarbons (e.g. ethane) in air were determined by a (destructive) hot wire thermal decomposition modulator, and phenol was determined in canal water with an electrochemical concentration modulator, a correlation HPLC system. The relatively slow dynamic behaviour of a chemical modulator can be the source of ghost peaks and correlation noise [24].

Discussion

Optimal signal processing and the application of correlation techniques can be very valuable in environmental analysis. The sometimes drastic decrease in the uncertainty in the determination and the corresponding lowering of the detection limit make these techniques very useful. However, hardware (injection system) for CC and the related techniques and software are not commercially available. The techniques are not easy to understand or easy to implement. This is probably the reason why the applications in practice have been restricted so far.

References

1. Smit HC (1983) Trends Anal Chem 2:1
2. Kaljurand M, Smit HC: (1994) Chromotographia 39: 210
3. Valentin JR, Carle GC, Phillips JB (1985) Anal Chem 57:1035
4. Laeven JM, Smit HC, Kraak JC (1987) Anal Chim Acta 194:11
5. Smit HC, Mars C, Kraak JC (1986) Anal Chim Acta 181:37
6. Louwerse DJ, Boelens HFM, Smit HC (1992) Anal Chim Acta 256:349
7. Phillips JB, Luu D, Pawliszyn JB, Carle GC (1985) Anal Chem 57:2779
8. Louwerse DJ, van der Moolen JN, Bruin GJM Poppe H, Smit HC (1992) presented at the 5th International Conference on Chemometrics in Analytical Chemistry (CAC '92), Montreal, Canada
9. Smit HC, Walg HL (1975) Walg Chromatographia 8:311
10. Scheeren PJH, Klous Z, Smit HC, Doornbos DA (1985) Anal Chim Acta 171:45
11. Bevington PR (1969) Data Reduction and Error Analysis for the Physical Sciences, McGraw Hill, New York
12. Scheeren PJH, Barna P, Smit HC (1985) Anal Chim Acta 167:65
13. Vaidya RA, Hester RD (1984) J Chromatogr 287:231
14. van den Heuvel EJ, van Malssen KF, Smit HC (1990) Anal Chim Acta 235:343
15. van den Heuvel EJ, van Malssen KF, Smit HC (1990) Anal Chim Acta 235:355
16. Shannon CE (1948) Bell System Tech J 28:379
17. Jaynes ET (1957) Phys Rev 106:620
18. Kaljurand M, Kullik E (1989) Computerized Multiple Input Chromatography, Ellis Horwood Ltd., Chichester
19. Laeven JM, Smit HC, Kraak JC (1983) Anal Chim Acta 150:253
20. Mars C, Smit HC (1990) Anal Chim Acta 228:193
21. Lovelock JE (1975) J Chromatogr 112:29
22. Engelsma M, de Graaf J, Smit HC (1991) Anal Chim Acta 252:187
23. Engelsma M, Smit HC (1991) Chromatographia 31:393
24. Engelsma M, Kok WTh, Smit HC (1990) J Chromatogr 506:201
25. Louwerse DJ, Smit HC (1994) Chromatographia 38:62

Information Theory of Signal Resolution – Precision of Measurements

Yuruzu Hayashi and Rieko Matsuda*

National Institute of Health Sciences 1-18-1 Kami-Yoga, Setagaya, Tokyo 158, Japan

Summary

Even a sophisticated signal processing suffers more or less from strong peak overlap or low signal-to-noise ratio. This discussion emphasizes not only the practical use of the Kalman filter but also the statistical and probabilistic aspects of it. The precision or relative standard deviation (RSD) of the estimates obtained from the Kalman filter is considered in a simple model where white noise is the only randomness. The RSD of the filter estimates is shown to be predicted from the degree of peak overlap and from the peak width and area with accuracy. The reliability of quantitative data can be evaluated with the predicted precision as a standard without repeated simulations. Mutual information has the equivalent meaning to the precision and is a useful concept for the optimization of instrumental conditions, especially for multi-peak output. The Kalman filter is selected as signal processing here on account of its mathematical simplicity and relevance to all the subjects discussed here.

List of Symbols and Abbreviations

A_j	area of peak j
$E[.]$	the ensemble mean of a random variable in the square brackets over the repeated measurements ($i = 1,\ldots,n$)
σ_j	width of Gaussian peak j (standard deviation)
ΔT	sampling intervals of an analog-to-digital converter
FUMI	Function of Mutual Information
RSD	relative standard deviation
F_k	signal intensity at data point k
L_k	Kalman gain at data point k
P_k	error variance at data point k
W_k	value of white noise at data point k (random variable)
$\widetilde{W}_k$	standard deviation of white noise W_k
X	quantity of target material to be estimated (e.g., concentration)
Y_k	observed data at data point k
k	data point ($k = 1,\ldots,N$)
n	the number of repeated measurements or experiments
N	the number of data points
$X(i)$	the estimate for X at the i-th measurement ($i = 1,\ldots,n$)
$\widehat{X}_k$	the estimate for X at data point k
$\kappa_c(j)$	cutoff point for peak j (the starting point of Kalman filtering)
$\kappa_f(j)$	filter-off point for peak j (the ending point of Kalman filtering)
Rs	resolution often used in chromatography
$\phi(j)$	FUMI for peak j
Φ	FUMI for all the peaks in a data set (total information)
$\psi(j)$	intact information for peak j (information free from peak overlap)
$\delta\phi(j)$	information loss (caused by peak overlap)
τ_j	position or retentiont time of peak j

Introduction

Various mathematical methods have been developed to deconvolute overlapped signals into their individual signals [1]. Such techniques are certainly useful in the situations where further separation is impossible or difficult to carry out through HPLC (high performance liquid chromatography) or other analytical instruments in separation science. However, even sophisticated signal processing is neither omnipotent nor can be relied upon too much. In general, the more strongly the signals overlap the more error the mathematical technique is accompanied by. Therefore, this discussion not only explains the practical use of signal processing but also emphasizes the statistical aspect of it.

Kalman filtering is taken here as signal processing, because: (i) the relative standard deviation (RSD) for the filter estimates is theoretically predictable from the peak shape, noise level and overlap without resorting to the repeated simulations in a simple measurement model; (ii) the precision of the filter (1/RSD) is equivalent to Shannon's information.

This information theory of signal processing will have wide applicability in environmental chemistry. The environmental levels and human exposure of substances which are detected only at low levels can be evaluated from both the observed quantity and its predicted precision. Various data from different instruments or from diverse environments will be compared with the universal quality criterion (precision). Governmental action for chemical safety programs can also be planned according to the universal judgement of environmental data.

This discussion is devoted to the improvement in the precision of trace and subtrace analyses, but also considers the judgement of the statistical reliability of the data provided by the signal processing.

Measurement Model

Figure 1 illustrates the measurement model used throughout this discussion. The data set of measurement 1 contains Gaussian peak signal, $F_1, \ldots, F_N$, in N data points. A random noise, called white noise, $W_1, \ldots, W_N$, is superimposed over the signal. The white noise is characterized by zero mean and normal distribution. An observed value, Y_k, at data point k is described as $Y_k = F_k + W_k\,(k = 1, \ldots, N)$.

Signal processing of the data set of measurement 1 provides an estimate, $\widehat{X}(1)$, for the analytical quantity such as concentration. A recursive signal processing such as Kalman filtering provides an estimate, $\widehat{X}_k$, at every data point (see below) as it proceeds from $k = 1$. Any one of the N estimates, $\widehat{X}_1, \ldots, \widehat{X}_N$, obtained from measurement 1 can be the result, $\widehat{X}(1)$. For example, the last estimate, $\widehat{X}_N$, will often be selected because of its stochastic reliability ($\widehat{X}(1) = \widehat{X}_N$; see below). The repetition of the above measurement produces n estimates, $\widehat{X}(1), \ldots, \widehat{X}(n)$. The RSD of the n estimates is the concern of this discussion.

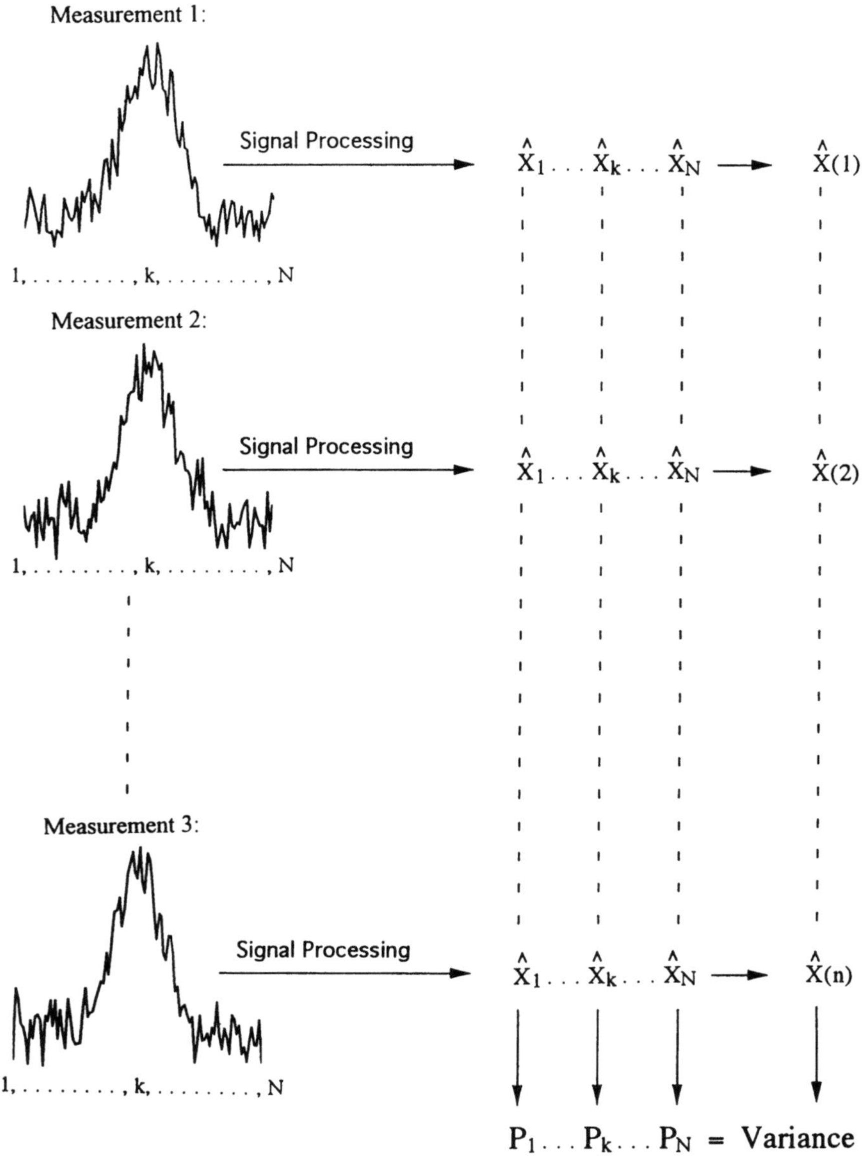

Fig. 1. Data and estimates in measurement model. The estimate, $\widehat{X}(i)$, of the i-th data set or measurement i is one of the estimates, $\widehat{X}_1,\ldots,\widehat{X}_N$, obtained by the signal processing of the data set. The S/N which is defined as the height of the peak maximum divided by the SD of the white noise ($\widetilde{W} = \widetilde{W}_k$) is 10 in this figure. The relationship shown in Eq. (5) is indicated

The n data sets differ only in the noise appearance from each other, while the entire areas, widths (standard deviation) and positions (mean) of the Gaussian peaks and standard deviation (SD), $\widetilde{W}_k$, of the white noise are kept constant. That is, the peak shape, $F_1,\ldots,F_N$ and noise level, $\widetilde{W}_k$, are invariant during all the measurements from *1* to n. The estimates, $\widehat{X}(1),\ldots,\widehat{X}(n)$, will vary from each other due to the random noise. Thus, the mean and variance of the estimates should result from a large number of data sets, n. The statistical study based on the artificial randomness is called a Monte Carlo technique [2]. The situation of Fig. 1 mimics the repeated measurements on the same samples under the same operating conditions of an analytical instrument.

In a linear measurement model, the amplitude of the signal for a target material is directly proportional to concentration, X, at every data point except for the white noise:

$$Y_k = F_k X + W_k \quad \text{(measurement model)}. \tag{1}$$

The aim of signal processing is to estimate the concentration, X, from the observed noisy data, $Y_1,\ldots, Y_k,\ldots, Y_N$. For example, a value near unit concentration ($X = 1$) will result from the data of the model signal, $F_1,\ldots,F_N$, and noise, W_k.

A BASIC program for generating the white noise of zero mean and unit *SD* from the uniform random number ranging between 0 and 1 (RND) is surprisingly simple:

```
10 AVE = 0
20 FOR J = 1 TO 12 : AVE = AVE + RND : NEXT
30 NOISE = AVE − 6
```

The white noise (NOISE) can be produced by adding 12 uniform random numbers and by drawing 6 from the sum. A product, b*NOISE, is identical to the white noise with zero mean and *SD* of b. In addition, the variance of the uniform distribution is $1/12$ (see line 20 of the program). The central limit theorem [3] should be referred to.

Kalman Filter

Algorithm

Detailed formulation of the Kalman filter is available in the chemical [4–7], engineering [8, 9] and mathematical [10] literature. As the Kalman filter proceeds from data point 1 to k of a data set in the measurement model (Eq.(1) and Fig. 1), it provides an estimate, $\widehat{X}_k$, of the concentration at data point k:

$$\widehat{X}_k = \widehat{X}_{k-1} + L_k\,(Y_k - F_k\widehat{X}_{k-1}) \tag{2}$$

where L_k is called Kalman gain. The update estimate, $\widehat{X}_k$, is supplied by correcting the last estimate, $\widehat{X}_{k-1}$, alone without referring to any previous estimate, $\widehat{X}_1,\ldots,\widehat{X}_{k-2}$ (recursive property). Innovation, $Y_k - F_k\widehat{X}_{k-1}$, represents the discrepancy between the update measurement, Y_k, and the last estimate, $F_k\widehat{X}_{k-1}$., in the scale of Y_k (not X). Thus, the Kalman gain, L_k, serves as a weighting factor for the correction.

In the Kalman filter analysis, the model signal, F_k, and noise level, $\widetilde{W}_k$, are assumed to be known at every data point before the signal processing is started. In Eq. (2), Y_k is observable at data point k and the last estimate, $\widehat{X}_{k-1}$, can be known recursively from the last estimation. Therefore the Kalman gain, L_k, is the only requirement for achieving the update estimation. An arbitrary value of $\widehat{X}_0$ should be assumed at the estimation at $k = 1$.

The Kalman gain at data point k can be obtained through error variance, P_k:

$$L_k = \frac{P_{k-1}F_k}{\widetilde{W}_{k2} + P_{k-1}F_{k2}} \tag{3}$$

$$P_k = P_{k-1} - L_k F_k P_{k-1}\,. \tag{4}$$

Given an arbitrary value of P_0, the Kalman gain can be calculated from Eqs. (3) and (4) as follows: $P_0 \rightarrow L_1 \rightarrow P_1 \rightarrow L_2 \rightarrow \ldots \rightarrow P_k \rightarrow \ldots \rightarrow P_N$. In practice, a large initial value of P_0 is preferable for successful filtering [4]. The arbitrariness of P_0 and $\widehat{X}_0$ vanishes as the filtering advances [8].

The observed value, Y_k, estimate, $\widehat{X}_k$, and white noise, W_k, are random variables because of the randomness. On the other hand, the Kalman gain and error variance involve only the deterministic value, F_k, and statistical constant, $\widetilde{W}_k$, and are not random variables. The clear distinction between the random variables and usual variables is a key to understanding the theory of Kalman filtering.

Time-Course of Filtering

Figure 2 shows the time-courses of the noise-contaminated data, Y_k (a), estimate, $\widehat{X}_k$ (b), error variance, P_k (c), and Kalman gain, L_k (d) [11]. As the signal of the Gaussian peak begins to predominate in the noisy data, Y_k, the error variance, P_k, abruptly decreases. In this region A, the filtering process exhibits conspicuous activity which is characterized by a large value of Kalman gain, L_k. This activity is reflected on the abrupt change in estimate, $\widehat{X}_k$: the last estimate, $\widehat{X}_{k-1}$, is improved more and more as new information, Y_k, is collected during region A.

After this active region A, the estimate, $\widehat{X}_k$, is no longer altered appreciably by the further acquisition of the information, Y_k. Now the filter does nothing, even if fully supplied with the raw data, Y_k, containing the useful information on the desired quantity, X. The error variance is a monotone-decreasing function with data point, k. This means that the estimation error never increases, as the information concerning the analyte quantity is accumulated increasingly.

The commonly used least squares method is similar to the Kalman filter in that the model peak, F_k, is fitted to the real data, Y_k. However, one of the most important differences is that the common least squares method works on the real data with equal weight at every data point or with preset weights in the algorithm. The Kalman gain which corresponds to the weight is out of control and uniquely

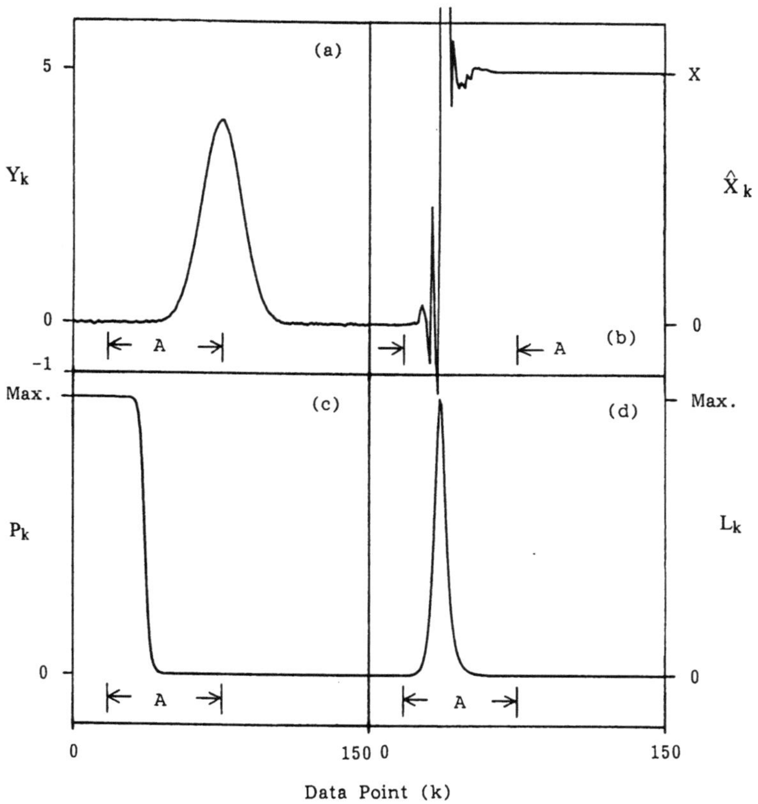

Fig. 2a-d. Time-course of the observed value: **a** Y_k; **b** estimate, $\widehat{X}_k$; **c** error variance, P_k; **d** Kalman gain, L_k (From [11])

determined from the peak shape and noise level as shown in Eqs. (3) and (4). So is the active region of the filter.

Precision of Estimates

We consider the precision of the filter estimate, $\widehat{X}_k$, at data point k in the simple measurement model of Fig. 1. The Kalman filter has the notable property that the observed RSD of the n estimates, $\widehat{X}(1),\ldots,\widehat{X}(n)$ $(= \widehat{X}_k)$, is equal to the error variance and is also predictable from the peak shape and noise level [12]:

$$\mathrm{RSD}^2 = \frac{1}{\mathrm{n}-1}\sum_{\mathrm{i}=1}^{\mathrm{n}}(\widehat{X}(i) - E[\widehat{X}(i)])^2 \tag{5A}$$

$$= P_k \tag{5B}$$

$$= \frac{\widetilde{W}^2}{F_1^2 + F_2^2 + \ldots + F_k^2} \tag{5C}$$

where the SD of the white noise is assumed to be constant at any data point ($\widetilde{W}_k = \widetilde{W}$) and the true concentration of the target material is unity: $X = 1$ and $E[\widehat{X}(i)] = 1$. Equation (5C) is derived from Eqs. (3) and (4) [12].

We should note that the statistic is defined over the series of measurements as shown in Eq. (5A). By analogy with the definition in spectral analysis [13], the variance, P_k, may be referred to as "ensemble" variance. The ensemble mean and ensemble variance of the white noise, $E[W_k]$ and $E[W_k^2]$, are equal to the time mean and time variance, respectively, $\frac{1}{N}\sum_{k=1}^{N} W_k$ and $\frac{1}{N-1}\sum_{k=1}^{N} W_k^2$. This property is called ergodic [13]. The time mean is carried out over time N with measurement i kept constant and the ensemble mean over the n measurements with time remaining constant. Of course, the time mean and variance of the filter estimates make no sense in discussing the statistical reliability of the filter.

The three ways to assess the precision of the filter are: (i) statistical method using Eq.(5A); (ii) numerical method using Eq.(5B); (iii) probability theory using Eq. (5C). These different approaches are in good agreement statistically [14]. The statistical method cannot be started, until all the n estimates are secured. On the other hand, all the demands of the theoretical approach (iii) are the model peak, F_k, and noise level, $\widetilde{W}$. The numerical method and theoretical approach dispense with the repeated simulations to perform the error prediction, but the error variance, P_k, at data point k results from the consecutive calculations from $k = 1$.

Influence of Peak Area and Width on Precision

The Kalman filtering of a Gaussian peak j with area, A_j, and width, σ_j (SD), over the infinite region (Eq. 5) takes another useful form of the precision [12, 14]:

$$\mathrm{RSD}^2 = \frac{2\pi^{1/2}\sigma j\, \Delta T\, \widetilde{W}^2}{Aj^2} \tag{6}$$

where ΔT denotes the sampling interval for the analog-to-digital conversion. As the number, k, of raw data processed by the Kalman filter increases, the estimation becomes more precise and the error variance converges to the lower limit shown in Eq. (6).

The effects of the entire area and width of a Gaussian peak on the filter precision can be seen from Eq. (6). The RSD decreases with increasing area, A_j, but increases with increasing peak width, σ_j. That is, a larger (larger A), sharper (smaller σ) signal provides higher precision. If the peak area is fixed, the highest precision can be obtained from the sharpest signal, called delta function, which takes all the signal intensities exclusively at a single data point.

Information Retrieved from Overlapped Signals

The stochastic properties of signals are summarized:
(A) as the peak broadens, the precision decreases (i.e., RSD increases);
(B) as the peak area increases, the precision increases;
(C) as the noise decreases, the precision increases;
(D) as the peak overlaps with another, the precision decreases.
Factors *A* and *B* refer to the property indigenous to a peak or signal-to-noise ratio (S/N), but *D* the topological relationship between neighboring peaks. The word, precision, can be replaced by "information" or "FUMI" without altering the meaning (see below).

The multi-dimensional filter is often employed in tackling the resolution of multi peaks. However, a one-dimensional filter, called a scalar filter [11], can be considered for this problem on account of the following advantages: (i) theoretical simplicity for the error prediction of the filter estimates; (ii) applicability of the error prediction to another peakresolving methods such as the adaptive Kalman filter (a version of the regular filter [7]).

Signal Overlap and Scalar Kalman Filter

The scalar Kalman filter has its own time region over which it works on a single peak. The shaded regions in Fig. 3 illustrate the filtering for peaks j and $j+1$. The complete isolation of these regions enables the scalar calculation of each precision of multi peaks by Eq. (5C); the squared signal intensities are summed up only in the region.

From the filtering region for peak j, the signal of the adjacent peak $j+1$, higher than 0.05% of the apex of peak j is excluded. The signal of the adjacent peak $j+1$ first reaches the highest acceptable signal at the filter-off point, $\kappa_f(j)$,

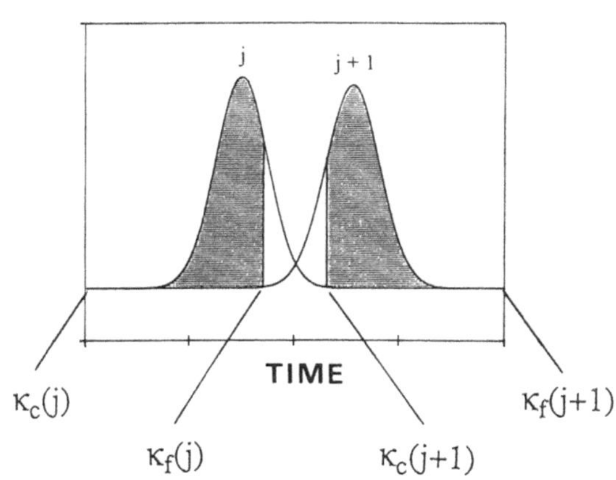

Fig. 3. Filtering regions of scalar Kalman filter for peak resolution. ($\kappa_c(j)$, $\kappa_f(j)$) denotes the filtering region of peak j and ($\kappa_c(j+1)$, $\kappa_f(j+1)$) that of peak $j+1$. (From [12])

of peak j. The cutoff point, $\kappa_c(j)$, is so far from the peak center that the sum of the signal intensities beyond the cutoff point is negligibly small.

If peak j overlaps with one more peak $j-1$ from the left, the shaded Gaussian peak lacking both the edges appears over the filtering region. The cutoff point, $\kappa_c(j)$, of peak j coincides with the highest acceptable signal limit of the adjacent peak $j-1$. The width of the region $(\kappa_c(j), \kappa_f(j))$ is governed by the signals of adjacent peaks and not by peak j itself. This definition of overlap is quite natural and useful for evaluating the quantification of a small peak adjacent to a large disturbing peak, e.g., an HPLC assay of a small impurity of an optical isomer present only in a small amount [14].

The error of the multi-dimensional Kalman filter for q peaks can be described by the determinant of the q-dimensional error variance, P_k. The calculation of this precision by a personal computer becomes time-consuming, as the dimension or the number of peaks increases. Moreover, the simple mathematical expression of the error like Eqs. (5C) or (6) will be more difficult to derive from the multi-dimensional error variance.

Scalar Kalman Filter for Signal Resolution

In Fig. 4, the RSD of the filter estimates is plotted against the resolution, Rs, for two Gaussian peaks obtained from HPLC [14]. The resolution is defined as $Rs = [(t_{j+1} - t_j)/(t_j + t_{j+1})]\sqrt{N}/2$ where N denotes the number of theoretical plates (constant) and t_j the position (mean) of peak j. The first eluted peak is fixed in position and the second eluted peak is moved with its width changed according to the fundamental equation: $N = (t_j/\sigma_j)^2$.

As the resolution decreases from $Rs = 1.0$, the RSD for the moving peak increases abruptly, because the peak overlap narrows the filtering region $(\kappa_c(j), \kappa_f(j))$ and reduces the denominator of Eq. (5C). The slight decrease in the S/N arising from the peak broadening, as Rs increases from 1.0, is the direct cause of

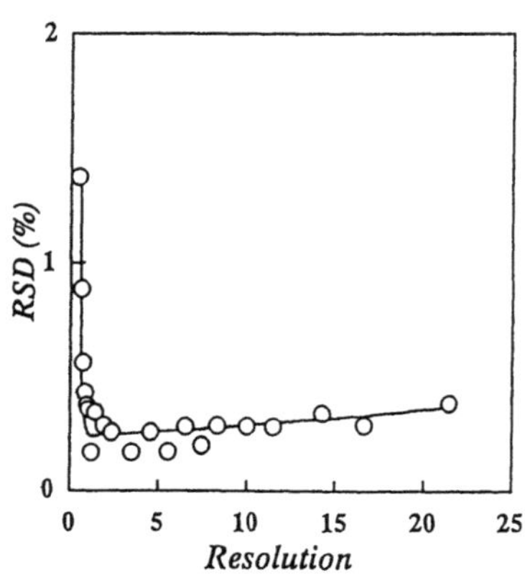

Fig. 4. Effect of overlap on precision of scalar Kalman filter applied to chromatography. o: the results from the Monte Carlo simulation and common least squares fitting;– : theoretical curve from Eq. (9) (RSD $= \exp[-\phi(j)]$). (From [14])

Table 1. The filter estimates from five experiments for the naphthalene and diphenyl mixture.

No.	Naphthalene		Diphenyl	
	Scalar Filter	2-D Filter	Scalar Filter	2-D Filter
1	99.8	100.4	100.2	100.1
2	98.0	99.6	101.2	101.1
3	100.1	100.5	99.6	99.6
4	99.7	99.9	100.4	100.4
5	101.0	100.4	100.3	99.5
Mean	99.7	100.2	100.3	100.1
RSD (%)	1.09	0.39	0.57	0.65

(From Ref. [11])

the gradual increase in the RSD. The resolution at the minimum RSD of Fig. 4 is called the optimal resolution [12].

Below $Rs = 1.0$, the RSD for the peak is more than that calculated from the entire peak shape without overlap (Eq. 6). There exists information loss caused by overlap (see below). Above $Rs = 1.0$, however, the active region of the peak, e.g., region A in Fig. 2, is completely covered by the filtering region ($\kappa_C(j)$, $\kappa_f(j)$) and the peak can successfully be resolved by the scalar filter with the overlap-free precision (Eq. 6), even if the peak appears to be overlapped in the time scale.

The precision of the Kalman filter (—) is identical to that of the common least squares curve fitting (o) in the example and Eq. (5C) can also be an exact description of the RSD for the latter [14]. This coincidence is not surprising, because they have an equivalent mathematical structure [11].

Table 1 lists the estimates of the scalar and two-dimensional Kalman filters in an HPLC determination for a mixture of naphthalene and diphenyl ($Rs \approx 1$ and $S/N \approx 10^6$) [11]. The observed RSD values for the filters are both satisfactory and indistinctive within experimental error; since the reproducibility of the HPLC system itself was 0.24%. The bias of the filters is at most 0.3% (the true value of the estimates is 100) and is much better than that of the well-known perpendicular dropping in many situations. The theoretical values calculated from Eq. (5C) fall below the observed ones, but are corrigible (see final section).

Adaptive Kalman Filter for Signal Resolution

Situations often arise where an "unknown" peak interferes with a target "known" peak to be quantified. The adaptive Kalman filter has more flexibility and ability to remove the adverse contributions of the unknown peaks than the regular Kalman filter [7, 15]. The regular filter can only analyze the known peaks exactly, because the peak shape, F_k, should be known or modeled before the measurement and filtering. Despite a slightly different algorithm, the precision of the adaptive filter can also be predicted in the similar way to the scalar Kalman filter [15].

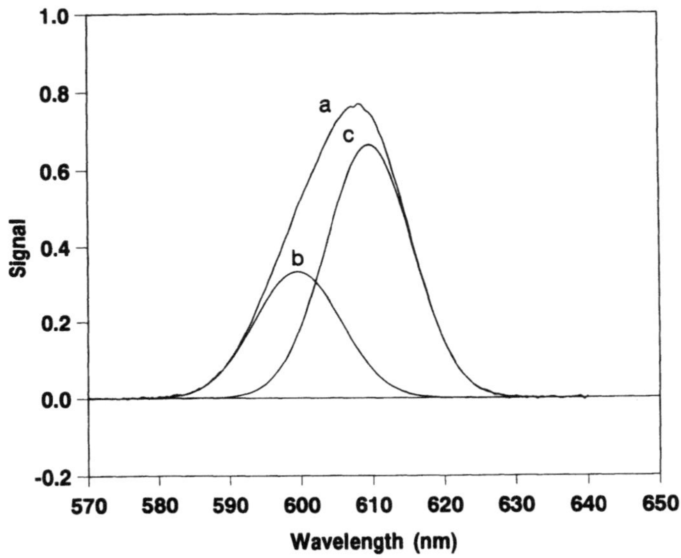

Fig. 5. Overlapped Gaussian signals. (a) mixture spectrum ($\tilde{W} = 2 \times 10^{-3}$); (b) unknown peak (peak maximum height 0.332); (c) known peak (peak maximum height 0.663). The widths of both the peaks are 6 nm. The S/N is 665 for the known peak. The area ratio of the unknown to known peaks is 1/2. (From [15])

Figure 5 illustrates the mixture spectrum (a) of the unknown Gaussian peak (b) and known Gaussian peak (c) with added white noise [15]. The overlap with the unknown peak would critically deteriorate the precision and accuracy of the regular filter. The adaptive filter, however, corrects the estimates by referring to the innovation which represents the gap between the estimated spectrum and actual spectrum.

Figure 6 shows the influence of the peak overlap on the precision of the estimates of the known peak (lower line) and unknown peak (upper line). The estimates of the target known peak are obtained by the adaptive filter and the area of the unknown peak is estimated by the subtraction of the estimated contribution of the target peak from the mixture spectrum. The known peak is ten times the unknown peak in area and has even higher precision. As the known peak approaches the fixed unknown peak, the RSD of the estimates increases especially below 620 nm ($Rs = 0.83$). In the region of the severe overlap below 606 nm, the adaptive filter mistakes the fused peaks for a single large peak, which causes a decrease in the RSD values.

The error prediction based on Eq. (5C) (*dotted lines*) is also excellent in this example. The observed statistics were obtained from 500 repeated simulations for each degree of separation, but the predicted RSD values resulted from a single, noisy spectrum. This is why the predicted RSD curves in Fig. 6 are noisier than the observed curves. As is expected, a larger area of the unknown peak reduces the quality of the error prediction of the adaptive filter. The precision for the overlap-free signal is constant irrespective of the peak position in this spectroscopic situation (see the chromatographic situation shown in Fig. 4).

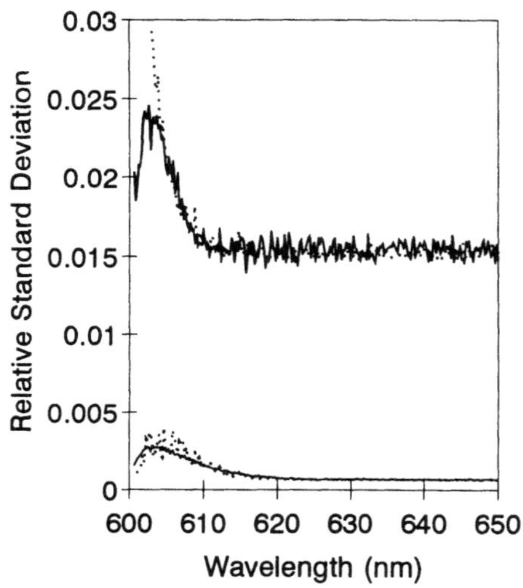

Fig. 6. Influence of overlap on precision of adaptive Kalman filter in the situation of spectroscopy. The abscissa denotes the position of the known peak and the unknown peak is fixed at 600 nm. *Solid lines* (from Monte Carlo simulation): *upper*, the RSD for the unknown peak; *lower*, the RSD for the known peak. *Dotted lines* are theoretically obtained. The area ratio of the unknown and known peaks is 1/10. The moving known peak has the same shape irrespective of its position

Information Theory of Measurement

Until now, we have paid attention only to a single peak. Now, a criterion for the precision of multi peaks is defined based on information theory. The combination with Shannon's theory might only seem to imply the mathematical transformation of formalism. On the contrary, the information theory of signal processing lends itself materially to chemical analysis, e.g., optimization of HPLC analysis [12, 14].

A major aim in analytical chemistry is to elaborate a method through which more information can be effectively transmitted from analytes of interest in a given sample. The total amount of information involved originally in the sample, however, cannot be elicited from the observed data because of inevitable noise contamination, interference, etc. The available information on the analyte concentration through measurement and signal processing is formulated as mutual information [12, 14].

Information, Precision and FUMI

According to the theory of the Kalman filter, the precision is simply related to the mutual information [12]:

$$\phi(j) = \log(1/\mathrm{RSD}(j)) \tag{7}$$

where RSD(j) denotes the RSD of the estimates for peak j shown in Eq. (5).

If there are multi peaks in an output of an analytical instrument and if the peaks are independent of each other, the total information, Φ, can be given as the sum of the individual peak information, $\phi(j)$ [12]:

$$\Phi = \sum_{j=1}^{q} \phi(j). \tag{8}$$

Equations (7)–(11) are abbreviated as FUMI (Function of Mutual Information).

Let the precision be the reciprocal of RSD. FUMI and precision have the same tendency against peak overlap and shape as pointed out by properties A-D. The mutual information, precision and FUMI are essentially equivalent concepts.

Calculation of Information

The Tailor series expansion of Eqs. (5C) or (6) around the peak center leads to a useful expression of FUMI [12]:

$$\phi(j) = \psi(j) - \delta\phi(j) \quad (\geq 0). \tag{9}$$

The first term, called intact information, denotes the information indigenous to the peak shape itself and free from overlap (see Eq. 6) [12]:

$$\psi(j) = \tfrac{1}{2} \log \left(\frac{A_j^2}{2\pi^{1/2}\sigma_j \Delta T\, \widetilde{W}^2} \right) \tag{10}$$

where A_j and σ_j denote the entire area and width, respectively, of Gaussian peak j. The second term of Eq. (9), called information loss, takes the form [12]:

$$\delta\phi(j) = -\tfrac{1}{2} \log \left(\frac{\kappa_f(j) - \tau_j}{\pi^{1/2}\sigma_j} + \frac{\tau_j - \kappa_c(j)}{\pi^{1/2}\sigma_j} \right). \tag{11}$$

All the functions, $\phi(j)$, $\psi(j)$ and $\delta\phi(j)$, are non-negative. The cutoff and filter-off point, $\kappa_c(j)$ and $\kappa_f(j)$, are indicated in Fig.3. The shaded region corresponds, though roughly, to the mutual information, $\phi(j)$, the white region the information loss, $\delta\phi(j)$, and the entire peak shape the intact information, $\psi(j)$.

Without peak overlap, the information loss, $\delta\phi(j)$, takes the lowest value (zero) and the information, $\phi(j)$, relies only on the peak shape (area and width) and noise variance, $\widetilde{W}^2$. If another peak is interfering with peak j, the loss, $\delta\phi(j)$, begins to increase from zero and spoils the precision with increasing overlap. The signal properties A-D can easily be recognized from Eqs. (9)–(11).

Attention should be paid to the following restrictions on the information loss: $1 \geq 1/2 + (\kappa_f(j) - \tau_j)/(\pi^{1/2}\sigma_j) \geq (2\pi^{1/2}\sigma_j \Delta T\, \widetilde{W}^2)/A_j^2$; $1 \geq 1/2 + (\tau_j - \kappa_c(j))/(\pi^{1/2}\sigma_j) \geq (2\pi^{1/2}\sigma_j \Delta T\, \widetilde{W}^2)/A_j^2$. All these restrictions come from the definition that FUMI, $\phi(j)$, should take the lower and upper limits: $\psi(j) \geq \phi(j) \geq 0$.

FUMI and Resolution

It is mathematically proved that more peak separation (increase in resolution, Rs) is equivalent to an increase in FUMI, $\phi(j)$, of peak j, only if peak j strongly overlaps another peak [12].

$$\frac{\partial \phi(j)}{\partial Z} = C\frac{\partial Rs}{\partial Z} \tag{12}$$

where Z denotes a chromatographic variable (e.g., mobile phase composition) and coefficient C (> 0) is a function of Rs. ∂Rs denotes a slight change in Rs and $\partial \phi(j)$ a slight change in FUMI. If the change, ∂Rs, is positive, then the change, $\partial \phi(j)$, should also be positive because of the positive coefficient C ($\partial \phi(j) = C\partial Rs$) irrespective of the sign of ∂Z. This means that the more separated the peaks ($\partial Rs > 0$), the more precise ($\partial \phi(j) > 0$) the analysis in the presence of peak overlap.

Further but excessive separation ($\partial Rs > 0$) causes peak broadening and spoils the precision ($\partial \phi(j) < 0$) in the chromatographic situations. Then, the relationship does not hold true in the case of sufficient peak separation, because the signs of the changes, $\partial \phi(j)$ and ∂Rs, become opposite. Equation (12) is the information-theoretical interpretation of the most commonly used separation function, Rs.

Probability Theory of Measurement in Environmental Analysis

The real RSD on HPLC and capillary electrophoresis is more or less underestimated by FUMI [14]. A major reason for this gap is the existence of some noises in the instrumental output other than the white noise included in the FUMI theory. The real RSD, however, can also be predicted with accuracy by regarding the baseline drift as the mixed random process of the white noise and Brownian motion [14]. The Kalman filter and information theory described here lay the foundations of this probability theory of measurement and its applications in quantitative analysis.

Chemical analysis for many environmental problems can be grouped into trace analysis. In a macro analysis on HPLC, however, the major cause of the measurement error is the injection volume error and any condition will yield nearly the same precision as long as the peaks are sharp and sufficiently separated. In a micro analysis, the precision varies substantially from peak to peak and from condition to condition [14]. Therefore, the precision should be consulted more carefully in selecting the optimum from among all the operating conditions examined. Mobile phase composition, column length, flow rate, detection wavelength, amount of internal standard and choice of the optimal internal standard material have been totally optimized in the determination for pesticides and drugs with the predicted RSD or FUMI as a criterion [14].

Acknowledgement. Hayashi would like to thank Prof. Sarah C. Rutan for her useful suggestions about the adaptive Kalman filter. Figure 6 was created as a part pf their work [15] in Richmond, Virginia, USA, in 1992.

References

1. Brown SD, Bear RS Jr, Blank TB (1992) Anal Chem 64:22R
2. Güell OA, Holcombe JA (1990) Anal Chem 62:529A
3. Nishio M (1985) Probability Theory. Jikkyo Shuppan, Tokyo
4. Brown SD (1986) Anal Chim Acta 181:1
5. Rutan SC (1989) Chemom Intel Lab Sys 6:191
6. Rutan SC (1990) J Chemometr 4:103
7. Rutan SC (1991) Anal Chem 63:1103A
8. Arimoto S (1976) Kalman Filter, Sangyo Tosho, Tokyo
9. Jazwinski AH (1970) Stochastic Processes and Filtering Theory. Academic Press, New York
10. Kunita H (1976) Estimation of Random Process. Sangyo Tosho, Japan
11. Hayashi Y, Yoshioka S, Takeda Y (1988) Anal Chim Acta 212:81
12. Hayashi Y, Matsuda R (1993) Chemom Intel Lab Sys 18:1
13. Hino M (1982) Spectral Analysis. Asakura Shoten, Tokyo
14. Hayashi Y, Matsuda R (1994) Advance in Chromatogr 34:347
15. Hayashi Y, Helburn RS, Pompano JM, Rutan SC (1993) Chemom Intel Lab Sys 20:163

Robust and Non-parametric Methods in Multiple Regressions of Environmental Data

Yvette L. Mallet[1], *Danny H. Coomans*[1] *and Olivier Y. de Vel*[2]

[1]Department of Mathematics and Statistics, James Cook University, Townsville QLD 4811, Australia
[2]Department of Computer Science, James Cook University, Townsville QLD 4811, Australia

Summary

Statistical regression methods in environmental chemistry are of vital importance. Regression techniques provide environmental chemical analysts with the ability to calibrate instruments and model large environmental systems.

It has become apparent that ordinary least-squares regression is not well suited to modeling data that contains outliers or strong nonlinearities. In the presence of outlying data robust regression methods prove to be a useful tool, while various non-parametric regression models are useful should the data possess nonlinearities or high levels of noise.

Robust techniques have the ability to detect outliers and dampen their effect on the modeling procedure. Several robust regression methods have been proposed but this article focuses on the least median of squares method and reweighted least squares regression.

The non-parametric models to be discussed include the ACE model, the PI model and the MARS model. Unlike ordinary least squares, these methods have evolved only recently, hence there is only limited documentation available on these methods.

List of Symbols and Abbreviations

LS	ordinary least-squares
ACE	alternating conditional expectations
PI	pi implementation
MARS	multivariate adaptive regression splines
LMS	least median squares
RLS	reweighted least-squares
PE_{gcv}	generalized cross-validation estimate of the prediction error
lof	lack-of-fit
mi	maximum level of interaction
RSS	residual sum of squares
mad	median absolute deviation
CV	cross-validating
NP	number of estimated parameters
GCV	generalized cross-validation
MSE	mean square error
AIC	Akaike's information criterion
y	response variable
x_i	i-th predictor variable; $i = 1, \ldots, m$
ε	residual component
β_i	i-th regression coefficient
m	number of predictor variables
$N(\cdot)$	normal distribution
n	number of observational units
Z	sample of data vectors
Z'	corrupted sample of data vectors
T	regression estimator
$(\hat{\cdot})$	estimate of $(\cdot)$
$\Vert(\cdot)\Vert$	magnitude of $(\cdot)$
Γ_n	finite sample breakdown point
σ^2	population variance
σ	population standard deviation
$\hat{\sigma}_{LS}$	population standard deviation estimated by LS
$\hat{\sigma}_{LMS}$	population standard deviation estimated by LMS
z	standardized residual component
$\vert(\cdot)\vert$	modulus of $(\cdot)$
w_i	weight assigned to the i-th observational unit
R^2	coefficient of determination
R^2_{cv}	cross-validated coefficient of determination
R^2_{adj}	adjusted coefficient of determination
s	scatterplot smoother
c_i	i-th cutpoint
R_k	k-th set of observations
$\eta((\cdot))$	neighborhood of $(\cdot)$

a	constant
$d((\cdot))$	an even function that decreases with $\lvert(\cdot)\rvert$
S_{0j}	weight given to y_j in producing a smoothing estimate of the observation x_0
k	number of observations in a symmetric neighborhood
r^{-i}	cross-validated residual component
s^{-i}	fitted smooth with the point x_i removed
$\hat{\xi}$	local error estimate
κ^*	initial span
$\varrho(\cdot)$	truncated power function
q	spline degree
t	knot
$\theta(\cdot)$	transformed response
$g(\cdot)$	transformed predictor
e^2	unexplained variance
$\theta^*(\cdot)$	optimal transformed response
$g^*(\cdot)$	optimal transformed predictor
$\prod_j$	j-th product in a PI model
$\phi(\cdot)$	cubic spline basis function
J	number of products in a PI model
K	number of knots in each cubic spline basis function
J^*	optimal number of products in a PI model
$U(\cdot)$	uniform distribution
B_p	p-th multivariate spline basis function
P	number of multivariate spline basis functions
V	number of groups the data is divided into for cross-validation
$(\bar{\cdot})$	mean value of $(\cdot)$
f^{-v}	estimated function when the v-th group has been removed.

Introduction

The application of statistical regression methods in environmental chemistry is important in many ways. Applications range from the small scale of calibration in environmental analytical chemistry to large scale applications such as modeling environmental systems.

A wide variety of regression models have been developed to cater for different situations. Factors such as outliers, the amount of noise, the presence of nonlinearities, correlations among the variables and the observation/variable ratio are just a few factors which strongly influence the path along which the regression analysis should follow. For example, when there is significant evidence of high correlations among the variables of a regression system, principal component regression analysis would be an obvious candidate for the regression process [1].

The regression model consists of a systematic component and a residual component. The systematic component contains information about the underlying dependencies between a set of predictor variables and a response. The aim of statistical regression analyses is to model the mechanisms that produce the systematic component. This model should contain a minimal amount of noise belonging to the residual component. Additionally, the same model should contain a maximal amount of information contained in the systematic component.

If the aim of the regression procedure is to predict, it is necessary for the model to fit the data closely. Any model which produces a *good* fit can be used to predict new response values from a set of new predictor values. For example one may wish to predict ozone concentrations from another set of variables which may be easier to measure such as wind speed or temperature.

If, on the other hand, the aim of the regression procedure is to gain a greater understanding about the true mechanisms of the system, a stricter approach is required to determine the best model. In this case, the model should have some inherent characteristics such as interpretability [2].

In some instances the underlying mechanism may be known. That is, the functional form of the model involving the relationship among the predictors has been established. This usually requires prior knowledge about the system. Unfortunately, in practice, this is often not the case and the functional form has to be estimated.

Traditionally, ordinary least-squares (LS) has been used to represent the underlying function of a regression system. This fact can be attributed to the ease in implementing and interpreting the ordinary least-squares model. Ordinary least-squares is best suited to data which can adequately be represented by linear or polynomial functions.

Ordinary least-squares is a parametric regression model but, due to the advancement in statistical theory and technology, several non-parametric regression models have evolved. A distinction between parametric and non-parametric regression models can be made. Parameteric models assume a particular functional form to represent the systematic component of the regression model. For example, the LS model has the form

$$y = \beta_0 + \beta_1 x_1 + \cdots + \beta_m x_m + \epsilon \,. \tag{1}$$

Here the response (y) consists of the error component ϵ, and the systematic component $\beta_0 + \beta_1 x_1 + \cdots + \beta_m x_m$. The systematic component is a linear combination of the m predictor variables $x_1, \ldots, x_m$. The parameters $\beta_1, \ldots, \beta_m$ are the regression coefficients and β_0 is the y-intercept. The coefficients are solutions to

$$\underset{\beta_0,\ldots,\beta_m}{\text{minimize}} \sum_{i=1}^{n} \hat{\epsilon}_i^2 \tag{2}$$

where n is the number of observational units and the estimated residual is given by

$$\begin{aligned}\hat{\epsilon}_i &= y_i - \hat{y}_i \\ &= y_i - \hat{\beta}_0 - \hat{\beta}_1 x_{1i} - \cdots - \hat{\beta}_m x_{mi} \,.\end{aligned}$$

Parametric models also make assumptions about the distribution of the residuals. In the case of LS it is assumed the residuals follow a normal distribution with mean equal to zero and a constant variance, that is $\epsilon \sim N(0, \sigma^2)$.

Whilst LS is a linear parametric model, nonlinear parametric models also exist. Nonlinear parametric models are typically used when the functional form of the systematic component is known. Such models often arise in the field of chemical kinetics. For instance, the relationship between the concentration of available dissolved organic substrate (x) and the rate of uptake (y) of that substrate is considered to be nonlinear [3].

Non-parametric models make no assumption about the functional form of the systematic component. Given a set of data, non-parametric methods estimate the systematic component by using smoothing techniques. Smoothing techniques provide a flexible approach to data analysis, and, because there is no prespecified model, the use of smoothers can detect extra information from a set of data that may well have gone undetected had a prespecified model been used. Additionally, non-parametric models place no restriction on the distribution of the residuals. Of course, one would hope the residuals have a mean equal to zero and are spread with constant variance.

Some examples of non-parametric models include ACE [5–9], PI[10], MARS [2, 11, 12], Additive Models [13–15] and Artificial Neural Networks [16]. These methods not only provide a flexible approach to data modeling, but are also capable of modeling data that contains nonlinearities mixed with high levels of noise.

Ordinary least-squares does not fare so well with data containing high levels of noise, nor does it produce accurate results in the presence of outliers. Robust regression methods [17, 24] are usually employed when outliers are contained in the data. Robust regression methods replace general estimators, namely the regression coefficients with robust estimators. In most cases robust regression techniques use a linear function to model the systematic component. Whilst these methods assume a particular function for the modeling procedure, no constraints are usually placed on the distribution of the regression residuals.

The prementioned regression techniques are alternatives to the ordinary least-squares model. Each method may produce varying results depending on the adaptability of the methods across different situations.

Other alternatives to ordinary least-squares which have not been mentioned include biased regression techniques. These methods reduce the number of parameters that need to be estimated to overcome the low observation/variable ratio. Such methods include partial least-squares [1, 18–21, 23], principal component analysis [1, 18] and ridge regression [1, 18].

Hybrid regression methods can also be produced by combining several regression techniques. For example, a biased regression technique such as partial least-squares could be modified by making the procedure robust [22]. Another hybrid method which combines smoothers and partial least-squares to form a nonlinear partial least-squares model is discussed by Frank [23].

In recent years there has been extensive amounts of literature in chemometrics about biased regression techniques, much more than appears about robust and non-parametric techniques. In order to provide more information about these methods, we will focus on robust and non-parametric methods for multiple regression analysis. Since many data sets from environmental chemistry have the potential to contain significant levels of noise, nonlinearities and outliers, these techniques are very important in this discipline.

Robust Multiple Regression Models

The LS technique for estimating the unknown regression coefficients in multiple linear regression models is very sensitive to the presence of outlying data vectors. However, the occurrence of naturally outlying data (that is data which involves no discrepancy) is quite common in data from environmental chemistry. For example, when dealing with contaminants, an atypically high concentration value in a data set may be an indication of an extreme pollution situation. Consequently, the detection of natural outliers is of major importance in observational environmental studies.

In multivariate data sets the identification of such atypical data vectors becomes difficult. This is because it is not easy to detect multivariate influential data vectors in two dimensional scatter plots. Consequently, tools are needed to identify these outliers through models not influenced by these aberrant data. One approach is to use robust regression methods which are able to model the majority of the data and automatically dampen the influence of atypical data vectors. Atypical data lying far from the cluster of *good data*, will have large residuals from the robust regression fit. Since robust regression estimates are insensitive to outliers, the detection of influential data is simplified. In contrast, the residuals from LS methods are not reliable for outlier detection. Indeed, the atypical data vectors may have small residuals since the least-squares fit is pulled too much in the direction of these outliers.

A number of robust regression methods have been proposed for simple and multiple linear regression [24–26]. As an example we will concentrate on the least median of squares (LMS) method proposed by Rousseeuw and Leroy [26]. This method was further introduced by Massart et al. [27] and Rousseeuw [17] in analytical chemistry and chemometrics.

The basic idea behind the method is to use the median as a criterion in the least-squares method. The LMS estimator which replaces the least sum of squares in Eq. (2) by the least median of squares, is given by

$$\underset{\hat{\beta}_0,\ldots,\hat{\beta}_m}{\text{minimize}} \ \underset{i=1,\ldots,n}{\text{median}}\, \hat{\epsilon}_i^2 \ , \tag{3}$$

where the median is the $(n/2+1)$th ranked squared residual.

A typical values can occur in the response variable as well as in the predictor variables. The worst case is when leverage data vectors are present in the data set, that is data vectors for which $(x_{k1},\ldots,x_{km})$ lies far from the cluster of the $(x_{i1},\ldots,x_{im})$ $i \in [1, n]$, $i \neq k$ data in the measurement space. Rousseeuw [17] explains this is of major concern because of the following.

1. LS estimators are vulnerable more to leverage points than to outliers purely in the response variable.
2. Leverage outliers are more likely to occur than data merely aberrant in the response variable, as there are m predictor variables compared to one response variable.
3. Due to the higher dimensionality of the predictor variables space, identifying leverage outliers can be a difficult task.

In order to be able to identify *several* outliers at a time, we need to model the main cluster of data vectors regardless, of the presence of multiple outliers. It means a high-breakdown estimator such as LMS is required [28]. The breakdown point of an estimator is the smallest fraction of observations that have to be replaced by corrupted ones to make the estimator unbounded. More formally, the notion of the breakdown point [26] in regression can be described as follows. Take any sample Z of n data vectors $(x_{11},\ldots,x_{1m},y_1),\ldots,(x_{n1},\ldots,x_{nm},y_n)$ and a regression estimator T (e.g. T being LS, LMS etc). Applying T to Z yields the regression estimate vector $\hat{\beta}$. Consider all possible corrupted samples Z' that are obtained by replacing any ℓ of the original data vectors by arbitrary data vectors, such as those that produce outliers. Let bias$(\ell; T, Z)$; be the supremum of $\| T(Z')-T(Z) \|$ for all corrupted samples Z', then the finite sample breakdown point can be defined as

$$\Gamma_n(T,Z) = \min\{\ell \mid n; \text{bias}(\ell; T, Z) \text{ is infinite}\} \ . \tag{4}$$

The finite sample breakdown point for the LMS estimator is

$$\frac{n/2 - m + 2}{n} \ . \tag{5}$$

The breakdown point depends little on n. To get a single breakdown point value for fixed m, one often prefers to report the infinite sample breakdown point. The infinite sample LS breakdown point is 0%, whereas the breakdown point for the LMS technique is as high as 50%, the best that can be achieved, as shown by Rousseeuw and Leroy [26].

For outlier detection in regression, one has to compare each residual from robust regression with the spread of all the resulting residuals. When using least-squares regression the residuals are assumed to have a normal distribution with an unknown scale parameter denoted by σ. This is the population standard deviation, which is estimated by LS as

$$\hat{\sigma}_{\mathrm{LS}} = \left(\frac{1}{n - m - 1} \sum_{\mathrm{i}=1}^{\mathrm{n}} \hat{\epsilon}_{\mathrm{i}}^{2} \right)^{\frac{1}{2}} . \tag{6}$$

For LMS regression, the appropriate scale estimate is

$$\hat{\sigma}_{\mathrm{LMS}} = 1.483 \left(\underset{\mathrm{i}=1,\dots,\mathrm{n}}{\mathrm{median}}\, \hat{\epsilon}_{\mathrm{i}}^{2} \right)^{\frac{1}{2}} \tag{7}$$

where the $\hat{\epsilon}_{\mathrm{i}}$s are the LMS residuals and 1.483 is the constant to make $\hat{\sigma}_{\mathrm{LMS}}$ a consistent estimator of σ. The LMS scale estimate has a 50% breakdown point whereas $\hat{\sigma}_{\mathrm{LS}}$ has a breakdown point of 0% which is another downfall of LS.

Rousseeuw and Leroy [26] mention that the standardized residuals,

$$z_{\mathrm{i}} = \frac{\hat{\epsilon}_{\mathrm{i}}}{\hat{\sigma}} , \tag{8}$$

resulting from a robust fit (in analogy to the standardized residuals in LS regression diagnostics) can be employed to diagnose outliers. If $|z_{\mathrm{i}}|$ is larger than 2.5, the i-th observation is considered to be a regression outlier. It is not advisable to use both the LS residuals and the LS scale estimate in Eq. (8), since leverage data tend to have small LS residuals and will not be detected as outliers. Moreover, some large residuals (such as those in the dependent variable) will greatly increase and artificially expand the tolerance limits for the residuals.

Let us now demonstrate the applicability of the robust regression approach in a real example using environmental data.

Example

The data to be used in this example forms part of a larger observational (survey) study. The study involved determining levels of lead and cadmium present in the surface enamel of permanent incisors of a cohort of 370 Belgian schoolchildren [29, 30]. Samples were taken by means of an acid etch microbiopsy method [31]. Two layers of enamel, a few micrometers thick were removed, and the concentration of heavy metals was determined. The cohort consisted of children from selected areas in Belgium. The selected areas came from distinct parts of the country, and were chosen according to geographic and demographic differences corresponding to different kinds and levels of environmental pollution. Both rural and coastal areas were included in the study, as were localities in the vicinity of industries that caused heavy metal pollution of the environment. The heavy metal concentrations in surface enamel was related and decorrelated (calibrated) with respect to etched depth and age of the child. It was shown the two layers of enamel were a reflection of environmental exposure to lead and cadmium. This can be related to different sources of body burden [29, 30].

In this example we selected two small data subsets. The first data set consisted of 55 children aged from 6.2 to 9.5 years. The children were randomly selected from a number of small schools in 11 villages in the southern part of Belgium (Ardennes). The villages are situated in the same region of the Ardennes and are predominantly rural. The second data set consisted of 27 children aged from 6.2 to 9.5 years selected from a school in the town of Schelle which lies in the northern part of Belgium. The area is moderately urban.

In the ARDENNES example we examined the first etch biopsy, i.e. the first layer of incisor enamel removed and in the SCHELLE data set we analyzed the second etch biopsy. Both data sets contain two independent variables and one response. One independent variable is etched depth (*etchd* in micrometers). Etched depth is estimated from the amount of calcium removed during the etch biopsy. The second independent variable is the age of the child which has been transformed to the decimal system from years and months (see above).

The response variable ($\ln Pb$ with Pb in ppm) was regressed on *age* and *etchd*. We used the logarithmic transform of the lead concentrations in the biopsy, as it is known the profile of lead levels in the outermost enamel decreases exponentially from the surface towards the inner enamel. Additionally the distribution of lead in enamel is positively skewed.

The ARDENNES and SCHELLE data subsets are given in Tables 1 and 2 respectively. The full data set can be found in [30].

The regression analysis in the original study [29, 30] was performed for the following reasons.

1. To remove some of the variation in the lead levels due to non-environmental factors such as age and etched depth.
2. To detect leverage outliers that make least-squares regression models unreliable and biased and (positive) outliers of special importance (in response variable $\ln Pb$) which may indicate an extraordinarily high level of heavy metal in the tooth surface of the particular subject.
3. To use the information in the regression coefficients of age and etched depth in the comparison of the regions.

The 3D scatter plot for the ARDENNES data in Fig. 1 shows the relationship between $\ln Pb$, *etchd* and *age* is rather weak. Figure 1 also shows that most variation in lead levels is not due to differences in *age* and *etchd*. This means,

Table 1. ARDENNES data set.

id	*etchd* (x_1)	*age* (x_2)	*Pb*	ln *Pb* (y)
1	3.82	7.67	387	5.96
2	3.10	7.67	1236	7.12
3	2.88	8.42	1339	7.20
4	2.52	7.33	584	6.37
5	2.37	6.67	320	5.77
6	1.80	9.00	1199	7.09
7	3.53	8.50	473	6.16
8	3.31	8.50	589	6.38
9	2.38	8.50	2864	7.96
10	2.73	9.33	692	6.54
11	3.02	7.42	595	6.39
12	2.66	7.17	788	6.67
13	2.64	7.83	507	6.23
14	2.55	8.00	1141	7.04
15	3.15	8.25	1012	6.92
16	2.86	8.92	2164	7.68
17	2.86	9.08	1669	7.42
18	2.42	8.42	1152	7.05
19	2.28	8.17	1919	7.56
20	3.02	7.67	1074	6.98
21	3.34	8.75	897	6.80
22	2.49	7.83	1236	7.12
23	3.52	7.00	1510	7.32
24	2.94	8.42	1299	7.17
25	2.59	8.75	1808	7.50
26	2.94	9.08	1510	7.32
27	2.59	8.67	1394	7.24
28	2.59	9.17	788	6.67
29	1.90	7.08	982	6.89
30	2.25	9.25	2275	7.73
31	2.59	8.42	1754	7.47
32	2.83	8.42	1495	7.31
33	2.77	7.17	915	6.82
34	2.77	8.67	1130	7.03
35	2.59	7.42	1043	6.95
36	2.94	8.67	1211	7.10
37	2.08	8.92	1436	7.27
38	3.06	7.08	720	6.58
39	4.17	8.92	699	6.55
40	3.03	9.00	720	6.58
41	3.81	8.50	607	6.41
42	3.58	8.25	566	6.34
43	4.03	7.83	550	6.31
44	2.82	8.00	720	6.58
45	2.82	7.67	735	6.60
46	2.99	7.92	1394	7.24
47	3.34	7.83	584	6.37
48	2.80	7.33	3165	8.06
49	3.03	8.25	1339	7.20
50	2.57	7.83	1571	7.36
51	2.68	7.67	1495	7.31
52	3.03	7.67	2617	7.87
53	3.15	8.17	1737	7.46
54	2.75	8.83	1380	7.23
55	2.85	8.75	1274	7.15

Table 2. SCHELLE data set.

id	*etchd* (x_1)	*age* (x_2)	*Pb*	ln *Pb* (y)
1	2.67	7.67	353	5.87
2	2.89	8.17	100	4.61
3	2.96	8.50	287	5.66
4	3.75	8.00	231	5.44
5	2.38	7.58	388	5.96
6	2.59	8.08	146	4.98
7	2.59	7.50	377	5.93
8	2.96	8.17	304	5.72
9	2.45	7.67	195	5.27
10	2.74	8.17	441	6.09
11	3.10	7.33	52	3.95
12	2.81	7.33	180	5.19
13	2.81	7.33	235	5.46
14	3.10	7.75	318	5.76
15	2.74	8.08	332	5.81
16	2.96	7.67	351	5.86
17	3.32	7.83	268	5.59
18	2.67	7.92	211	5.35
19	3.10	7.92	349	5.86
20	2.89	8.17	268	5.59
21	2.71	7.92	479	6.17
22	2.71	7.67	164	5.10
23	3.31	8.58	468	6.15
24	2.64	8.58	117	4.76
25	3.39	8.00	218	5.38
26	3.10	8.25	318	5.76
27	2.96	7.83	227	5.42

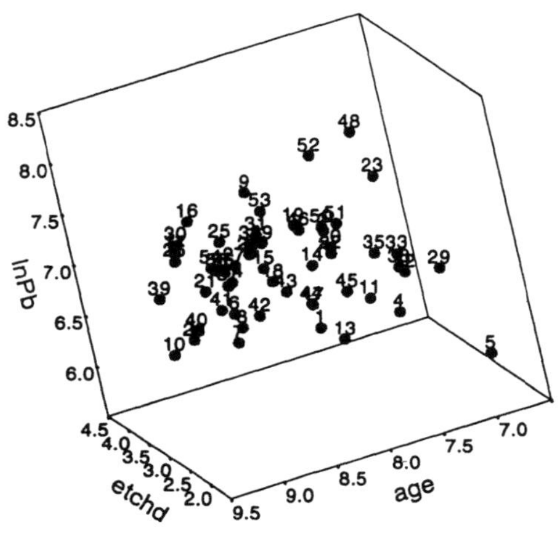

Fig. 1. ARDENNES data set. A 3D plot of the regression variables

from a calibration point of view, we are dealing with a noisy data set. Moreover, it is clear there are three points (23, 48, and 52) which lie slightly further from the main data cluster. These points do not seem to be severe outliers given the large amount of variation in the data set. The data points 23, 48 and 52 represent children with high lead levels in enamel (outlying in the dependent variable ln *Pb*) and thus a high past body burden. They don't have extreme values in the independent variables. It is important to detect these cases from a toxicological point of view. Moreover, these cases may affect the LS regression model, although they do not have extreme values in the independent variables and we do not expect them to be strong leverage points. The cases 5 and 10 deviate in a different way, namely somewhat extreme in the independent variables with a rather negative deviation in ln *Pb*,

The LS regression function for this data set is

$$\hat{y} = 6.65 - (0.407)x_1 + (0.183)x_2 \tag{9}$$

whereas the LMS model gives

$$\hat{y} = 5.42 - (0.581)x_1 + (0.395)x_2 \, . \tag{10}$$

Although there is no dramatic change in the regression coefficients when going from LS to LMS, it is clear in this case that LMS gives steeper slopes. The reason for this is LMS detects four outliers and LS detects only one. The three outliers in the response, ln *Pb* which were obvious from the 3D plot in Fig. 1 were all detected by LMS but only one was detected by LS. The unstandardised and standardized residuals are given in Table 3. One can see both LS and LMS identify case five as borderline within the tolerance region of the regression residuals ($|z_i| < 2.5$) and LMS detects case ten as an outlier but LS does not.

In a further step, after the LMS model was developed, a least-squares multiple regression analysis was performed with the weights of the outliers detected by LMS reduced to zero. This means we are making use of the following weights,

$$w_i = \begin{cases} 1 & \text{for } |z_i| \leq 2.5 \\ 0 & \text{for } |z_i| > 2.5. \end{cases} \tag{11}$$

This simply means case *i* will be retained in the weighted LS if its LMS residual is small to moderate, but disregarded if it is an outlier. The bound 2.5 is

Table 3. ARDENNES data set. Regression residuals of selected cases.

	LS		LMS		RLS	
Index (i)	$\hat{\epsilon}_i$	z_i	$\hat{\epsilon}_i$	z_i	$\hat{\epsilon}_i$	z_i
5	−1.31	−2.45	−0.915	−2.40	−0.857	−2.32
10	−0.701	−1.52	−0.987	−2.59	−0.856	−2.31
23	0.827	1.79	1.173	3.07	1.125	3.04
48	1.213	2.63	1.364	3.58	1.405	3.80
52	1.055	2.29	1.173	3.08	1.205	3.26

arbitrary, but quite reasonable, since in the case where the residuals follow a normal distribution there will be very few residuals larger than 2.5. This way the destabilizing influence of outliers on the LS model is eliminated. This procedure is called reweighted least-squares regression (RLS). The estimator can be written

$$\underset{\beta_0,\ldots,\beta_m}{\text{minimize}} \sum_{i=1}^{n} w_i \hat{\epsilon}_i^2 \,. \tag{12}$$

This estimator still possesses the high breakdown point, but it is more efficient in a statistical sense under the Gaussian assumptions. An additional advantage of using RLS instead of LMS for the final model is it yields the classical output such as the parametric coefficient of determination (see section entitled Model Selection Criteria), confidence intervals on the regression parameters and hypothesis tests. This result, despite the distribution theory valid for LS, is no longer exact for the RLS regression.

Table 4 summarizes the regression models LS, LMS and RLS of the ARDENNES data. The value ln *Pb* is the estimated ln *Pb* value for the reference etched depth of 3 μm and age equal to 7 years. The LMS and RLS estimates are very similar. The associated standard error is an estimate of the scale of the ln *Pb* residuals after regression. The standard error is smaller for LMS and RLS when compared to LS. The coefficient of determination (R^2) is also displayed in Table 4. This value gives an idea of the strength of the linear relationship between the response variable and the predictor variables. Additionally it measures the proportion of total variability explained by the regression model. For LS

Table 4. ARDENNES data set. Regression estimates and model fitting statistics.

	LS	LMS	RLS
No of children	55	55	51
No of outliers	1	4	4
ln *PB*; $E(y \mid x_1 = 3, x_2 = 7)$	6.710	6.448	6.443
$\hat{\sigma}$	0.462	0.395	0.370
R^2	0.208	0.542	0.435
p-value	0.0023		0.0000
intercept ($\hat{\beta}_0$)	6.65	5.42	5.40
standard error	0.87		0.74
t-value	7.65		7.36
p-value	0.0000		0.0000
etch depth ($\hat{\beta}_1$)	−0.407	−0.581	−0.478
standard error	0.131		0.107
t-value	−3.11		−4.45
p-value	0.0030		0.00005
age ($\hat{\beta}_2$)	0.183	0.395	0.354
standard error	0.094		0.082
t-value	1.94		4.29
p-value	0.058		0.00009

regression, one can test the hypothesis that R^2 equals zero, as R^2 under the null hypothesis can be related to an F-distribution if the $\hat{\epsilon}_i$s are normally distributed. The null hypothesis is accepted if the calculated F-statistic for R^2 is less than the $(1-\alpha)$th quantile of the associated F-distribution, that is, if the p-value is the larger than α (say $\alpha = 0.05$). The R^2 for the RLS is defined in an analogous way to LS, except all the observations are now multiplied by their corresponding weight w_i.

As for the LMS model, the RLS model explains a larger proportion of the variability in the ARDENNES data than LS. The result is a more significant R^2 for RLS (p-value for RLS is much smaller). For LMS the measure R^2 is defined in a robust way (see Model Selection Criteria section). No simple parametric test for R^2 can be formulated here. Nevertheless, the R^2 values indicate that the LMS and RLS models linearly fit the ARDENNES data better than LS. This is not necessarily the case for other examples, where the LMS and RLS models may result in substantially smaller R^2 values and less significant values for RLS. The reason is that the R^2 for LS is very sensitive to outliers and leverage outliers may cause R^2 to be artificially high. It all depends on the particular characteristics of the contaminated data set whether the R^2 of LS will be smaller or larger than the R^2 for LMS or RLS. In the ARDENNES example the leverage effect of the outliers is not consistent and strong enough to increase the LS R^2 value greatly.

Table 4 also reveals that a steeper slope of etched depth and age (i.e. larger modulus of the regression coefficients) after outlier deletion (RLS) in conjunction with smaller standard errors of the coefficients gives rise to higher significance (i.e smaller p-values) for the t-test on the regression coefficients than the original LS model. This indicates a better fit of the model.

The 3-D scatter plot for the SCHELLE data in Fig. 2 shows a small cluster of points $\{2, 6, 9, 18, 22, 24\}$ below the main cluster and an isolated outlier $\{11\}$. The proportion of outliers present here is rather large (7/27). It was found these outliers were not natural ones in the sense that they do not indicate abnormally high heavy metal body burden levels. We discovered these children had a high caries status and very poor oral hygiene which may have resulted in deeper

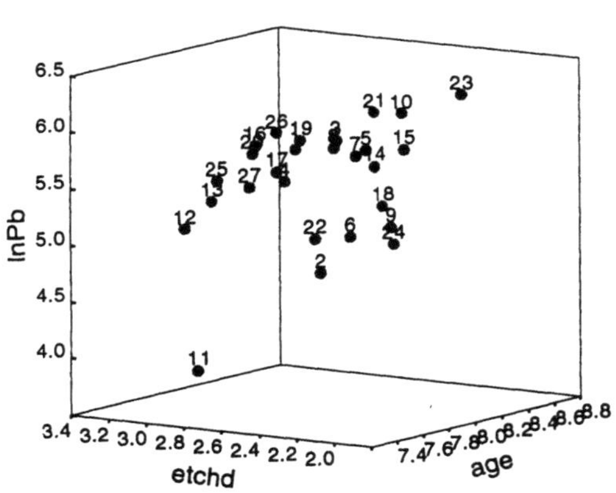

Fig. 2. SCHELLE data set. A 3D plot of the regression variables

etching and a post eruptive dissolution of lead into saliva. The SCHELLE data were collected very early in the survey and after this finding it was decided to enrol only children with a reasonable caries status into the study.

LS was unable to detect the outliers since no very large standardized residuals were found (see Table 5). On the contrary, LMS identified all the outliers and labeled them as strong ($|z_i| > 3.0$). RLS confirmed the LMS results.

As can be seen from Table 6, the R^2 was severely affected by the presence of outliers. The deletion of the outliers had a strong effect on the regression coefficient of *etchd*, on the standard errors of the coefficients and on the significance of the associated t-test.

Table 5. SCHELLE data set. Regression residuals of selected cases.

	LS		LMS		RLS	
Index(i)	$\hat{\epsilon}_i$	z_i	$\hat{\epsilon}_i$	z_i	$\hat{\epsilon}_i$	z_i
2	−0.893	−1.83	−1.87	−6.73	−1.32	−7.26
6	−0.611	−1.25	−0.946	−5.37	−0.866	−5.56
9	−0.462	−0.95	−0.883	−5.00	−7.51	−4.81
11	−1.253	−2.56	−1.490	−8.44	01.474	−9.46
18	−0.344	−0.70	−0.732	−4.15	−0.619	−3.97
22	−0.415	−0.85	−0.751	−4.26	−0.672	−4.31
24	−0.924	−1.89	−1.260	−7.14	−1.177	−7.55

Table 6. SCHELLE data set. Regression estimates and model fitting statistics.

	LS	LMS	RLS
No of children	27	27	20
No of outliers	1	7	(7)
ln *PB*; $E(y \mid x_1 = 3, x_2 = 7)$	4.997	4.500	5.198
$\hat{\sigma}$	0.489	0.176	0.156
R^2	0.145	0.819	0.697
p-value	0.152		0.00004
intercept ($\hat{\beta}_0$)	4.39	4.84	4.72
standard error	2.19		0.83
t-value	2.01		5.65
p-value	0.056		0.00003
etch depth ($\hat{\beta}_1$)	−0.528	−0.736	−0.606
standard error	0.292		0.104
t-value	−1.81		−5.81
p-value	0.082		0.00002
age ($\hat{\beta}_2$)	0.313	0.364	0.328
standard error	0.276		0.105
t-value	1.15		3.14
p-value	0.263		0.0060

Transformations

Sometimes in regression analysis it may be useful to transform the data. Transformations can be as simple as converting data from some unit such as centimeters to meters or, as in the previous section, converting years and months into a decimal system. These sorts of manipulations should have no effect on the data analysis. Other transformations exist which can result in a regression model producing better fits or improved predictions.

An infinite number of transformations exist, and the reader should refer to [33–35] for more information. Some common transformations involve taking the logarithm, power or square root of a variable(s).

Because there are so many transformations available it is necessary for the analyst to determine which transformation would be appropriate. By examining scatterplots of the response vs the predictors, or of the residuals vs the regression variables, suitable transformations can be suggested.

Besides improving fits and predictions, transformations can also be used to get the data variables to follow a certain distribution, since in some regression procedures it is assumed the data is sampled from a particular distribution. For example, should the data be required to follow a normal distribution, it could be worthwhile to perform a Box-Cox transformation [32].

Sometimes the type of transformation needed may not be obvious by simply viewing scatterplots. Large amounts of noise, for instance, can distract the naked eye from perceiving a suitable transformation. Should this be the case, smoothing techniques can prove to be a useful modeling method to apply.

Since smoothing techniques are often incorporated in algorithms of non-parametric models, a selection of smoothing techniques will be briefly mentioned in the next section.

Smoothing Techniques

Smoothers enhance visual aid by discovering trends in the data that may have gone undetected if the use of a smoother were not employed. Besides being a visual aid, smoothers also provide estimative information about the trend of one or more predictors. Smoothers are non-parametric since there is no assumption involving the form of the dependence of the response on its predictors.

For notational purposes, only smoothing techniques involving one predictor will be described; the notation for multiple predictors is simply an extension of the univariate case.

Smoothing Notation and Definitions

Given one has a data set with n response measurements y_i, $i = 1, \ldots, n$ and n predictor values x_i, $i = 1, \ldots, n$ such that $x_1 < \cdots < x_n$, the scatterplot smoother is defined as

$$s = \mathcal{S}(y|x) \,. \tag{13}$$

The function s has the same domain as x. The scatterplot smooth estimate of some value $x_0 \in \{x_1,\ldots,x_n\}$ is $s(x_0)$ which is calculated by evaluating the function $\mathcal{S}(y|x)$ at x_0. If x_0 is not an element of $x_1,\ldots,x_n$ then some interpolation strategy is required.

Bin Smoother

The bin smoother groups the independent variable into disjoint clusters each having the same number of elements. Usually about five groups are formed and the response in each region is averaged. The bin smoother has jump discontinuities at the cut points and is therefore not very smooth.

A set of indices defining the cut points $c_0, c_1, \ldots, c_K$ is

$$R_k = \{i : c_k \leq x_i \leq c_{k+1}\} \,.$$

Provided $0 \in R_k$,

$$s(x_0) = \underset{i \in R_k}{\text{mean}} \; y_i \,.$$

Running Mean Smoother

The running mean is similar to the bin smoother in that the response variable is averaged. The main difference between the two is that for the bin smoother an average is taken for every group, while with the running mean an average is taken for every measurement in x.

A neighborhood $\{\eta(x_0)\}$ for each target value x_0, is specified by selecting a number of points on each side of x_0. For example a neighborhood of some point x_i could be

$$\eta(x_i) = \{i-2, i-1, i, i+1, i+2\} \,.$$

For each target value, average the response values whose corresponding predictor values lie in the neighborhood of the target. Formally, define the running mean smoother as

$$s\,(x_i) = \underset{j \in \eta(x_i)}{\text{mean}} \; y_j \,.$$

The running mean is often referred to as the *moving average*. It is popular for equispaced time series data but tends to be too variable and can be very biased since the running mean smoother tends to flatten out trends near the endpoints.

Running Line Smoother

Instead of averaging in the neighborhood of some point x_0, the running line smoother performs a least-squares regression so that

$$s\,(x_0) = \hat{\beta}_0 + \hat{\beta}_1 \; x_0$$

where $\hat{\beta}_0$ and $\hat{\beta}_1$ are least-squares coefficients calculated from the data points in the neighborhood of x_0.

Unlike the running mean smoother, the running line smoother captures the trend at the endpoints and therefore reduces the bias. Cleveland [36] proposed a locally weighted running line smoother which performs a weighted least-squares regression to calculate $\hat{\beta}_0$ and $\hat{\beta}_1$.

Kernel Smoothers

Kernel smoothers use a sequence of weights, S_{0j}, to estimate each target value. S_{0j} is the weight given to y_j in producing an estimate for x_0. An expression for these weights is given by Hastie and Tibshirani [13] as

$$S_{0j} = \frac{a}{\lambda} d\left(\left|\frac{x_0 - x_j}{\lambda}\right|\right)$$

where a is a constant, appropriately chosen so the weights sum to unity, λ is the bandwidth and $d(\cdot)$ is an even function that decreases with $|\cdot|$. The sequence of weights forms a weight function whose shape is called a *kernel.* An example of a kernel is the Gaussian density function. The weight sequence for a kernel decreases the further you move from the target value, i.e., for a point x_j that is a large distance from x_0, S_{0j} will be small. Computationally, $s(x_0)$ is calculated as

$$s(x_0) = \frac{\sum_{j=1}^{n} d\left(\frac{x_0 - x_j}{\lambda}\right) y_j}{\sum_{j=1}^{n} d\left(\frac{x_0 - x_j}{\lambda}\right)}.$$

Kernel smoothers have a tendency to show biased behavior at the end points [13]. Incidentally, a smoother is said to be *linear* if it can be represented in the form $s(x_0) = \sum_{j=1}^{n} S_{0j} y_j$.

Supersmoother

The supersmoother is a very advanced running line smoother developed by Friedman [37]. The supersmoother uses a symmetric k-nearest neighborhood where k is estimated by a cross-validation routine (see Model Selection Criteria).

For each x_i a linear least-squares fit is constructed as done for the running line smoother. This is done three times, producing three smooths $s_1(x)$, $s_2(x)$ and $s_3(x)$. For s_1, $k = k_1 = 0.5n$; for s_2, $k = k_2 = 0.2n$; and for s_3, $k = k_3 = 0.02n$. A cross-validated residual for each point x_i is computed. Denote the cross-validated residuals for x_i by $r_1^{-i}(x_i), r_2^{-i}(x_i)$ and $r_3^{-i}(x_i)$ where

$$r^{-i}(x_i) = y_i - s^{-i}(x_i)$$

and s^{-i} is the smooth fitted when the point x_i has been removed.

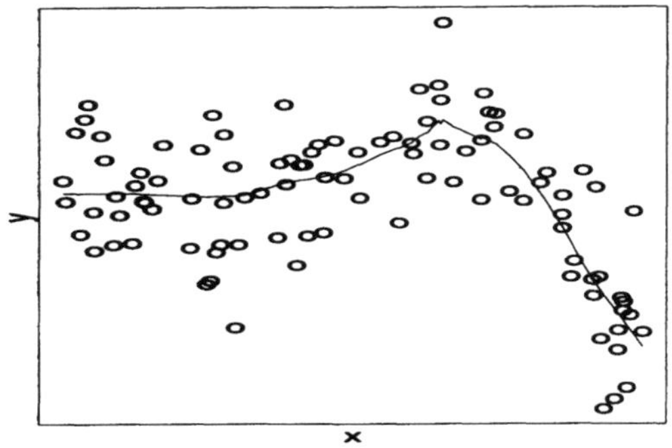

Fig. 3. A supersmoother applied to the scatterplot of y vs x

Next, take the absolute values for each of the elements in the cross-validated residual components, and smooth $|r_1^{-i}|$, $|r_2^{-i}|$ and $|r_3^{-i}|$ with $k = k_2$. Here $r = |r^{-i}|$. This produces three local error estimates $\hat{\xi}_1(x_i)$, $\hat{\xi}_2(x_i)$ and $\hat{\xi}_3(x_i)$ for each x_i. Here, $\hat{\xi}_j(x_i) = s(r_j(x_i))$.

An initial span $\kappa*$ for each x_i is chosen as the one that produces the minimum $\hat{\xi}_1(x_i)$, $\hat{\xi}_2(x_i)$ or $\hat{\xi}_3(x_i)$. For instance, if for the point x_0, $\hat{\xi}_3(x_0)$ was smaller than both $\hat{\xi}_1(x_0)$ and $\hat{\xi}_2(x_0)$, the span of x_0 would be $\kappa_3^* = 0.02n$.

The spans $\kappa*$ are then smoothed against x_i which produces an estimated span $\hat{\kappa}*$ for each x_i. The final smooth is constructed by interpolating between two of the three smooths s_1, s_2 and s_3. For example if $\hat{\kappa}*(x_0) \in (\kappa_1, \kappa_2)$ then the supersmooth $s(x_0)$ is an interpolation of $s_1(x_0)$ and $s_2(x_0)$.

The supersmoother can be computed very quickly, hence the prefix *super*. Supersmoothers are also highly adaptable to changes in the curvature of the underlying function, more so than ordinary running lines and locally weighted smoothers. Figure 3 displays an example of data being smoothed using a supersmoother.

Splines

Splines are non-parametric functions that can be used to explore the relationship between some response and its predictor(s). Splines, and in particular piecewise polynomial splines, have many advantages compared to polynomial regression. A major attraction of splines is the local nature to their fit. This is achieved by fitting several piecewise polynomials to the data. The polynomials are separated by a set of breakpoints $t_1, \ldots, t_K$ commonly called knots. The polynomials are usually constrained to join smoothly at the knots.

Several choices need to be made when dealing with splines. Such choices include:

1. number and positioning of knots;
2. type of basis functions; and
3. degree of the spline function.

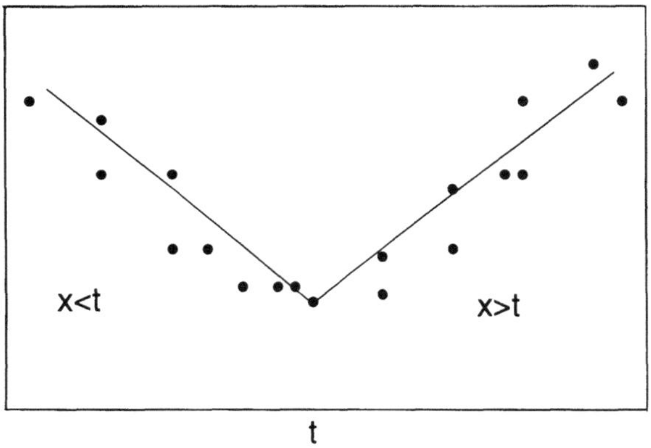

Fig. 4. A graph showing the division of the univariate scatterplot into two regions separated by a knot at t

The MARS model (see section entitled The MARS Model) has an elaborate approach to choose the number of knots and their locations. The basis functions MARS employs are the truncated power functions:

$$\varrho_+^{(q)}(x|t) = +(x-t)_+^q \, , \tag{14}$$

$$\varrho_-^{(q)}(x|t) = -(x-t)_+^q \, . \tag{15}$$

The power q is the degree of the spline which controls the amount of smoothness of the approximated function. For the MARS procedure $q = 1$. Generally for $q = 1$ the univariate spline basis functions do not have continuous derivatives, but Friedman [2] actually makes a modification to the basis functions so they can have continuous derivatives and still resemble $q = 1$ splines.

The $+$ sign on the right-hand side of the parenthesis in Eqs. (14) and (15) indicates the non-negative part. For example $+(x-t)_+$ will be positive for $x > t$ and similarly, $-(x-t)_+$ will be positive for $x < t$. Thus, Eqs. (14) and (15) split a scatterplot into two regions by a knot located at t. Figure 4 demonstrates this subdivision.

The PI model (see section entitled The PI Model) uses the cubic spline basis functions 1, x and $(x-t)_+^3$ to represent some response. In choosing the number of knots the reader is referred to that section. Breiman [15] explains how the position of the knots is chosen.

Non-parametric Multiple Regression Models

The ACE Model

As mentioned in the section entitled Transformations, transformations can be very effective in enhancing model fitting and predicting performances. By examining scatterplots and residual plots, suitable transformations can be suggested. In many instances however, such transformations are not so clear and this is where ACE can be very useful. The ACE algorithm proposed by Breiman and Friedman [4] essentially estimates optimal transformations for a set of predictors and a response. Sometimes these transformations can suggest a closed expression for a transformation such as the logarithm or square root.

ACE can be used on both continuous and categorical data, and can be very useful when applied in conjunction with LS. Unfortunately, ACE can sometimes produce misleading results. This can occur if there are sharp changes in the transformations [4].

The ACE model represents the transformed response $\theta(y)$ as a sum of transformed predictors $g(x)$

$$\theta(y) = \sum_{i=1}^{m} g_i(x_i) + \epsilon . \tag{16}$$

The transformed functions $\theta(y)$ and $g_1(x_1), \ldots, g_m(x_m)$ are smooth functions that are not required to have any particular form [5]. The error component, ϵ, is assumed to follow a normal distribution with mean zero and constant variance and, additionally, ϵ is to be independent of $x_1, \ldots, x_m$ [6].

Defining e^2 to be the fraction of unexplained variance obtained by regressing $\theta(y)$ on $g_1(x_1), \ldots, g_m(x_m)$, a data-driven expression for e^2 is

$$e^2 = \frac{\sum_{i=1}^{n} \left(\hat{\theta}(y_i) - \sum_{k=1}^{m} \hat{g}_k(x_k) \right)^2}{\sum_{i=1}^{n} \hat{\theta}^2(y_i)} . \tag{17}$$

For convenience, it is useful if the mean value of the response and predictor variables is zero and, additionally, if the variance of the response equals one. The optimal transformations $\theta^*(y)$ and $g_1^*(x_1), \ldots, g_m^*(x_m)$ are determined by a procedure which minimizes e^2. This is equivalent to maximizing the correlation between $\theta(y)$ and $g_1(x_1), \ldots, g_m(x_m)$.

Initial guesses for θ and g_i are to be $\theta^{(0)} = y / \| y \|$ and $g^{(0)} = 0$. These guesses are continually updated (until e^2 fails to decrease) by the expressions

$$\theta^{(k+1)}(y) = \mathcal{S} \left(\sum_{i=1}^{m} g_i^{(k)}(x_i) \,|\, y \right) / \| \mathcal{S} \left(\sum_{i=1}^{m} g_i^{(k)}(x_i) \,|\, y \right) \| \tag{18}$$

and

$$g_j^{(k+1)}(x_j) = \mathcal{S} \left(\theta(y) - \sum_{\substack{i=1 \\ i \neq j}}^{m} g_i^{(k)}(x_i) \,|\, x_j \right) . \tag{19}$$

(The smoothing operator $\mathcal{S}$ is discussed in the section entitled Smoothing Techniques.)

The following example provides an application of the ACE algorithm, which demonstrates ACE's ability to suggest possible transformations.

Example

An artificial data set

$$y_i = e^{(x_i^2 + \epsilon_i)} \quad i = 1, \ldots, 200 \tag{20}$$

is constructed so that we have some idea about the form of the transformations. Here x_i and ϵ_i are drawn independently from the standard normal distribution $N(0,1)$.

Figure 5 displays the original data and the transformed data resulting from the ACE procedure. In the first frame it is clearly seen that the relationship between the original variables is by no means linear. The plot of $g(x)$ vs x suggests the transformation $g(x) = x^2$ and the plot of $\theta(y)$ vs y suggests the transformation $\theta(y) = \ln(y)$.

The bottom right corner of Fig. 5 indicates that the relationship between the transformed variables is somewhat more linear than the relationship between the original variables.

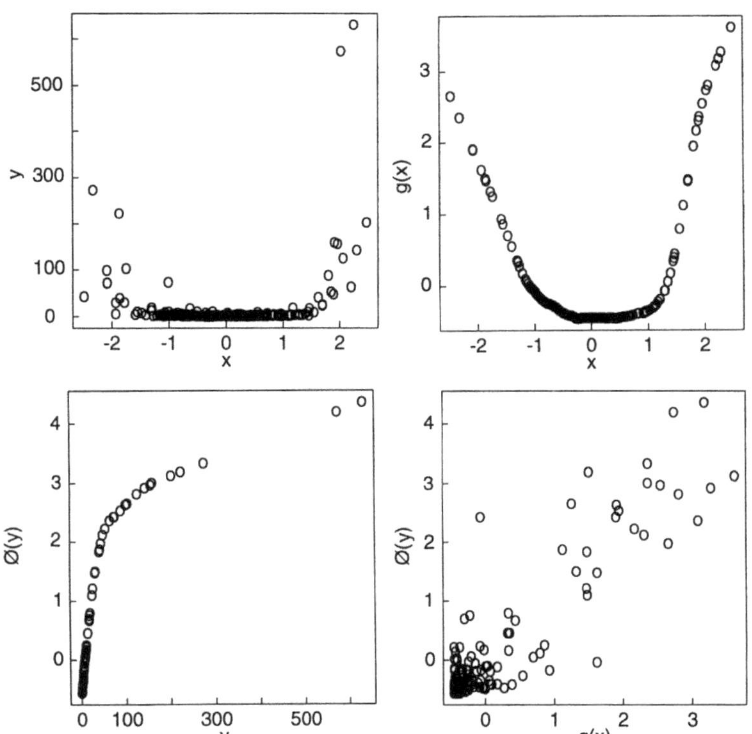

Fig. 5. Original data and transformations produced from the ACE algorithm

ACE and Ordinary Least-Squares

ACE can provide assistance to the least-squares procedure in two ways.

1. The transformations suggested by the ACE algorithm can be performed on the original variables. The transformed variables can then be used as the regression variables in LS. Sometimes ACE does not suggest a simple transformation to be made on a variable(s). Should this be the case, the ACE algorithm can force the appropriate transformations to be linear, and the original variables can form part of the LS model.
2. The ACE procedure can be used to assess the choice of a least-squares model. It is hoped, for a suitable least-squares model, the ACE transformations will be linear.

The PI Model

The PI model proposed by Breiman [10] approximates the response y by a sum of products:

$$y = \prod_{1} + \cdots + \prod_{J} + \epsilon \,. \tag{21}$$

Each $\prod$ consists of a product of univariate functions of the predictor variables,

$$\prod_{j} = \prod_{\ell=1}^{m} \phi_{j\ell}(x_\ell) \,. \tag{22}$$

The univariate functions ϕ are cubic spline basis functions (see section entitled Smoothing Techniques) having the representation

$$\phi_{j\ell}(x_\ell) = \hat{\beta}_{j\ell 0} + \hat{\beta}_{j\ell 1} x_\ell + \sum_{k=1}^{K} \hat{\beta}'_{j\ell k}(x_\ell - t_{j\ell k})^3_+ \,. \tag{23}$$

The subscript j ranging from 1 to J is a product label, indicating the product to which each univariate function belongs. The subscript ℓ is simply a variable label. The subscript k ranging from 1 to K indicates the number of knots in each univariate function ϕ.

The PI model assumes the underlying function is smooth and the error component has a zero mean. The PI model can produce superior results when modeling data consisting of a few terms involving high order interactions [10]. It is also claimed by Breiman [10] that the PI model is more than capable of performing well in situations when the ratio of the standard deviation of the function to the standard deviation of the noise is quite low.

The development of the model consists of two main algorithms. The first can be called a forward algorithm and the second a backward algorithm. During the forward algorithm the number of products J, and the number of basis elements ($K+2$) are determined. From this model the backward algorithm begins to remove basis elements that do not contribute to the fit, irrespective of what product they belong to.

In the PI model, Breiman [10] uses a score involving the residual sum of squares to determine the number of products and knots in the fit. The score is called a generalized cross-validation estimate of the prediction error (PE_{gcv}). A lower PE_{gcv} score is preferred to a higher PE_{gcv} score (see section entitled Model Selection Criteria).

Initially, the forward algorithm fixes the number of knots (K) in each of the products and attempts to fit one product, two products, three products and so on. If the PE_{gcv} score for three products were greater than the PE_{gcv} score for two products, then, for that particular K, two products would be used in the fit. Generalizing the problem for K fixed, if $PE_{gcv}(J) \leq PE_{gcv}(J+1)$, J products would be used in modeling the dependent variable. For the optimal number of products J^* with K fixed denote the generalized cross validation estimate of the prediction error as $PE_{gcv}(K, J^*)$.

In applying the algorithm, initially no knots are used in the fit, so each ϕ is linear. Then in each ϕ the forward search uses two knots up to the maximum number of knots (specified by the user). This means that during the forward algorithm the minimum number of knots in each factor ϕ is two. A sequence of PE_{gcv} scores $PE_{gcv}(2, J^*), \ldots, PE_{gcv}(K, J^*)$ is formed. Let $PE_{gcv}(k^*)$ denote the minimum of these scores.

The backward algorithm then begins operating on the model that produced $PE_{gcv}(k^*)$. Basis elements causing a minor increase in the residual sum of squares are gradually removed from the model. This will cause a decrease in degrees of freedom in the model. Again, a sequence of PE_{gcv} scores are formed.

The final model is not necessarily the model with the smallest PE_{gcv} score as seen in the application of the OZONE data (see section entitled PI Model, under Applications). The reasoning for this is explained by Breiman [10].

Let us now present an example, which illustrates the ability the PI algorithm possesses, to reproduce an original function from a noisy function.

Example

Simulated data will be used to compare an underlying function with the function produced by the PI model. The simulated data consists of two independent variables which were sampled from the uniform distribution $U(0, 4\pi)$. Noise was drawn from the normal distribution $N(0, 10)$ and added to the function,

$$y_i = x_{i1} \cos(x_{i2}) + \epsilon_i \quad i = 1, \ldots, 200 . \tag{24}$$

The upper frames in Fig. 6 shows plots of the response vs the predictor variables. It is clear that the relationship between the predictors and the response

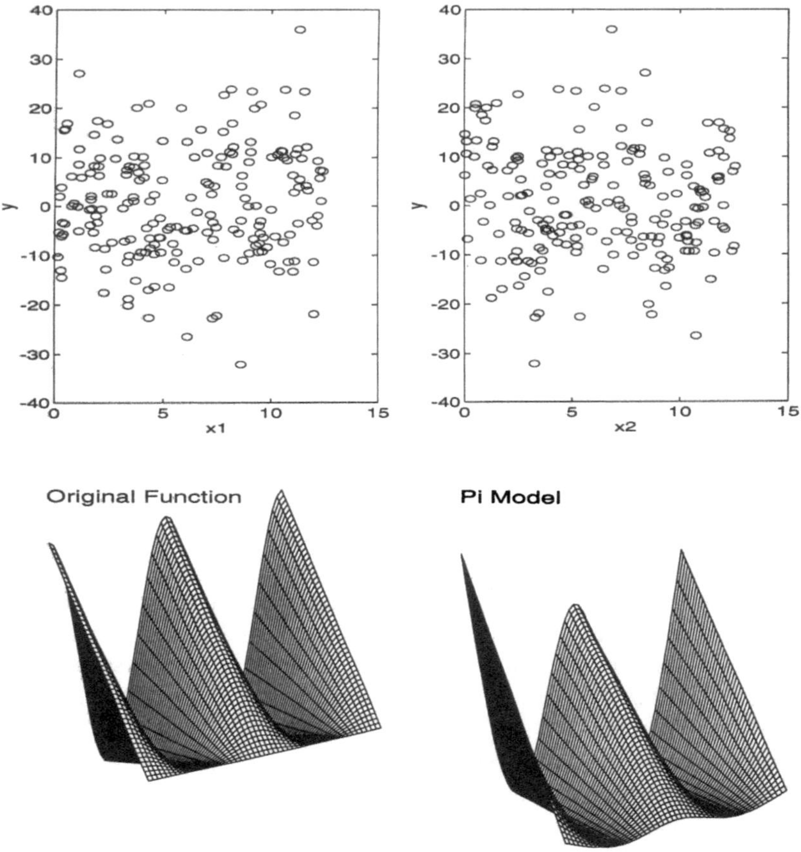

Fig. 6. The upper frame shows scatterplots of the response y vs x_1 and x_2. The lower frames display a surface plot of the original model (no noise) and the model produced by the PI algorithm (with noise). There is only one minor difference between the two surface plots – the foreground in the middle section of the PI plot is slightly lifted

contains an extreme level of noise. The lower frames display the surface of the original function $x_1 \cos(x_2)$ in the absence of noise and the surface plot produced by the PI model. Despite the addition of noise, the plot produced by the PI model is nearly identical to the original function.

The MARS Model

As in the case of the PI algorithm, the MARS algorithm developed by Friedman [2] consists of a forward and backward algorithm. The forward algorithm develops the full model and the backward algorithm removes terms that do not contribute significantly to the model. The MARS model is very similar to the PI model but the algorithms differ mainly in their forward searches. One could perhaps say that the forward algorithm for MARS is slightly more elaborate than that of the PI model.

The forward algorithm builds the MARS model by gradually incorporating subregions of the predictor variables that contribute to the fit. To explain the idea of subregions in context to the MARS algorithm consider the development of a MARS model, in representing some response y, by two independent variables – x and w. The MARS algorithm will initially incorporate one of the predictors into the model. The algorithm will try entering both x and w into the MARS model. The predictor variable that minimizes a score, called the lack-of-fit (*lof*) criterion (see Model Selection Criteria), will enter the model. As each term is being entered into the model all the data points in that term will be tested as possible knot locations. At this stage there can only be one knot in each variable. The knot position is also chosen by the lack-of-fit criterion.

Let us say the *lof* score is minimized when w enters the model with a knot placed at t_1. An estimate of the response is now expressed as a sum of two domains of w:

$$\hat{y} = \hat{\beta}_0 + \hat{\beta}_1[+(w - t_1)_+] + \hat{\beta}_2[-(w - t_1)_+] . \tag{25}$$

The algorithm now has the choice of entering x into the model (either additively or interactively) or adding another knot to w in which case one of the existing regions of w would be split. If the next greatest contribution is achieved when x enters the model additively, with a knot placed at t_2, the estimate of the response becomes

$$\begin{aligned} \hat{y} = \hat{\beta}_0 &+ \hat{\beta}_1[+(w - t_1)_+] + \hat{\beta}_2[-(w - t_1)_+] \\ &+ \hat{\beta}_3[+(x - t_2)_+] + \hat{\beta}_4[-(x - t_2)_+] . \end{aligned} \tag{26}$$

The algorithm can now test for further splitting of the specified regions in Eq. (26) by adding another knot to either x or w or search for interactions. If the region defined by $[+(w - t_1)_+]$ is found to interact with x, the estimate of the response becomes

$$\begin{aligned} \hat{y} = \hat{\beta}_0 &+ \hat{\beta}_1[+(w - t_1)_+] + \hat{\beta}_2[-(w - t_1)_+] \\ &+ \hat{\beta}_3[+(x - t_2)_+] + \hat{\beta}_4[-(x - t_2)_+] \\ &+ \hat{\beta}_5[+(w - t_1)_+][+(x - t_2)_+] + \hat{\beta}_6[+(w - t_1)_+][-(x - t_2)_+] . \end{aligned}$$

In summary, when updating the MARS model, the algorithm has the choice of the following.

1. Introducing a new variable to the model.
2. Splitting an existing region by placing an additional knot in a variable already present in the model.

3. Introducing an interaction term in the model. (This can only be done if one of the interaction terms has previously been defined.)

The operator has the option of specifying the level of interaction (*mi*) that may occur in the model. If $mi = 1$ the MARS model would be additive, if $mi = 2$ the maximum level of interaction would be of first order and so on.

Once the forward model has been established, elements such as $[+(x - t_2)_+]$ may be deleted by the backward algorithm. The elements that cause the smallest increase in the *lof* score or degrade the fit the least are subject to deletion.

Formally, the MARS model used to estimate a response y from m independent variables $x_1, \ldots, x_m$ is written,

$$\hat{y} = \sum_{\mathrm{p}=0}^{\mathrm{P}} \hat{\beta}_{\mathrm{p}} B_{\mathrm{p}}(x_1, \ldots, x_{\mathrm{m}}) \tag{27}$$

where

$$B_{\mathrm{p}}(x_1, \ldots, x_{\mathrm{m}}) = \prod_{\mathrm{k}=1}^{k_{\mathrm{p}}} \varrho(x_{\nu(\mathrm{k,p})} \mid t_{\mathrm{kp}}) \tag{28}$$

and

$$B_0(x_1, \ldots, x_{\mathrm{m}}) = 1 \, . \tag{29}$$

The subscript p in Eq. (27) is an index ranging from 1 to P (the number of regions in the model). The B_{p} are called multivariate spline basis functions which contain products of univariate spline basis functions, with a predictor $x_{\nu(\mathrm{k,p})}$ and associated knot t_{kp}. The functions used as a basis for the univariate splines are the truncated power functions (see Splines). The symbol k_{p} specifies the number of products in each B_{p}, while the subscript $\nu(k, p)$ simply labels the predictors in the k-th univariate spline basis function of the p-th multivariate spline basis function.

Another way of expressing the multivariate spline basis function in Eq. (28) could be,

$$B_{\mathrm{p}}(x_1, \ldots, x_{\mathrm{m}}) = B_{\tau}(x_1, \ldots, x_{\mathrm{m}}) \, \varrho \, (x_{\nu} \mid t) \, , \tag{30}$$

where $\tau \in [1, \mathrm{p} - 1]$. The variable x_ν cannot appear in the selected B_τ.

A simpler representation of the MARS model exists called the *ANOVA decomposition*. This model groups the terms having the same level of interaction. As Friedman [2] explains, the first sum in the ANOVA decomposition (Eq. (31)) contains the functions $g_{\mathrm{i}}(x_{\mathrm{i}})$ that consist purely of additive terms, i.e., the terms which are not involved in interactions. The second sum contains the functions $g_{\mathrm{ij}}(x_{\mathrm{i}}, x_{\mathrm{j}})$ involving first order interactions. The third summation contains the functions $g_{\mathrm{ijk}}(x_{\mathrm{i}}, x_{\mathrm{j}}, x_{\mathrm{k}})$ that consist of second order interactions and so on. Thus,

$$\hat{y} = \hat{\beta}_0 + \sum_{\mathrm{k_p}=1} g_{\mathrm{i}}(x_{\mathrm{i}}) + \sum_{\mathrm{k_p}=2} g_{\mathrm{ij}}(x_{\mathrm{i}}, x_{\mathrm{j}}) + \sum_{\mathrm{k_p}=3} g_{\mathrm{ijk}}(x_{\mathrm{i}}, x_{\mathrm{j}}, x_{\mathrm{k}}) + \cdots \, . \tag{31}$$

The following example provides an application of MARS.

Example

The MARS algorithm was applied to a simulated data set consisting of three independent variables, x_1, x_2 and x_3, each drawn from the uniform distribution $U(0, 10)$. Residuals were drawn from the standard normal distribution $N(0, 1)$ and were added to the model,

$$y_i = -2(x_{i1} - 5)^2 + 2x_{i2}x_{i3} + \epsilon_i \quad i = 1, \ldots, 100\,. \tag{32}$$

For illustrative purposes the data in this example were not standardized, but for stability purposes it is good practice to standardize the data before running MARS [2].

The model MARS constructed to represent the response consisted of linear terms in x_1 and first order interactions between x_2 and x_3. The ANOVA decomposition for this example can be written

$$y = g_1(x_1) + g_{23}(x_2, x_3)\,. \tag{33}$$

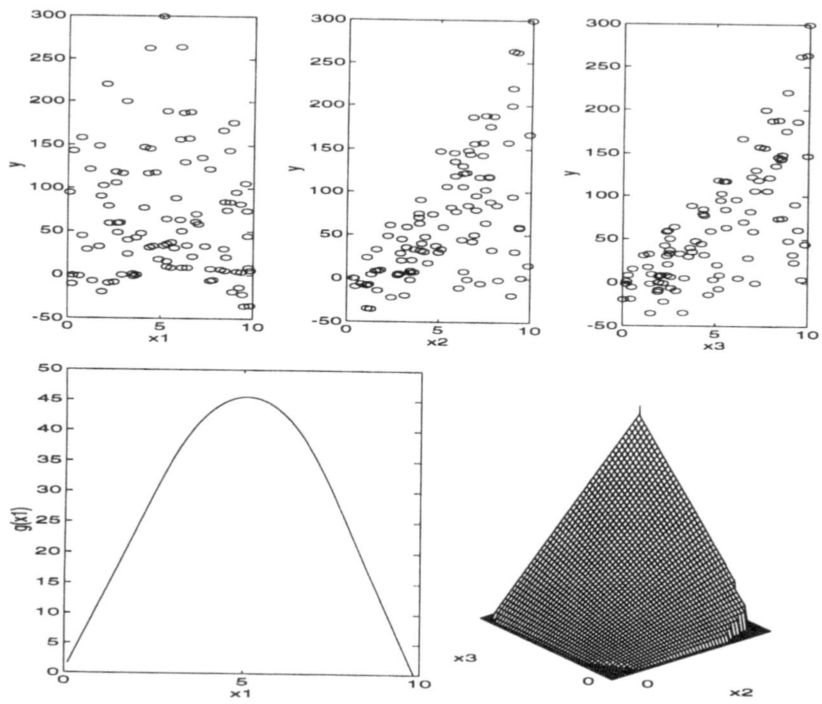

Fig. 7. Scatterplots of the simulated data (above) and the ANOVA functions (below) produced by MARS

In addition to scatterplots of the original response vs the predictors, Fig. 7 also displays the ANOVA functions of Eq. (33). The lower left frame is simply the additive contribution of $g_1(x_1)$ and the lower right frame is a perspective mesh plot of the bivariate function $g_{23}(x_2, x_3)$. The surface plot represents the joint contribution of x_2 and x_3 to a smooth of y on the variables x_1, x_2 and x_3. The symbol '0' designates on the axes smaller values of the variables. From this plot y is seen to be monotonically increasing for increasing x_2 and x_3.

Model Selection Criteria

In the previous examples, each of the non-parametric models was applied to simulated data. Whilst noise was added to the data, the original function was known. This knowledge allows one to assess subjectively how well the regression procedure modeled the data. When the underlying function is unknown, the quality of the model must be measured using some other criteria. Some of these criteria only measure how well the model fits the data, while other criteria can measure the accuracy of a model to predict future observations.

Model Fitting Criteria

The residual sum of squares (RSS) and the multiple coefficient of determination (R^2) both measure how well the model fits the data. Here

$$RSS = \sum_{i=1}^{n}(y_i - \hat{y}_i)^2 \tag{34}$$

where $\hat{y}$ is the estimated response. The coefficient of determination is defined as

$$R^2 = 1 - \frac{\sum_{i=1}^{n}(y_i - \hat{y})^2}{\sum_{i=1}^{n}(y_i - \bar{y})^2} . \tag{35}$$

R^2 is the squared correlation between y and $\hat{y}$. The symbol $\bar{y}$ is the mean average of the response values. $R^2 \in [0, 1]$. An R^2 value equal to 1 indicates a perfect fit of the estimated function and the data, while an R^2 value close to 0 indicates a poor fit. It is important to note that a high R^2 value could simply be a reflection of overfitting, since unnecessary terms incorporated into a model capture a larger portion of the error component.

The coefficient of determination, R^2, can also be determined in a robust way when dealing with robust regression. For LMS (Rousseeuw and Leroy [26]) it is defined as follows (in the case of regression with a constant term):

$$R^2 = 1 - \left(\frac{\text{median } |\hat{\epsilon}_i|}{\text{mad } (y_i)}\right)^2 . \tag{36}$$

The abbreviation "mad" stands for median absolute deviation,

$$\text{mad}\ (y_i) = \underset{i}{\text{median}} \left\{ |\ y_i - \underset{i}{\text{median}}\ (y_i)\ | \right\}. \tag{37}$$

Whilst the RSS and R^2 measure the quality of fit, they cannot be used to compare fits produced by different models of differing degrees of freedom. Other criteria exist which can be used to compare how well models of different degrees of freedom fit the data. Such measures include the mean square error (MSE), adjusted R^2 (R^2_{adj}), Akaike's information criterion (AIC) and Mallow's C_P.

Model Predicting Criteria

The cross-validating (CV) score provides a measure that reflects how well a model can predict. In determining the CV score for assessing the quality of future predictions of some function $\hat{f}(x)$ used to model a response y, the observations are randomly divided into V distinct groups (usually of the same size) $G_1, \ldots, G_V$. The number of groups will vary depending on the number of observational units in the data set and the accuracy the data analyst desires. One approach is to have ten groups with each group containing 10% of the data. Each group is then removed from the data set one at a time and the model $\hat{f}(x)$ is constructed in the absence of a particular group. Denote the model constructed in absence of the v-th group as $f^{-v}(x)$. The observations in the group removed from the data set are then predicted using $\hat{f}^{-v}(x)$. The predicted values are compared to the observed values and a measure of the squared difference is taken. Once each of the groups has been removed, V models will have been created and n squared differences will have been measured. The CV score as displayed in Eq. (38) is simply the average of all the squared distances. This equation has been taken from Breiman [15];

$$CV = \frac{1}{n} \sum_{v=1}^{V} \sum_{(x_i, y_i) \in G_v} (y_i - \hat{f}^{-v}(x_i))^2. \tag{38}$$

This kind of cross-validation is usually referred to as *V-fold* cross-validation. If the number of groups equals the number of observations then this form of cross-validation is called the *leave-out-one method.* Frank [18] converts the CV score to a cross-validated R^2 score (R^2_{cv}).

Cross-validating is very time consuming. This has led to the introduction of generalized cross-validating methods which are computationally quicker to calculate. Generalized cross-validation (GCV) scores provide only an estimate of the prediction error.

A general form of the GCV estimate is

$$GCV = \frac{1}{n} \left(\frac{RSS}{1 - NP/n} \right)^2 \tag{39}$$

where NP is the number of parameters that need to be estimated. Equation (39) is of the same form that Friedman [2] used to estimate a lack-of-fit score for the MARS model (see The MARS Model).

Breiman [10] removes the $1/n$ in Eq. (39) to produce what he calls the generalized cross-validation estimate of the prediction error (see The PI Model),

$$GCV = \left(\frac{RSS}{1 - NP/n}\right)^2 . \quad (40)$$

Applications

The regression models discussed so far will be applied to two data sets. Firstly, the WATER QUALITY data set will be used, followed by the OZONE data set. Both data sets have been standardized.

WATER QUALITY Data

The WATER QUALITY data [38] is a chemical data set consisting of 76 observational units. There are five independent variables – *sal*, *din*, *tip*, *chl*, *ss* and one response – *pha*. The variables are abbreviated to represent salinity, dissolved inorganic nitrogen, total inorganic phosphorus, chlorophyll, suspended solids and phytoplankton, respectively. Samples were taken from eight reefs located off the east coast of Australia. Figure 8 is a matrix plot showing the relationships between each combination of variables belonging to the WATER QUALITY data. Some plots between the variables reflect some degree of linearity. For instance, some linearity is evident between *pha* and *ss*, *pha* and *chl*, and *chl* and *ss*. Grouping of the data can be seen in plots involving *tip*, especially between *tip* and *din*.

ACE Model

The ACE model was applied to the WATER QUALITY data and gave the following equation:

$$\theta(pha) = g(sal) + g(din) + g(tip) + g(chl) + g(ss) . \quad (41)$$

Each of the transformations $\theta(\cdot)$ and $g(\cdot)$ is simply a transformation of the original variable. So, like the original variables, the transformed variables contain $n = 76$ data points. Figure 9 shows plots of the transformed variables vs the original variables. The transformations of *sal* and *din* seem peculiar since they

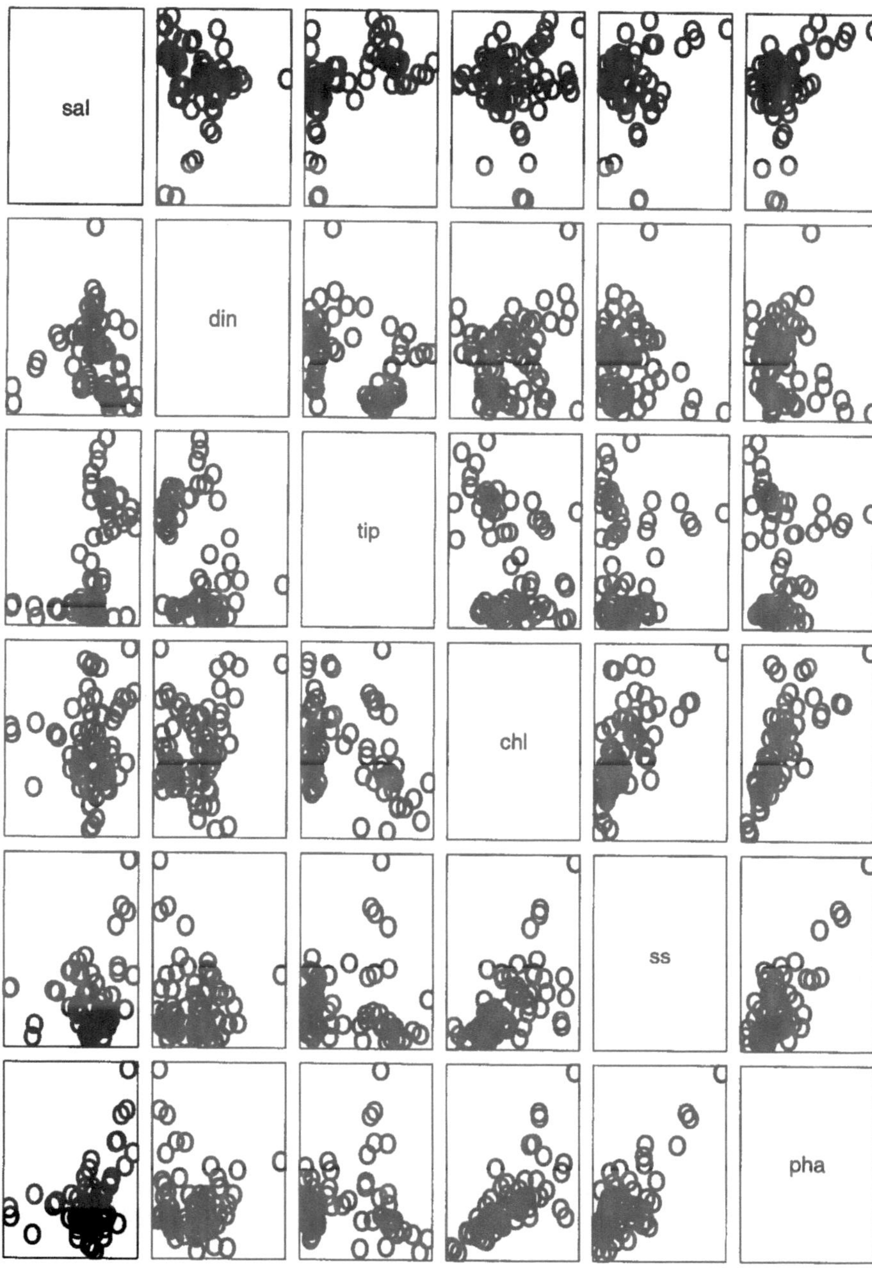

Fig. 8. WATER QUALITY data set. A matrix of scatterplots showing the relationship between each pair of variables

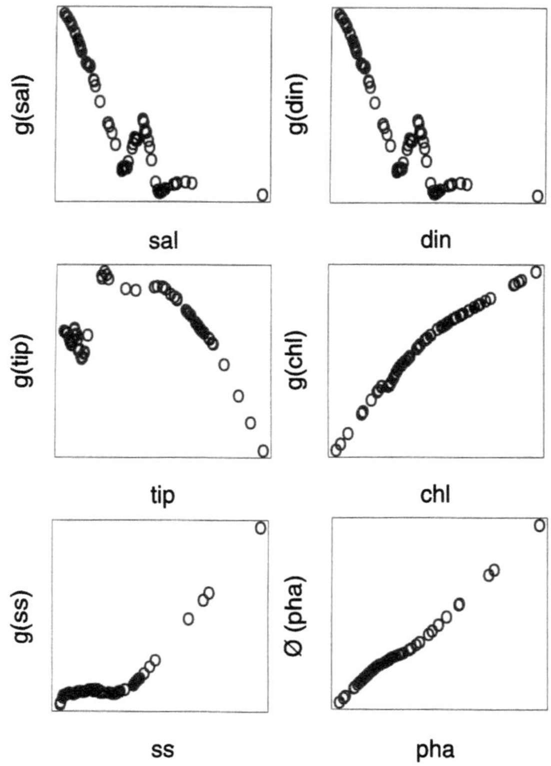

Fig. 9. WATER QUALITY data set. Optimal transformations produced by ACE

possess sharp changes in direction. This could possibly be attributed to the original data being clustered. When the transformations have a strange appearance it could be worthwhile to force these transformations to be linear, or alternatively remove them from the ACE analysis.

The transformation of *tip* possesses more structure than the transformations of *sal* and *din*. While some roughness is present in the beginning stages of the transformation, it does seem to possess a quadratic trend. The transformation of *chl* could be seen to have a very slight curvature. The transformation of *ss* increases linearly for larger values of *ss*; for smaller values of *ss* the transformation possesses hints of curvature. From an overall perspective the transformation could reflect a cubic. The transformation of the response is mostly linear.

The ACE procedure was re-run forcing the transformations of *sal* and *din* to be linear. This increased the R^2 score from 0.947 to 0.952. Figure 10 shows the transformations produced from the second ACE procedure. With the restriction being placed on *sal* and *din*, the transformations of *tip* and *ss* have changed slightly. If the two points in the upper left corner of the transformation of *tip* are ignored, the plot again seems to possess some quadratic nature; alternatively this could be represented by two piecewise linear functions. The transformation of *ss* now tends to have a quadratic appearance. The transformations of *chl* and the response *pha* remain virtually unchanged from that in Fig. 9.

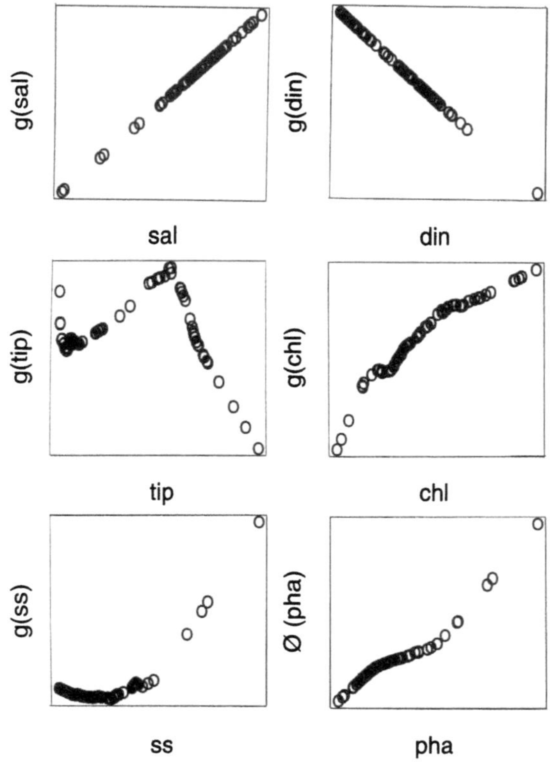

Fig. 10. WATER QUALITY data set. Optimal transformations produced by ACE with *sal* and *din* forced to be linear

LS was performed taking into consideration the transformations suggested by ACE. The LS model,

$$pha = -0.0272 + (0.229)\,sal - (0.112)\,din - (0.141)\,tip^2 + (0.551)\,chl + (0.169)ss^2, \quad (42)$$

produced an R^2 score of 0.848. When each of the predictors appear linearly in the LS model, the coefficient of determination is 0.792, approximately 5% lower than the R^2 produced with the aid of the ACE transformations.

PI Model

Two tables summarize the development of the PI model. Table 7 shows results of the forward stepwise strategy. This table consists of three columns indicating the optimal number of products J^*, for a fixed number of knots K. The final column shows the associated estimate of the generalized cross-validation estimate of the prediction error $PE_{gcv}(K,J^*)$. Table 8 shows results of the backward elimination process. The two columns shown in this table identifies the degrees of freedom

Table 7. WATER QUALITY data set. Results of the forward stepwise procedure of the PI model.

No. Knots K	No. Prods. J^*	$PE_{\text{gcv}}(K, J^*)$
2	2	9.58
3	2	7.36
4	2	8.30
5	2	14.49

Table 8. WATER QUALITY data set. Results of the backward stepwise procedure of the PI model.
No. of initial knots = 3; No. of products = 2; Final Df = 29

Df	PE_{gcv}
32	7.36
30	6.51
29	6.26
28	6.24
27	6.28

and the PE_{gcv} score of the corresponding model. The decrease in degrees of freedom reflects the deletion of basis elements from the regression model.

By examining Table 7, it is seen that when there are two, three, four and five knots in each factor, the optimal number of products is two. Upon completion of the forward algorithm a model consisting of three knots in each of the five (= number of predictors) factors, for both products, gave the minimum generalized estimate of the prediction error $PE_{\text{gcv}}(k^*) = 7.36$. It is this model that undergoes backward deletion of the basis elements.

Table 8 displays the results of sequential deletion of basis elements, which in turn reduces the degrees of freedom in the model. The PI model with 29 degrees of freedom,

$$\begin{aligned} pha = {} & \phi_{1\,\text{sal}}(sal)\, \phi_{1\,\text{din}}(din)\, \phi_{1\,\text{tip}}(tip)\, \phi_{1\,\text{chl}}(chl)\, \phi_{1\,\text{ss}}(ss) \\ & + \phi_{2\,\text{sal}}(sal)\, \phi_{2\,\text{din}}(din)\, \phi_{2\,\text{tip}}(tip)\, \phi_{2\,\text{chl}}(chl)\, \phi_{2\,\text{ss}}(ss)\,, \end{aligned} \tag{43}$$

was produced to represent *pha*. The R^2 value for this model is 0.968. Each factor in each of the products need not have the same amount of knots, since a different number of basis elements have been deleted from different factors.

Plots of the functions representing each of the factors for the two products are displayed in Fig. 11. The first column of graphs are the functions belonging to the first product, while the second column of graphs are the functions belonging to the second product.

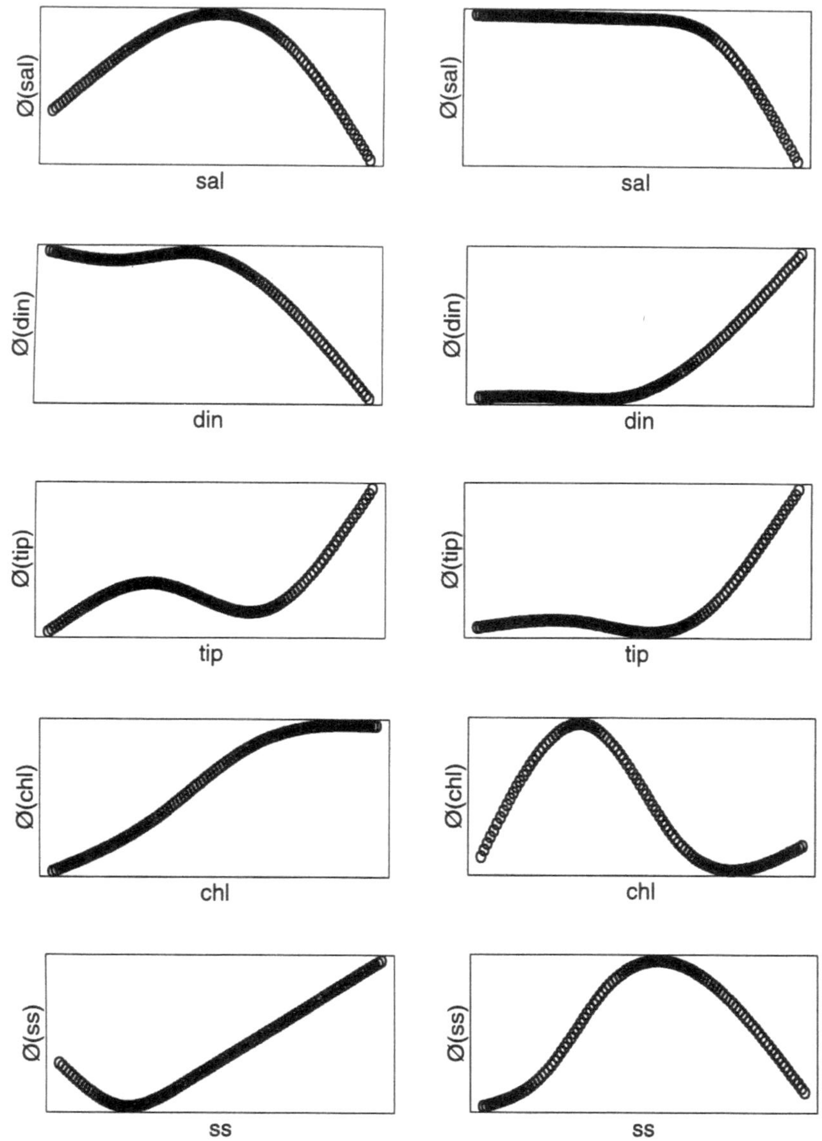

Fig. 11. WATER QUALITY data set. Graphical presentation of the univariate functions in the first (left) and second (right) products of the PI model

MARS Model

Two tables summarize the development of the MARS model. Table 9 consists of four columns and shows the forward development of the MARS model. The first column simply shows the iteration level of the MARS algorithm. The second

Table 9. WATER QUALITY data set. Results of the forward stepwise procedure of the MARS model.

Iteration	Basis Function		Variable	Parent
0	0			
1	1	2	ss	0
2	3	4	chl	0
3	5	6	sal	0
4	7	8	din	6
5	9	10	tip	5
6	11	12	din	6
7	13	14	tip	0
8	15		chl	14
9	16	17	ss	5
10	18	19	sal	14
11	20		din	2

Table 10. WATER QUALITY data set. Results of the backward stepwise procedure of the MARS model.

Basis Fn.	0	1	2	3	4	5	6	7	8	9	10
	√	√	×	×	√	×	×	√	×	√	√
Basis Fn.	11	12	13	14	15	16	17	18	19	20	
	×	×	×	×	√	×	×	×	×	×	

column shows which basis functions have entered the model. The third column identifies the variable that belongs to its corresponding basis function(s). The final column shows the parent basis function. This is the basis function which the latest basis functions at the current iteration are multiplied with. For example during the fourth iteration basis functions B_7 and B_8 are multiplied with basis function B_6, producing an interaction term between *sal* and *din*. The zeroth basis function is unity.

Table 10 identifies which basis elements have been deleted during the backward stepwise deletion process. If the basis function has been deleted from the model a cross appears while a tick appears if the basis function remains. From Table 10 the second and third basis elements have been deleted, along with the fifth, sixth and eighth, eleventh to fourteenth and all the basis elements after and including the sixteenth.

The MARS model for the WATER QUALITY data had the following ANOVA decomposition:

$$pha = g(ss) + g(chl) + g(din, sal) + g(sal, tip) + g(tip, chl) \,. \tag{44}$$

This model with an $R^2 = 0.968$ consists of one univariate function and three bivariate functions. Figure 12 displays plots of these ANOVA functions. The dependence of *pha* on *ss* is mostly linear, while the joint dependence of *pha* on *sal* and *din* is quite strong. The dependence of *pha* on *sal* and *tip* increases

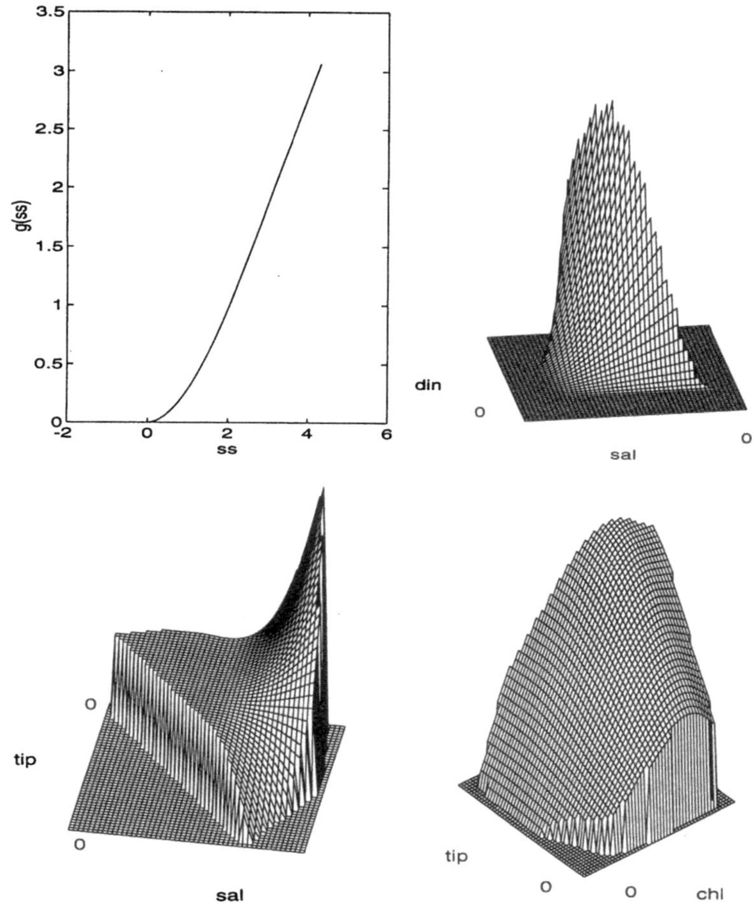

Fig. 12. WATER QUALITY data set. Graphical presentation of the ANOVA functions of the MARS model

sharply for increasing values of *sal* and decreasing values of *tip*. In the final frame, *pha* monotonically increases as *chl* and *tip* become larger.

OZONE Data

The OZONE data [6] is an environmental data set consisting of 111 observational units and four variables. Data was sampled from New York City between May 1 and September 30, 1973. The independent variables are *rad, temp* and *wind* each abbreviated to represent radiation, temperature and wind speeds. The dependent variable is *oz* which represents ozone levels. Figure 13 shows a matrix

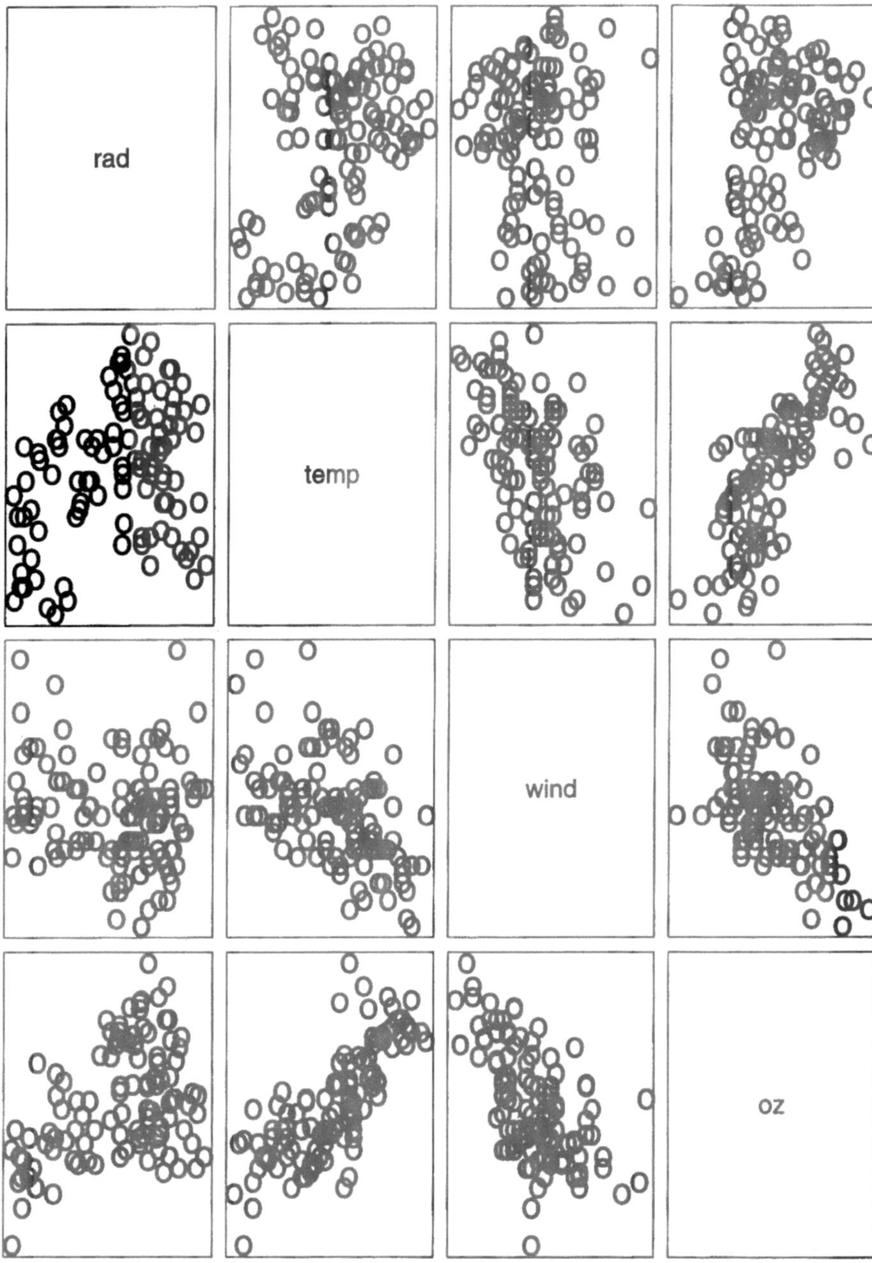

Fig. 13. OZONE data set. A matrix of scatterplots showing the relationship between each pair of variables

plot for the OZONE data. Negative correlations are seen to exist between *temp* and *wind*, and also between *wind* and *oz*. A strong positive correlation is evident between the variables *temp* and *oz*. Less structure appears in the remaining plots.

ACE Model

ACE was applied to the OZONE data, producing the regression model

$$\theta(oz) = g(rad) + g(temp) + g(wind) \,. \tag{45}$$

Equation 45 has an R^2 score equal to 0.890. The transformations in Eq. 45 can be seen in Fig. 14. The transformation of *rad* looks to consists of two piecewise linear fits. The plot of $g(temp)$ against temp has an overall increasing trend, with high temperatures corresponding to increased ozone levels. This transformation is not very smooth so it may be worthwhile to force this transformation to be linear. The transformation of wind is not completely smooth either with some roughness

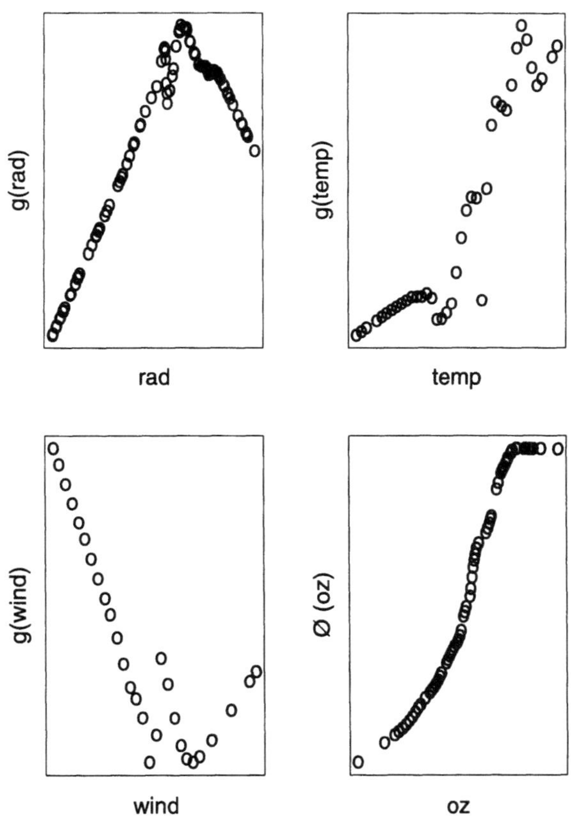

Fig. 14. OZONE data set. Optimal transformations produced by ACE

occurring for the mid-ranged values of wind speed. Generally, however, this transformation has mostly a negative trend. The transformation of the response is initially increasing but then reaches a plateau for larger ozone levels.

PI Model

Tables 11 and 12 summarize the development of the PI model for the OZONE data. As seen in Table 11, upon completion of the forward PI algorithm the model chosen to undergo backward stepwise deletion consisted of two products with each factor in the products containing four knots. This model is expressed as follows:

$$oz = \phi_{1\mathrm{rad}}(rad)\phi_{1\mathrm{temp}}(temp)\phi_{1\mathrm{wind}}(wind) + \phi_{2\mathrm{rad}}(rad)\phi_{2\mathrm{temp}}(temp)\phi_{2\mathrm{wind}}(wind)\,.$$

Initially the PE_{gcv} score for this model was 27.77, but was lowered to 22.62 after some knots were removed during the backward stepwise deletion strategy. These results can be seen in Table 12. The final model has $R^2 = 0.850$.

Figure 15 displays the univariate functions in the first and second products. In the third frame of the first column the function involving wind is constant giving the resulting PI model the following representation:

$$oz = \phi_{1\mathrm{rad}}(rad)\phi_{1\mathrm{temp}}(temp) + \phi_{2\mathrm{rad}}(rad)\phi_{2\mathrm{temp}}(temp)\phi_{2\mathrm{wind}}(wind)\,. \quad (46)$$

Table 11. OZONE data set. Results of the forward stepwise procedure of the PI model.

No. Knots K	No. Prods. J^*	$PE_{\mathrm{gcv}}(K,J^*)$
2	2	35.69
3	2	33.37
4	2	27.77
5	2	30.09
6	2	30.84

Table 12. OZONE data set. Results of the backward stepwise procedure of the PI model.
No. of initial knots = 4; No. of products = 2; Final Df = 16

Df	PE_{gcv}
26	27.77
22	24.68
21	24.34
19	23.80
16	22.62
14	23.46

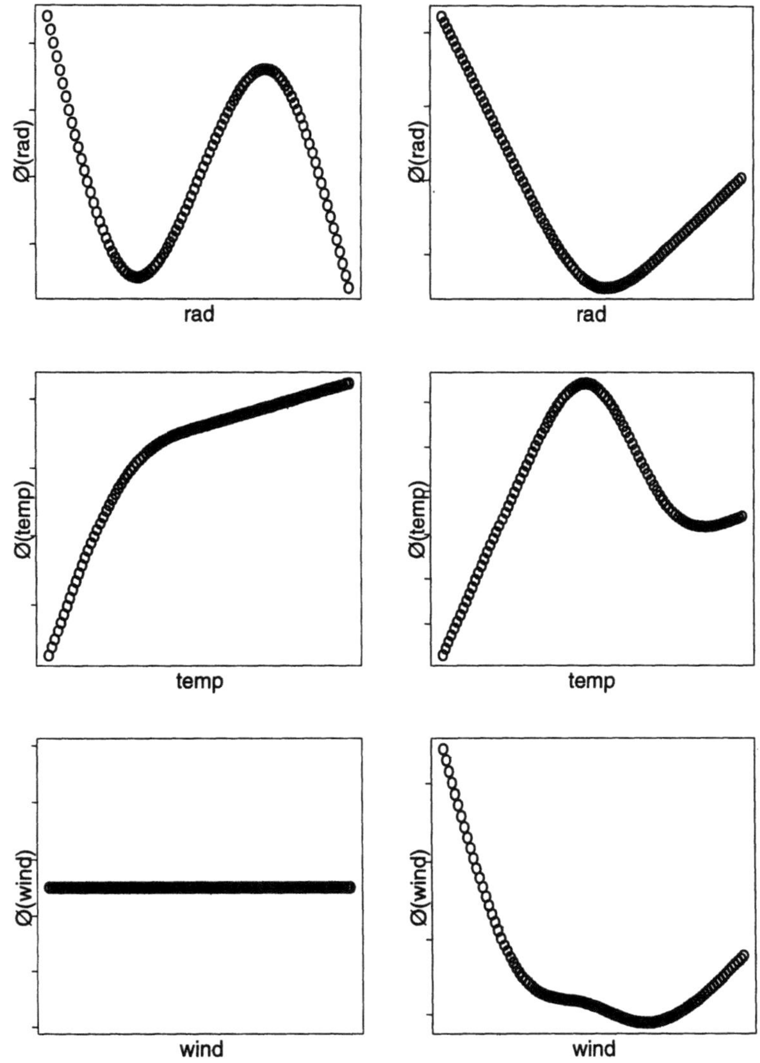

Fig. 15. OZONE data set. Graphical presentation of the univariate function in the first (left) and second (right) products of the PI model

MARS Model

Tables 13 and 14 show the forward and backward development of the MARS model for the OZONE data. The model chosen to represent *oz* is written

$$oz = g(temp) + g(wind) + g(rad, temp) + g(temp, wind)\ . \tag{47}$$

This model consists of two bivariate functions and has $R^2 = 0.954$.

Table 13. OZONE data set. Results of the forward stepwise procedure of the MARS model.

Iteration	Basis Function		Variable	Parent
0	0			
1	1	2	temp	0
2	3	4	wind	0
3	5	6	rad	0
4	7	8	wind	1
5	9	10	temp	6

Table 14. OZONE data set. Results of the backward stepwise procedure of the MARS model.

Basis Fn.	0	1	2	3	4	5	6	7	8	9	10
	√	√	×	×	√	×	×	×	√	√	√

Table 13 shows the first variable to enter the model was *temp* followed by *wind* and *rad*. During the fourth iteration an interaction term involving *wind* and *temp* entered the MARS model followed by another interaction between the variables *temp* and *rad* during the fifth iteration. Upon completion of the backward stepwise procedure the basis functions to remain in the model were B_0, B_1, B_4, B_8, B_9 and B_{10}, as displayed in Table 14.

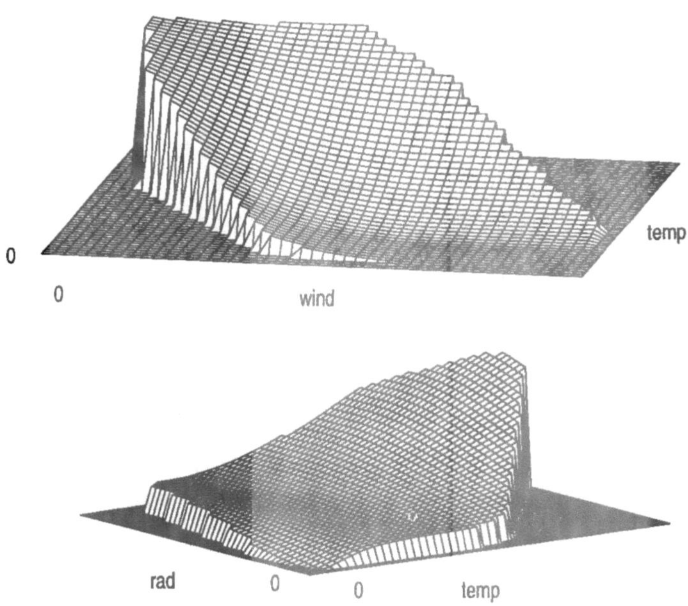

Fig. 16. OZONE data set. Graphical presentation of the ANOVA functions of the MARS model

Perspective plots in Fig. 16 display the joint bivariate contributions of *rad* and *temp* and of *temp* and *wind*. The perspective plot involving *temp* and *wind* is almost symmetrical. It seems that higher values of *temp* and lower values of *wind* have a monotonically increasing effect on ozone levels. The joint dependence of *oz* on *rad* and *temp* is reasonable with larger values of each variable corresponding to a rise in ozone levels.

Concluding Remarks on Non-parametric Regression Models

Equations (41)–(44) display the models resulting for each non-parametric application on the WATER QUALITY data. The equations produced to model the OZONE data are seen in Eqs. (45)–(47). Each of the models are seen to be quite different from each other.

It is important to remember there is no justification for comparing these models based on the R^2 value since each of the models has been constructed using a different number of degrees of freedom. Additionally, each of the models presented so far are not necessarily the best to represent each set of data, since each model was proposed purely to help gain a better understanding in the selected non-parametric methods.

In conclusion there is no non-parametric regression model which is preferable across all situations, since each of the methods have special characteristics that enhance performance given suitable circumstances. As a final word of caution, each of the methods have evolved fairly recently especially when compared to LS, so all results should be interpreted with care.

Software

The program PROGRESS [26] was used to present applications of robust regression methods on the ARDENNES and SCHELLE data sets. The 3D scatterplots of the ARDENNES and SCHELLE data were produced using the statistical program SPSS for MS-Windows, release 6.00 [39]. ACE was applied using the statistical package S-plus for MS-Windows version 3.1 [6]. The plots presented in the ACE examples were also produced using the same package. The applications of the PI model were accomplished using the FORTRAN code developed by Breiman [10]. The surface plots seen in the PI examples were produced using MATLAB version 4.1. A FORTRAN code developed by Friedman [2] was implemented to present the MARS model. The illustrations presented in the MARS applications were produced using MATLAB. All remaining plots were created in S-plus.

Acknowledgements. The authors wish to thank Jerome Friedman of Stanford University and Leo Breiman of Berkely University for sending their FORTRAN codes of the MARS and PI models respectively. Additional thanks to P. Rousseeuw for the program PROGRESS used in the applications of robust regression.

Appreciation is also extended to Glenn De'ath and A.I.M.S for allowing us the use of the WATER QUALITY data, and finally to Shaun Belward for constructive comments on the manuscript.

References

1. Draper N, Smith H (1981) Applied regression analysis. John Wiley and Sons, California
2. Friedman J (1991) Multivariate adaptive regression splines (with discussion). Annals of Statistics 19:1–141
3. Myers H (1990) Classical and modern regressions with applications. PWS-KENT, Boston
4. Breiman L, Friedman J (1985) Estimating optimal transformations for multiple regression and correlation (with discussion). J American Statistical Association 80:580–619
5. Frank I (1988) ACE: a non-linear regression model. Chemometrics and Intelligent Laboratory Systems 3:301–313
6. (1993) S-PLUS for windows user's manual volume 2 Version 3.1, Seattle: Statistical Sciences, Inc.
7. Fox J, Long J (1990) Modern methods of data analysis. Sage Publications, Inc. California
8. Härdle W (1990) Applied nonparametric regression. Cambridge University Press
9. Clare B (1993) Structure-activity correlations for psychotomimetics. 2. Phenylalkylamines: a treatment on nonlinearity using the alternating conditional expectations technique. Chemometrics and Intelligent Laboratory Systems 18:71–92
10. Breiman L (1991) The Π method for estimating multivariate functions from noisy data. Technometrics 33:125–160
11. Sekulic S, Kowalski B (1992) Mars: A tutorial. J Chemometrics 6:199–216
12. Friedman J (1988) Fitting functions to noisy data in high dimensions. Technical Report 101, Statistics Department, Stanford University
13. Hastie T, Tibshirani R (1991) Generalized additive models. Chapman and Hall, London
14. Hastie T, Tibshirani R (1986) Generalized additive models (with discussion). Statist Sci 1:297–318
15. Breiman L (1993) Fitting additive models to regression data. Computational statistics and data analysis 15:13–46
16. Blank T, Brown S (1993) Nonlinear multivariate mapping of chemical data using feedforward neural networks. Analytical chemistry 65:3081–3089
17. Rousseeuw P (1991) Tutorial to robust statistics. J Chemometrics 5:1–20
18. Frank I (1989) Comparative Monte Carlo study of biased regression techniques. Technical Report 105, Statistics Department, Stanford University
19. Höskuldsson A (1988) PLS regression methods. J Chemometrics 2:211–228
20. Manne R (1987) Analysis of two partial-least squares algorithms for multivariate calibration. Chemometrics and Intelligent Systems 2:187–197
21. de Jong S (1993) SIMPLS: an alternative approach to partial least squares regression. Chemometrics and Intelligent Systems 18:251–263
22. Wakeling I, Macfie H (1992) A robust PLS procedure. J Chemometrics 6(4):189–198
23. Frank I (1990) A nonlinear PLS model. Chemometrics and Intelligent Systems 8:109–119
24. Siegel A (1982) Robust regression using repeated medians. Biometricka 69:242–244
25. Rousseeuw P (1983) Regression techniques with high breakdown point. IMS Bull, 12 155
26. Rousseeuw P, Leroy A (1987) Robust regression and outlier detection. Wiley-Interscience, New York
27. Massart D, Kaufman L, Rousseeuw P, Leroy A (1986) Least median of squares: a robust method for outlier and model eror detection in regression and calibration. Anal Chim Acta 187:171–179
28. Rousseeuw P, van Zomeren B (1990) Unmasking multivariate outliers and leverage points. J AM Stat Assoc 85:633–639
29. Cleymaet R (1991) Lead and Cadmium in tooth enamel: measurement via acid etch biopsies. PhD thesis, Free University of Brussels
30. Coomans D, Slop D, Cleymaet R (1991) Lead and cadmium content in tooth surface enamel of Belgian schoolchildren from different geographic areas. Technical Report, Dept of Mathematics

and Statistics, James Cook University and Eenheid Prothetische Tandheelkunde, Vrije Universiteit Brussel
31. Cleymaet R, Quartier E, Retief D, Slop D, Coomans D (1991) Reappraisal of an in vitro and in vivo acid etch microbiopsy method applied to human tooth surfaces. Trace El Med 8:74–82
32. Box G, Cox D (1964) An analysis of transformations. J Royal Statistical Society B26:211–252
33. Bartlett M (1947) The use of transformations. Biometrics 3:39–52
34. Bennett C, Franklin N (1954) Statistical analysis in chemistry and the chemical industry. Wiley, New York
35. Bickel P (1981) An analysis of transformations revisted. J American Statistical Association 76:296–311
36. Cleveland W (1979) Robust locally-weighted regression and smoothing scatterplots. J American Statistical Association 74:829–836
37. Friedman J (1984) A variable span smoother. Technical report LCS5, Department of statistics, Stanford University
38. Australian Institute of Marine Science (1992) Long term monitoring of the Great Barrier Reef: dissolved and particulate nutrients. Australian Institute of Marine Science, Townsville, Australia
39. Norusis M (1993) SPSS for windows release 6.00. SPSS Inc.

Extension and Application of Univariate Figures of Merit to Multivariate Calibration

Karl S. Booksh and Ziyi Wang

University of Washington, Department of Chemistry BG-10, Seattle, Washington 98195, USA

Summary

Univariate "figures of merit" (i.e. sensitivity, selectivity, limit of detection, etc.) are common benchmarks employed in univariate calibration and instrument comparison. It is shown that the univariate figures of merit are easily transferable to multivariate calibration. Like their univariate brethren, the multivariate figures of merit and useful not only to compare and contrast calibration and instrumental performance, but to better understand and optimize multivariate calibration.

[1] Present address: Department of Chemistry and Biochemistry, University of South Carolina, Columbia, SC 29208

[2] Present address: Ohmeda Inc., 1315 West Century Dr., Louisville, CO 80027

List of Symbols and Abbreviations

$\|\| \|\|_2$	Euclidean norm
$\|\| \|\|_F$	Frobenius norm
ALS	alternating least squares
$\boldsymbol{A}^T$	transpose of $\boldsymbol{A}$
$\boldsymbol{A}^+$	generalized inverse of $\boldsymbol{A}$
$\boldsymbol{a}^*$	part of $\boldsymbol{a}$ that is unique to the analyte
$\bar{\boldsymbol{a}}$	mean of $\boldsymbol{a}$
$\boldsymbol{b}$	regression vector
c	analyte concentration in one sample
$\boldsymbol{c}$	vector of analyte concentrations
$\boldsymbol{C}$	matrix of component concentrations
CI	confidence interval
CLS	classical least squares
EBP	eigenvalue based problem
ε_r	univariate random instrumental errors
$\boldsymbol{\varepsilon}_\mathbf{r}$	vector of random instrumental errors
$\boldsymbol{E}$	matrix of random instrumental errors
$\boldsymbol{h}$	leverage of sample
$\boldsymbol{I}$	identity matrix
I	number of rows in a tensor (e.g. number of samples with first order data)
i	row index
ILS	inverse least squares
J	number of columns in a tensor (e.g. number of digitized wavelengths)
j	column index
K	number of slices in a tensor (e.g. number of samples with second order data)
k	slice index or arbitrary number of standard deviations
LOD	limit of determination
M	number of replicate unknown samples
$\boldsymbol{m}$	univariate regression vector (slope)
N	number of calibration samples
$\boldsymbol{N}$	second order instrument response of a pure analyte
NAR	net analyte rank
NAS	net analyte signal
$\boldsymbol{r}$	a first order tensor (a $1 \times I$ column vector)
$\boldsymbol{R}$	a second order tensor (a $I \times J$ matrix)
$\mathbb{R}$	a third order tensor (a $I \times J \times K$ cube)
$\boldsymbol{r}_i$	the i-th row (sample) of $\boldsymbol{R}$
$\boldsymbol{R}_k$	the slice (sample) of $\mathbb{R}$
σ_c	estimated prediction error
σ_r	estimated error about the regression line

$\boldsymbol{S}$	matrix of component sensitivities
SEL	selectivity
SEN	sensitivity
S/N	signal to noise ratio
var	variance
$\boldsymbol{X}$	estimated intrinsic profiles from the first instrument of the hyphenated pair
$\boldsymbol{x}_a$	estimated intrinsic profile of the analyte from the first instrument of the hyphenated pair
$\boldsymbol{Y}$	estimated intrinsic profiles from the second instrument of the hyphenated pair
$\boldsymbol{y}_a$	estimated intrinsic profile of the analyte from the second instrument of the hyphenated pair
$\boldsymbol{Z}$	estimated relative concentrations of all components from second order analysis
$\boldsymbol{z}_a$	estimated relative concentrations of the analyte from second order analysis

Introduction

In the broad field of multivariate calibration there are many vital and interesting subjects. An experienced practitioner of multivariate methods must be well versed in the areas on experimental design [1, 2], data pretreatment [3], sample and variable selection [4, 5], model selection and validation [6–8], outlier detection [9], and statistical interpretation and validation of the results [10]. These topics are covered in many chemometric textbooks [3, 10–12] and tutorials [13–17] and are used successfully by most application oriented analysts.

The above topics provide an excellent framework for creating and assessing multivariate calibration models and comparing the applicability of two or more different models to a particular calibration problem. However, this framework is not well suited for facilitating a comparison of different multivariate instrumental techniques. For univariate data, "figures of merit" such as the "limit of determination", "sensitivity", and "signal to noise ratio" are almost universally used to compare and contrast rapidly the applicability of two or more instrumental techniques to a particular problem. Univariate analytic methods are further characterized by the explicit propagation of instrumental errors through calibration and into the predicted analyte concentration. The figures of merit and error propagation also serve as a concise set of guiding principals to aid the analyst in the optimal implementation of the univariate analytical methods

Although the notions of figures of merit and calculation of prediction errors exist for multivariate analytic methods, these ideas have not been embraced by practicing analysts to the same degree as their univariate brethren have been embraced. Like the univariate figures of merit and error propagation equations these notions, when derived for multivariate data, serve to compare multivariate analytic methods and act as a guide for optimal implementation of multivariate methods.

Nomenclature

In this discussion the nomenclature of Sanchez and Kowalski will be used [18]. Instrumentation that produces a zero order tensor, a scalar, per sample is dubbed a "zero order instrument." The associated algorithms to calibrate this type of instrument are "zero order calibration" methods. Equivalently, "first order instruments" and "second order instruments" produce first order tensors, vectors, and second order tensors, matrices, per sample, respectively.

Zero order instrumentation, e.g. an electrode, generates a single datum, e.g. one voltage, for each sample analyzed. In mathematical tensor notation a single datum is a zero order tensor, a scalar, and represented by an italic, lower case letter, r. Analysis of multiple samples produces one scalar per sample. These measurements are distinguished by including a subscript, e.g. r_i is the measured voltage of the i-th sample. These I measurements can be collected into a I

dimensional column vector, $\boldsymbol{r}$, where the i-th entry is the signal associated with the i-th sample.

First order instrumentation collects multiple measurements per sample. These measurements can be highly correlated, e.g. digitized wavelengths on a multichannel spectrometer, or largely independent, e.g. an array of ion selective electrodes. The instrument response for each sample is dimensioned one sample by J variables, $\boldsymbol{r}^{\mathrm{T}}$. A collection of I samples forms the $I \times J$ matrix, $\boldsymbol{R}$, where $\boldsymbol{r}_{\mathrm{i}}$ represents the i-th row of $\boldsymbol{R}$ and R_{ij} is the signal of at the j-th sensors for the i-th sample.

Second order instrumentation produces a second order tensor, a matrix, of data per sample, for example, GC-MS in combination, are becoming increasingly common. The response from these hyphenated instruments is an $I \times J$ matrix, $\boldsymbol{R}$, where I is the number of discrete measurements taken at J variables. A collection of K samples forms the $I \times J \times K$ third order tensor, $\mathbb{R}$, where $\boldsymbol{R}_{\mathrm{k}}$ represents the instrument response from the k-th sample and R_{ijk} is the signal at the i-th measurement, j-th variable for the k-th sample.

First Order Calibration

First order, multivariate calibration has four major advantages over zero order, univariate calibration. With first order data, analysis can be accurately performed in the presence of spectral interferents [18, 19]. Multiple analytes can be quantitated simultaneously. Signal averaging with multiple highly correlated channels significantly reduces prediction errors in many applications. And more sophisticated outlier detection is possible [20].

First Order Calibration Models

First order calibration is encountered in two separate forms. In classical least squares (CLS) the instrument response is considered to be a function of the concentrations of the compounds in the sample [21–23]. Hence, the CLS model is

$$\boldsymbol{R} = \boldsymbol{C}\boldsymbol{S}^{\mathrm{T}} \tag{1}$$

where the I rows of $\boldsymbol{R}$ are the instrument responses of the I samples, the I rows of $\boldsymbol{C}$ are the concentrations of each of the K compound present in the I samples, and the K columns of $\boldsymbol{S}$ are the sensitivities of the J sensors to the K compounds. The matrix of sensitivities is found by least squares regression of $\boldsymbol{C}$ on $\boldsymbol{R}$ by calculating the generalized inverse of $\boldsymbol{C}$, $\boldsymbol{C}(\boldsymbol{C}^{\mathrm{T}}\boldsymbol{C})^{-1}$

$$(\hat{\boldsymbol{S}}) = \boldsymbol{R}^{\mathrm{T}}\boldsymbol{C}(\boldsymbol{C}^{\mathrm{T}}\boldsymbol{C})^{-1} \tag{2}$$

and the concentrations of all K compounds in the unknown sample is also estimated by least squares

$$\hat{c}_{un} = \boldsymbol{r}_{un}{}^{T}\hat{\boldsymbol{S}}(\hat{\boldsymbol{S}}^{T}\hat{\boldsymbol{S}})^{-1} \tag{3}$$

where the superscript^{-1} is the inverse of the matrix. As can be inferred from the model, the concentration of all compounds is required for regression and analysis.

The disadvantage of requiring that the concentration of every compound in each calibration sample be known is circumvented by employing the inverse least squares (ILS) model [24]. The ILS model assumes the concentration is a function of the instrument response,

$$\boldsymbol{c} = \boldsymbol{R}\boldsymbol{b} \tag{4}$$

where c is the column of $\boldsymbol{C}$ from Eq. (1) that corresponds to the analyte of interest, and $\boldsymbol{b}$ is the regression vector for the analyte of interest. The regression vector, $\boldsymbol{b}$, is found by least squares regression of $\boldsymbol{R}$ against $\boldsymbol{c}$,

$$\hat{\boldsymbol{b}} = (\boldsymbol{R}^{T}\boldsymbol{R})^{-1}\boldsymbol{R}^{T}\boldsymbol{c} \tag{5}$$

and the analyte concentration in an unknown sample is directly estimated as

$$\hat{c}_{un} = \boldsymbol{r}_{un}{}^{T}\boldsymbol{b}. \tag{6}$$

Inversion of the matrix $\boldsymbol{R}^{T}\boldsymbol{R}$ is required to estimate the ILS regression vector, $\boldsymbol{b}$. For a matrix to be invertable it must be square and full rank. This, however, is usually not the case in chemical applications. Here, $\boldsymbol{R}^{T}\boldsymbol{R}$ is often a nearly singular matrix dimensioned 1024 digitized channels by 1024 digitized channels. Problems associated with the inversion of a singular nonsquare matrix can be avoided by calculating the pseudoinverse, of $\boldsymbol{R}$ (designated $\boldsymbol{R}^{+}$) instead the generalized inverse of $\boldsymbol{R}$. Here,

$$\hat{\boldsymbol{b}} = \boldsymbol{R}^{+}\boldsymbol{c} \tag{7}$$

provides a least squares solution for $\boldsymbol{b}$ in Eq. (4).

Calibration and analysis is usually performed by multiple linear regression, (MLR) [25, 26], principal component regression (PCR) or partial least squares (PLS) [27–29], although other algorithms exist [30–32]. These algorithms differ only in the manner in which they estimate the pseudoinverse of $\boldsymbol{R}$. With data void of all instrumental errors each algorithm will perform identically. However, results from the disparate algorithms will differ in the presence of model and instrumental errors.

The ILS regression vector from Eq. (4), $\boldsymbol{b}$, is intimately related to the matrix of sensitivities in Eq. (1), $\boldsymbol{S}$. The regression vector, $\boldsymbol{b}$, is proportional to the part of the vector of analyte sensitivities that is orthogonal to the vectors of sensitivities of all other compounds present in the calibration set. That is

$$\boldsymbol{b} \propto (\boldsymbol{I} - \boldsymbol{S}_{a}\boldsymbol{S}_{a}^{+})\boldsymbol{s}_{a}{}^{T} \tag{8}$$

where $\boldsymbol{I}$ is the $J \times J$ identity matrix, $\boldsymbol{s}_{a}$ is the vector of sensitivities for the analyte, and $\boldsymbol{S}_a$ is the matrix of sensitivities for all other compounds present in the calibration set. It is important to note that the vectors of sensitivities for each compound is, in fact, the pure spectrum of each compound at unit concentration.

This leads to the well known fact that the ILS regression vector is correlated to the spectrum of the analyte of interest and orthogonal to the spectrum of the other compounds present in the calibration set [18, 19]. That is

$$\boldsymbol{s}_i{}^T\boldsymbol{b} = 0 \tag{9}$$

where $\boldsymbol{s}_i$ is the spectrum of any interferent in the calibration set. The orthogonal property of the regression vector is illustrated in Fig. 1 for a two variable, binary system. The black and light gray lines represent the spectra of the analyte and interferent in the two variable spaces. The dark gray line is the projection of the analyte signal that is orthogonal to the interferent's signal. This is termed the "net analyte signal" (NAS) and the length of the NAS is inversely proportional to the squared length of the regression vector [19]. That is,

$$\mathrm{NAS} = (\boldsymbol{I} - \boldsymbol{S}_a\boldsymbol{S}_a{}^+)\boldsymbol{s}_a{}^T \tag{10}$$

and

$$\|\boldsymbol{b}\|_2 = 1/\|\mathrm{NAS}\|_2 \tag{11}$$

where $\|\cdot\|_2$ is the Euclidean norm of the vector.

Figure 2 shows the absorbance spectra of trichloroethylene (TCE), chloroform ($CHCl_3$) and 1,1,1 trichloroethane (TCA), three common chlorinated hydrocarbons mandated for environmental monitoring by the EPA. There is complete spectral overlap of TCE by both $CHCl_3$ and TCA. The NAS of TCE in the presence of $CHCl_3$ and TCA (*black*), just $CHCl_3$ (*dark gray*), and just TCA (*light gray*) are given in Fig. 3. Note that the NAS depends on the interferents present. When both interferents are present in the calibration set the $\|\mathrm{NAS}\|_2$ of TCE is 0.084 Absorbance units, while when only $CHCl_3$ or TCA is present as an interferent, the $\|\mathrm{NAS}\|_2$ is 0.16 or 0.28 Absorbance units, respectively. Also

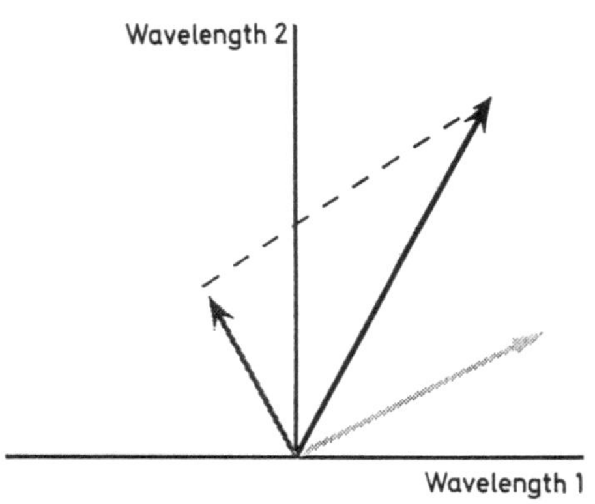

Fig. 1. Two wavelength analyte (*black*), and interferent (*light gray*), spectra with NAS for the analyte orthogonal to the interferent spectrum (*dark gray*)

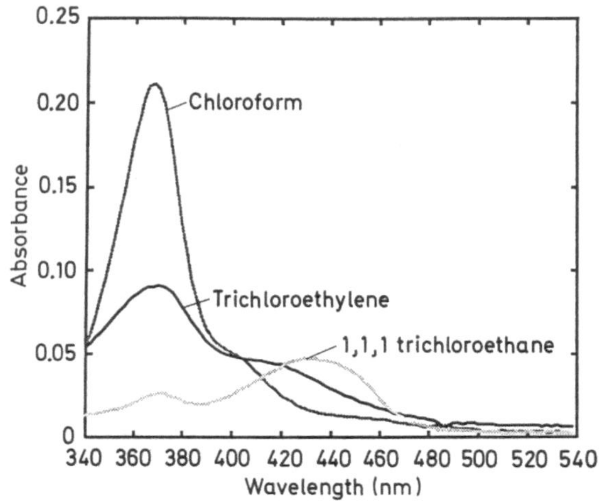

Fig. 2. Abosrbance spectra of pure trichloroethylene (*black*), chloroform (*dark gray*), and 1,1,1 trichloroethane (*light gray*)

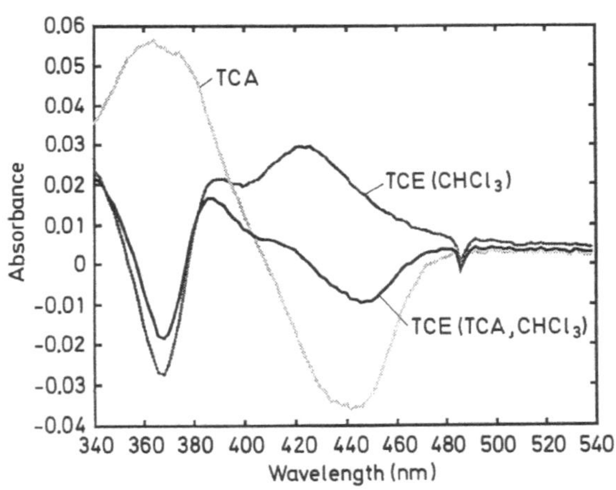

Fig. 3. Net analyte signal for TCE in the presence of TCA and $CHCl_3$ (*black*), TCE in the presence of $CHCl_3$ (*dark gray*), and TCE in the presence of TCA (*light gray*)

note that the form of the NAS changes with the presence of different spectral interferents. The NAS tends to be positively correlated with the spectral features of the analyte and negatively correlated with spectral features of the interferents. However, even in cases of complete spectral overlap, the NAS exists as long as the spectrum of the analyte is different from the spectra of the interferents.

Figures of Merit

Four figures of merit can be defined for first order data not by the total signal, as in zero order calibration, but by the NAS. Intuitively, it makes more sense to define figures of merit such as the selectivity, sensitivity, signal to noise ratio, and limit of detection by the part of the total signal that is used for calibration, the NAS, and not the total signal. On the surface, this appears to be a daunting task, and not very useful in practice, since calculation of the NAS requires measurement of the pure spectra of each interferent (Eq. 10). Fortunately, however, the calculation of the figures of merit require only the Euclidean norm of the NAS, $||\mathrm{NAS}||_2$, which can be calculated by Eq. (11) from the ILS regression vector. These figures of merit then, in turn, provide useful insights about optimizing the implementation of multivariate methods.

Any discussion about the relative merits of univariate methods begins with mentioning each methods selectivity,

$$\mathrm{SEL} = \frac{r^*}{r} \tag{12}$$

The selectivity is the fraction of the total signal, r, that is unique to the analyte, r^*. This dimensionless number varies from zero (cannot detect the analyte) to unity (complete selectivity to the analyte). For univariate calibration it is assumed that the selectivity is unity, else the estimated analyte concentration will be biased.

By the same logic, for first order analysis, the selectivity is defined as

$$\mathrm{SEL} = \frac{||\mathrm{NAS}||_2}{||\boldsymbol{s}_\mathrm{a}||_2} = \frac{1}{||\boldsymbol{b}||_2||\boldsymbol{s}_\mathrm{a}||_2} \tag{13}$$

recalling that sa is the pure analyte spectrum at unit concentration. The selectivity varies from 0 to 1 where a SEL of zero means that analysis is impossible from the analyte spectrum, and a SEL of 1 implies that there is no spectral overlap of the analyte and interferents (for positive spectra). Two points must be stressed. A selectivity of less than 1 does not mean that analysis will result in a biased concentration estimate as in zero order data. Estimation is accurate as long as SEL is greater that 0. That the SEL is proportional to the NAS implies that the selectivity changes, depending on the interferents in the calibration set. If an additional interferent is added the selectivity will usually decrease, while removing an interferent from the calibration set improves the selectivity. Therefore, the selectivity must be specified for each particular calibration problem. For example, the selectivity of spectroscopic calibration for TCE in the presence of TCA and $CHCl_3$ is less than the SEL when only TCA of $CHCl_3$ is present (Table 1).

Table 1. First order figures of merit for analysis of TCE

	0.49 μM TCE w/ TCA & $CHCl_3$	0.49 μTCE w/ $CHCl_3$	0.49 μM TCE w/ TCA
SEL	0.1869	0.3628	0.6283
SEN	0.1723 A/μM	0.3344 A/μM	0.5791 A/μM
S/N	84	164	284
LOD	0.017 μM	0.0052 μM	0.0030 μM

The definitions of selectivity for univariate calibration (Eq. 12) and for multivariate calibration (Eq. 13) do differ in one very important respect. For univariate calibration, the selectivity of analysis changes with the concentration of interferents present. Increase the concentration of a detectable interferent, and the univariate selectivity decreases. For multivariate calibration, the selectivity is independent of the interferent concentration. Regardless of the interferents' concentration, the NAS is the part of the analyte signal that is orthogonal to the interferents' spectra. Therefore, the magnitude of the univariate SEL is related to the magnitude of the bias associated with analysis, while the multivariate SEL means accurate calibration is impossible when the SEL is zero. It is assumed that unless the analysis is unbiased, the estimated concentration is worthless (i.e., what is the use of calculating the wrong answer!).

For univariate calibration, the sensitivity of a method is defined by the slope of the calibration line in the CLS model where the signal, r, regresses against the concentration, c. That is

$$\mathrm{SEN} = \frac{\Delta r}{\Delta c} = m \tag{14}$$

where m is the univariate regression vector and defined as the change in the instrument response with respect to a change in the analyte concentration. For first order analysis,

$$\mathrm{SEN} = \frac{1}{\|\boldsymbol{b}\|_2} = \|\mathrm{NAS}\|_2 \,. \tag{15}$$

As with univariate calibration, the SEN is given in units of signal/concentration (e.g. Absorbance units/Molarity).

Note that where the univariate sensitivity is proportional to the length of the regression vector, the first order sensitivity is inversely proportional to the length of the regression vector. This occurs because univariate calibration employs the CLS model where first order calibration methods employ the ILS model. Also, the SEN is proportional to the NAS. Therefore the SEN is a function of both the signal intensity at unit concentration and the uniqueness of the signal. For example, a spectroscopic technique would not be very sensitive to a highly absorbing species if the spectrum of the analyte is very similar to the spectrum of an interferent.

Also, like the SEL, the SEN must be defined for each calibration problem. This is evident in Table 1. The SEN of the spectroscopic method for TCE depends upon whether TCA, $CHCl_3$, or both are included in the calibration model.

The signal to noise ratio is an indication of the measurement precision of any instrumental technique. For univariate data this is defined as

$$\mathrm{S/N} = \frac{r}{\varepsilon_r} \tag{16}$$

where ε_r is the standard deviation of repeated measurements of r. The S/N for first order measurements can be equivalently defined as

$$\mathrm{S/N} = \frac{\|\boldsymbol{r}\|_2}{\|\boldsymbol{\varepsilon_r}\|_2} \tag{17}$$

where $\boldsymbol{r}$ is either the pure analyte spectrum or an analyte/interferent spectrum. Equation (17) does provide a statistic to compare the measurement precision. However, this definition of S/N is does not provide any insight to how the measurement precision affects the analysis (estimation) precision. A better definition, therefore, is

$$\mathrm{S/N} = \frac{c\|\mathrm{NAS}\|_2}{\|\varepsilon_{\mathbf{r}}\|_2} = \frac{c}{\|\mathrm{b}\|_2\|\varepsilon_{\mathbf{r}}\|_2} \tag{18}$$

where $c\|\mathrm{NAS}\|_2$ is the part of the total signal that is useful for quantitation. The $\|\mathrm{NAS}\|_2$ is multiplied by the concentration since NAS is defined at unit concentration. Note that the first order NAS is a function of the magnitude of the analyte signal on the instrument, the similarity between the spectrum of the analyte and the interferents, and the reproducibility of the measurement. Analysis for TCE in only TCA has the greatest S/N of the three examples shown (Table 1). This is due to TCA and TCE spectra being less similar that TCE and $CHCl_3$.

The limit of determination (LOD) is a measure of the analysis (estimation) precision. The International Union of Pure and Applied Chemists define the LOD for univariate analysis as

$$\mathrm{LOD} = \frac{k\varepsilon_{\mathrm{r}}}{m} \tag{19}$$

where k is an integer that defines the number of standard deviations of measurement error that constitutes "different" [19]. Usually k equals three which is nearly the 99% confidence level that two measurements are different.

This definition for the LOD can easily be translated for first order calibration. As with the SEN, $\|\mathrm{NAS}\|_2$, or $\frac{1}{\|\boldsymbol{b}\|_2}$ can be substituted for m, such that

$$\mathrm{LOD} = \frac{3\|\varepsilon_{\mathrm{r}}\|_2}{\|\mathrm{NAS}\|_2} = 3\|\varepsilon_{\mathrm{r}}\|_2\|\boldsymbol{b}\|_2 \tag{20}$$

As with the other first order figures of merit, the LOD differs between specific applications as the interferents in the calibration set change (Table 1).

Estimating Prediction Error

Although there is an exact translation of the figures of merit from univariate and multivariate calibration, no such exact translation exists for translating the estimation of prediction error from univariate to multivariate calibration. For univariate calibration the prediction error can be estimated directly by statistical propagation of errors through CLS [33, 34]. The estimated prediction error is

$$\sigma_{\mathrm{c}} = \frac{\sigma_{\mathrm{r}}}{m} * \sqrt{\frac{1}{M} + \frac{1}{N} + \frac{\bar{r}_{\mathrm{c}} - \bar{r})^2}{m^2 \sum_{\mathrm{i}=1}^{\mathrm{N}} (c_{\mathrm{i}} - \bar{c})^2}} \tag{21}$$

where m is the slope of the calibration curve, $\bar{r}_c$ is the mean of the M replicate unknown samples, $\bar{c}$ is the mean of the N calibration samples, and σ_r is the estimated error about the regression line. The error about the regression line is

$$\sigma_r = \sqrt{\frac{\sum_{i=1}^{N}(r_i - \bar{r})^2 - m^2 \sum_{i=1}^{N}(c_i - \bar{c})^2}{N-2}}. \tag{22}$$

Equations (21) and (22) provide insights to univariate calibration. The prediction error can be reduced by increasing the number of calibration samples, N, the number of replicates for each unknown, M, or the sensitivity of the measurement, m. The rate of improvement in the prediction error will be, at best, linear with improvements in the sensitivity and proportional to the square root of the increase in the number of calibration samples and replicates. Prediction error can also be improved by decreasing the magnitude of the last term under the radical in Eq. (21). This term decreases as the spread of the concentrations in the calibration set increases. Ideally half of the calibration samples would be at each extreme of the linear dynamic range of the instrument.

Unfortunately, no such exact equations can be derived for first order calibration. Malinowski provides an excellent foundation on the effects of instrumental errors on the decomposition and calculation of the pseudo inverse of the instrument response matrix, $\boldsymbol{R}$ [35]. However, it is impossible to propagate errors explicitly through the calculation of the pseudoinverse in Eq. (7). Nonetheless, many useful insights can still be derived based on the propagation of instrumental errors in first order calibration.

Lorber and Kowalski [8] derived an empirical equation to predict the confidence interval on estimation in multivariate analysis:

$$\mathrm{CI}_{\hat{c}_{\mathrm{unk}}} = \sqrt{\alpha^2 \sum_{j=1}^{j} b_j^2 \mathrm{var}(r_j) + \beta^2 \sum_{I=1}^{I} h_i{}^2 \mathrm{var}(\boldsymbol{c})} \tag{23}$$

where $\boldsymbol{r}$ is the unknown's instrument response, $\boldsymbol{h} = \boldsymbol{r}^{\mathrm{T}}\boldsymbol{R}^{+}$ is the leverage of the unknown sample relative to the calibration set, and α and β are the tabulated t-statistics with J-K and I-K (where K is the number of latent variables in the model) degrees of freedom, respectively. The variance of the unknown sample, var($\boldsymbol{r}$), and the variance of the calibration set var($\boldsymbol{c}$) can, be estimated as

$$\mathrm{var}(\boldsymbol{r}) = \frac{\boldsymbol{r}^{\mathrm{T}}(\boldsymbol{I} - \boldsymbol{R} + \boldsymbol{R})\boldsymbol{r}}{J-K} \tag{24}$$

and, assuming the variance in the calibration is constant at all concentrations of analyte,

$$\mathrm{var}(\boldsymbol{c}) = \frac{\boldsymbol{c}^{\mathrm{T}}(\boldsymbol{I} - \boldsymbol{R}\boldsymbol{R}^{+})\boldsymbol{c}}{I-K}. \tag{25}$$

Equations (23)–(25) can be used to show that many of the insights for optimizing zero order data gained from scrutinizing Eqs. (21) and (22) hold for first

Table 2. 95% confidence levels for prediction of TCE in different regions of the calibration model

TCE (units)	$CHCL_3$ (units)	TCA (units)	95% CI (±) (Equation 23)	First Term (Equation 23)	Second Term
2	2	2	0.0236	0.0230	0.0053
3	3	3	0.0244	0.0230	0.0080
2	4	0	0.0303	0.0230	0.0197
4	2	2	0.0264	0.0230	0.0129
4	4	0	0.0326	0.0230	0.0230

order calibration. Obviously the prediction error can be reduced by minimizing the variance of the errors in the unknown sample, var($\boldsymbol{r}$), and the calibration set, var($\boldsymbol{c}$). The confidence interval of prediction can also be reduced by increasing the NAS, equivalent to reducing $\|\boldsymbol{b}\|_2$. This implies that the more unique the spectrum of the analyte is from the interfering species, the better the expected results are from analysis. The leverage, $\boldsymbol{h}$, is a measure of the distance from the unknown sample to the center of the calibration set [10]. Therefore, prediction is best for samples in the center of the calibration set. Note that not all measurements and samples have the same effect on the prediction error. Variables with large weights on the regression vector and unique samples (high leverages) have a much greater effect on the calibration precision than measurements with small weights on the regression vector and samples close to the center of the calibration set. As with the figures of merit, the confidence interval on prediction is a function of the interferents in the calibration set. Change the interferents, and the prediction error changes.

Table 2 shows the estimated prediction error for five samples in the analysis of TCE in the presence of TCA and $CHCl_3$. The error estimate is based on 27 samples in a 3^3 experimental design where the component concentrations vary between 1, 2, and 3 units. Note that the error estimate is lowest when the unknown sample is at the center of the calibration set (2 units each of TCE, TCA, and $CHCl_3$). The estimated confidence limits increase when the sample is at the extreme of the calibration set (3 units each). The largest confidence intervals occur when both the analyte and interferent are outside of the calibration set. Note that the first term in Eq. (23) is independent of the sample position in the calibration set of the analyzed sample. The differences in the prediction error are only a function of the placement of the unknown relative to the calibration set. The 95% confidence intervals are much greater than the LOD. The LOD is a best case scenario that assumes an infinite calibration set centered around the sample.

Second Order Calibration

Second order calibration permits analysis in the presence of any unaccounted, uncalibrated species in the unknown sample. This is particularly applicable for environmental problems where is impossible or impractical to include every possible interferent in the calibration set.

Second order calibration can be viewed as a three step problem. In the first step the collection of standard and calibration samples, $\mathbb{R}$, is decomposed into sets of three intrinsic profiles, $\boldsymbol{X}$, $\boldsymbol{Y}$, and $\boldsymbol{Z}$. The second step entails determining which of the sets of profiles pertain to the analyte of interest, $\boldsymbol{x}_a$, $\boldsymbol{y}_a$, $\boldsymbol{z}_a$. One of the intrinsic profiles that correspond to the analyte is uniquely related to the analyte concentration in each sample, $\boldsymbol{z}_a$. Therefore, the third step is constructing an univariate calibration curve from the values of $\boldsymbol{z}_a$ that correspond standards and estimating the concentrations of the unknown samples on this curve.

Second Order Calibration Models

For the decomposition of $\mathbb{R}$ into sets of three intrinsic factors, the model

$$R_{ijk} = \sum_{n=1}^{N} X_{in} Y_{jn} Z_{kn} \tag{26}$$

where N is the number of factors, latent variables, required to describe $\mathbb{R}$ sufficiently. The number of factors in the model is the "rank" of the model. Wang et al. define the number of factors that uniquely pertain to the analyte as the "net analyte rank" (NAR) [36]. Formally,

$$\text{NAR} = \text{rank}(\boldsymbol{M}) - \text{rank}(\boldsymbol{M}|\boldsymbol{N}) \tag{27}$$

where rank ($\boldsymbol{M}$) is the number of factors required to model a mixture of an analyte and a number of interferents and rank ($\boldsymbol{M}|\boldsymbol{N}$) is the number of factors required to model the mixture without the analyte. For second order calibration to be successful, an analyte must have a NAR of at least one.

The third order tensor $\mathbb{R}$ can be decomposed by many different algorithms. Usually an alternating least square (ALS) [37] or eigenvalue based problem (EBP) [38, 39] is employed. For chemical species that have a spectrum that can be completely modeled by one factor, the decomposition of $\mathbb{R}$ is unique to a scalar multiple [40]. In this instance, the data is said to be "bilinear" and the columns of $\boldsymbol{X}$ and $\boldsymbol{Y}$ that correspond to the bilinear species are, in actuality, the estimates of the normalized instrumental profiles in the two orders (i.e. $\boldsymbol{x}$ is the chromatographic and $\boldsymbol{y}$ is the spectroscopic profile of the species if analyzed by LC-UVVis). If a species requires more than one factor to model the second order spectra, the decomposition of $\mathbb{R}$ is not unique. The true intrinsic profiles of these "nonbilinear" species can be formed from linear combinations of the columns of $\boldsymbol{X}$ and $\boldsymbol{Y}$. However, the proper linear combinations cannot be determined without a priori knowledge (e.g. unimodal or nonnegative spectral profiles).

Fortunately, whether the data is bilinear or nonbilinear, a number of columns of $\boldsymbol{X}$, $\boldsymbol{Y}$, and $\boldsymbol{Z}$ equal to the NAR are unique to the analyte. The respective columns of $\boldsymbol{Z}$ are uniquely related to the relative concentration of analyte in the sample. That is, if Z_{in} (where z_n is unique to the analyte of interest) was the instrumental response of a univariate sensor, the SEL would be unity. Therefore, univariate calibration can be performed on the z_n without fear of obtaining biased results.

The difficulty arises in determining which columns of $\boldsymbol{Z}$ are unique to the analyte. If the data is bilinear, positive identification can be obtained by observing the columns of $\boldsymbol{X}$ and $\boldsymbol{Y}$ (the chromatographic and spectroscopic profiles). If the n-th column of $\boldsymbol{X}$ and $\boldsymbol{Y}$ corresponds to the analyte of interest, the n-th column of $\boldsymbol{Z}$ would also. If the data is nonbilinear, a second order spectrum of pure analyte is needed for positive identification [36]. Mathematically doubling the intensity of the pure analyte spectrum decreases the relative concentration of analyte by a factor of two only in the column of $\boldsymbol{Z}$ that corresponds uniquely to the analyte.

Figures of Merit

The figures of merit for second order data are logical extensions of the figures of merit for zero and first order data. The zero and first order figures of merit are based on the part of the signal that is useful for calibration (i.e. the NAS). This is also the case for second order data. However, the definition of NAS differs from first to second order. For first order data, the NAS is orthogonal to all other spectra (see Eq. 10). In the second order, this orthogonality constraint is relaxed. As in the first order, the NAS is the part of the signal that is unique to the analyte. However, second order calibration does not require an orthogonal decomposition of $\mathbb{R}$. The NAS is instead the part of the signal that is related to the NAR. That is,

$$\mathrm{NAS}_{ijk} = \sum_{n=1}^{\mathrm{NAR}} X_{in} Y_{jn} Z_{kn} \tag{28}$$

where the first NAR factors are unique to the analyte. The second order NAS for the k-th sample is a matrix, not a vector. It is important to note that the second order NAS, like its first order counterpart, changes in intensity but not form from sample to sample in $\mathbb{R}$.

Hence, the second order figures of merit can be defined by the NAS:

$$\mathrm{SEL} = \frac{\|\mathrm{NAS}\|_F}{\|\boldsymbol{N}\|_F} \tag{29}$$

$$\mathrm{SEN} = \|\mathrm{NAS}\|_F \tag{30}$$

$$\mathrm{S/N} = \|\mathrm{NAS}\|_F \|\boldsymbol{E}\|_F \tag{31}$$

$$\mathrm{LOD} = \frac{3\|\boldsymbol{E}\|_F}{\|\mathrm{NAS}\|_F} \tag{32}$$

where $\|\cdot\|_F$ is the Frobenius norm, the square root of the sum of the squared elements in the matrix. The matrix $\boldsymbol{E}$ is the measurement errors. Malinowski details many schemes for accurately determining $\|\boldsymbol{E}\|_F$ [35].

For bilinear data the selectivity is one. As with first order calibration, analysis is unbiased whenever the selectivity is greater that zero. Furthermore, the selectivity will only be zero if there is no NAR (no NAR uniquely implies no NAS for second order calibration).

The advantages associated with second order calibration become evident when comparing the first and second analysis of Pb(II) in groundwater. Figure 4 shows the second order spectrum of 3.5 μM Pb(II) when analyzed by a dialysis-spectroscopic sensor developed by Lin and Burgess [41]. The dialysis membrane provides partial temporal separation of the Pb(II) from the Co(II), Ni(II), Mn(II), and Zn(II) interferents. The spectroscopic order simultaneously provides partial spectral discrimination between the Pb(II) and the four interferents. At no time or wavelength is the Pb(II) ever completely resolved from the other heavy metals. First order analysis can be performed by either the chromatographic or the spectroscopic method. The instrumental response for pure 3.5 μM Pb(II) and a 2.5 μM Pb(II) mixture along with the first order NAS is shown in Fig. 5 for the chromatographic technique and in Fig. 6 for the spectroscopic method. Note that the first order NAS for both methods is quite small compared to the pure Pb(II) signal. For second order analysis, the second order NAS is identical, except for noise reduction, to the pure 3.5 μM Pb(II) signal. Therefore the selectivity is, by definition, unity while the selectivity for the first order methods is much lower (Table 3). Similarly the SEN for the second order method is greater than the sensitivity of the first order methods due to the increased NAS.

The second order analysis also shows a small improvement in the S/N and LOD. These second order figures of merit could be further improved by eliminating some rows and columns of $\mathbf{R}$ that do not contain significant analyte infor-

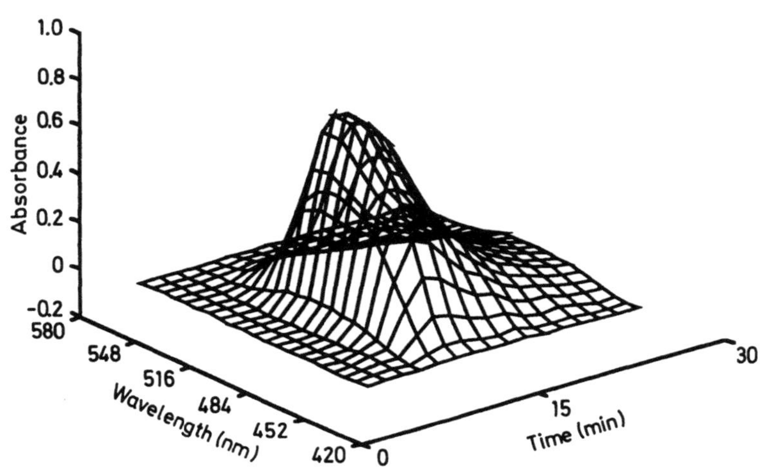

Fig. 4. Second order instrument response of 3.5 μM Pb(II)

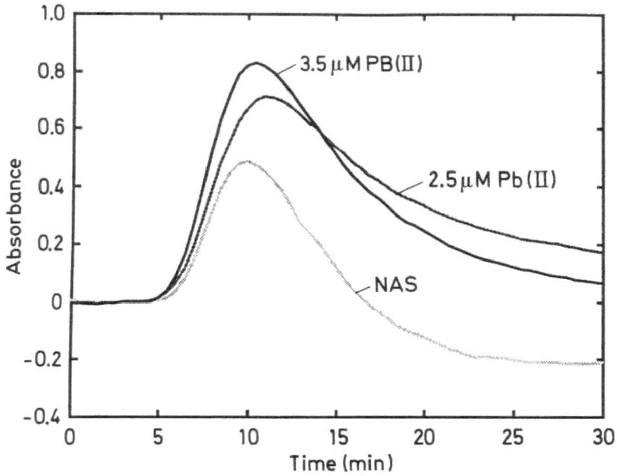

Fig. 5. Elution profile of 3.5 μM PB(II) (*black*), 2.5 μM Pb(II) with interferents (*dark gray*), and NAS of 3.5 μM Pb(II) (*light gray*)

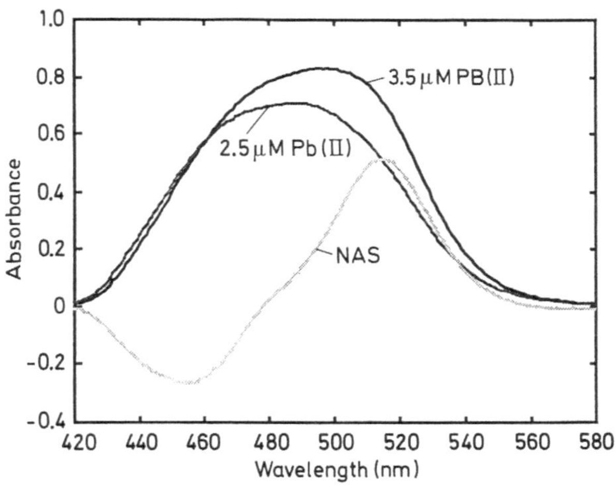

Fig. 6. Spectral profile of 3.5 μM PB(II) (*black*), 2.5 μM Pb(II) with interferents (*dark gray*), and NAS of 3.5 μM Pb(II) (*light gray*)

Table 3. Figures of Merit for analysis of 3.5 μM Pb(II)

	Temporal Domain	Spectral Domain	Second Order
SEL	0.6086	0.4558	1
SEN	0.5274 A/μM	0.8374 A/μM	6.8684 A/μM
S/N	283.02	294.03	319.06
LOD	0.0106 μM	0.0102 μM	0.0094 μM

mation. These rows and columns contribute noise without supplying additional signal related to the analyte.

Note that the second order figures of merit are not defined by a regression vector, only by the NAS. This is a function of the fact that, by nature, a regression vector is orthogonal to the interferent spectra. Second order calibration does not determine a space orthogonal to the interferents yet correlated to the analyte of interest; therefore, no regression vector, per say, exists for second order calibration.

Estimating Prediction Errors

Analysis of the effects of random instrumental errors is an active and relatively virgin area of research. To date no holistic set of equations has been postulated to predict the errors associated with analysis as Eqs. (21)–(23) do for zero and first order analysis. However a number of theoretical and empirical studies lead to important insights for second order calibration. The ALS and EBP algorithms are all converge to the correct unbiased estimate of analyte concentration in the absence of random instrumental errors or model errors. Model errors have a greater effect on the precision concentration of the estimates than on the precision of the estimated intrinsic profiles [42]. Random instrumental errors primarily affect the precision of the intrinsic profiles [42] but also result in biased concentration estimates [43]. This bias increases with the distance of the unknown sample from the center of the calibration set [43].

Studying the effects of error propagation helps chemists to optimize the analyzability of second order data. To achieve maximum quantitative precision, the first instrument in the hyphenated pair should be made as stable as possible to minimize model errors. The degree of discrimination between the analyte and interferent in either order is not essential as long as the first order NAS for both instruments is greater than the instrumental noise. Otherwise, the analyte is indistinguishable from the interferent in one order. The data could not be accurately analyzed by second order methods if the first order NAS in one domain was essentially zero. However first order analysis by the profile in the other order is possible. Also, as with zero and first order calibration, the prediction error is lowest when the unknown sample is centered in the calibration set.

References

1. Box GEP, Hunter WG, Hunter JS (1978) Statistics for experimenters: an introduction to design. Wiley, New York
2. Deming SN, Morgan SL (1987) Experimental design: a chemometric approach. Elsevier, New York

3. Williams P, Norris K (1987) Near-infrared technology in the agricultural and food industries. American Association of Cereal Chemists, St. Paul
4. Naes T (1987) J Chemom 1:121
5. Miller AJ (1990) Subset selection in regression. Chapman and Hall, London New York
6. Osten D (1988) J Chemom 2:39
7. Seasholtz MB, Kowalski BR (1993) Anal Chim Acta 277:165
8. Lorber A, Kowalski BR (1988) J. Chemom. 2:93
9. Rousseeuw PJ, Leroy AM (1987) Robust regression and outlier detection. Wiley, New York
10. Martens H, Naes T (1989) Multivariate Calibration. John Wiley & Sons, New York
11. Muhammad AS, Deborah LI, Kowalski BR (1986) Chemometrics. John Wiley & Sons, New York
12. Massart DL, Vandeginste BGM, Deming SN, Michotte Y, Kaufman L (1988) Chemometrics: A Textbook. Elsevier, New York
13. Geladi P, Kowalski BR (1986) Anal. Chim. Acta 185:1
14. Beebe KR, Kowalski BR (1987) Anal. Chem. 59:1007A
15. Haaland DM (1987) Spectrosc. 2:56
16. Kvalheim OM (1988) Chemom. Intell. Lab. Syst. 14:11
17. Stahle L, Wold S (1990) Chemom. Intell. Lab. Syst. 9:127
18. Sanchez E, Kowalski BR (1988) J. Chemom. 2:247
19. Lorber A, Kowalski BR (1988) J. Chemom. 2:265
20. Rousseeuw PJ (1991) J. Chemom. 5:1
21. Haaland DM, Easterling RG (1982) Appl. Spectrosc. 36:665
22. Haaland DM, Thomas EV (1988) Anal. Chem. 60:1193
23. Neter J, Wasserman W, Kuntner MH (1989) Applied linear regression models, 2nd edn. Irwin, Boston
24. Seasholtz MB, Kowalski BR (1991) J. Chemom. 5:129
25. Kalivas JH, Robberts M, Sutter JM (1989) Anal. Chem. 61:2024
26. Kelly JJ, Barlow CH, Jinguji TM, Callis JB (1989) Anal. Chem. 61:313
27. Haaland DM (1988) Anal. Chem. 60:1202
28. Pell RJ, Erickson BC, Hannah, RW, Callis JB, Kowalski, BR (1988) Anal. Chem. 60:2824
29. Wise BM, Ricker NL, Veltkamp DF, Kowalski BR (1990) Proc. Control Qual. 1:31
30. Sekulic S, Seasholtz MB, Wang Z, Kowalski BR, Lee S, Holt B (1993) 65:835A
31. Wang Z, Isaksson T, Kowalski BR (1994) Anal. Chem. 66:249
32. Isaksson T, Wang Z, Kowalski BR (1993) J. Near Infrared Spectrosc. 1:85
33. Skoog DA, West DM (1986) Analytical Chemistry: An Introduction, 4th edn. Saunders College Publishing, Philadelphia
34. Mendenhall W, Beaver R (1991) Introduction to Probability and Statistics, 8th edn. PWS-Kent Publishing Co., Boston
35. Malinowski, E. Howery D (1991) Factor Analysis in Chemistry, 2nd edn. John Wiley & Sons, New York
36. Wang Y, Borgen O, Kowalski BR, Gu M, Turecek F (1993) J. Chemom. 7:117.
37. Kroonenberg P (1983) Three-mode Princopal Component Analysis, DSWO Press, Leiden.
38. Wilson BE, Sanchez E, Kowalski BR (1989) J. Chemom. 3:493.
39. Sanchez E, Kowalski BR (1990) J. Chemom. 4:29.
40. Kruskal JB (1989) Multiway Data Analysis, Elsevier, Amsterdam.
41. Lin Z, Burgess LW (1994) Anal. Chem. 66:000
42. Booksh KS, Kowalski BR (1994) J. Chemom. 8:45.
43. Faber NM, Buydens LMC, Kateman G (1994) J. Chemom. 8:000.

Robust Procedures for the Identification of Multiple Outliers

Anita Singh[1] *and John M. Nocerino*[2]

[1] Lockheed Environmental Systems & Technologies Company, 980 Kelly Johnson Drive, Las Vegas NV 89119, USA
[2] United States Environmental Protection Agency, Environmental Monitoring Systems Laboratory-Las Vegas P.O. Box 93478, Las Vegas NV 89193–3478, USA

Summary

Classical and robust/resistant procedures for the estimation of population parameters and the identification of multiple outliers in univariate and multivariate populations are reviewed. The successful identification of anomalous observations depends on the statistical procedures employed. Commercial industries, local communities, and government agencies such as the United States Environmental Protection Agency (U.S. EPA), often need to assess the extent of contamination at polluted sites. Identification of these contaminants having potentially adverse effects on human health is especially important in various ecological and environmental applications. An environmental scientist typically generates and analyzes large amounts of multidimensional data. These practioners often need to identify experimental conditions and results which look suspicious and are significantly different from the rest of the data. The classical Mahalanobis distance (MD) and its variants (e.g., multivariate kurtosis) are routinely used to identify these anomalies. These test statistics depend upon the estimates of population location and scale. The presence of anomalous observations usually results in distorted and unreliable maximum likelihood estimates (MLEs) and ordinary least-squares (OLS) estimates of the population parameters. These in turn result in deflated and distorted classical MDs and lead to masking effects. This means that the results from statistical tests and inference based upon these classical estimates may be misleading. For example, in an environmental monitoring application, it is possible that the classification procedure based upon the distorted estimates may classify a contaminated sample as coming from the clean population and a clean sample as coming from the contaminated part of the site. This in turn can lead to incorrect remediation decisions.

It is well established among practioners that, for the identification of multiple outliers, one should use robust procedures with a high breakdown point. The estimates obtained using the robust procedures should be in close agreement with the corresponding classical OLS and MLEs when no discordant observations (from different population(s)) are present. Robust procedures for the identification of outliers and the estimation of population parameters of location and scale typically use an influence function. The robust procedure based upon a recently developed "proposed" influence function, called the PROP function, works quite effectively in estimating population parameters accurately, and correctly identifying multiple outliers in univariate and multivariate populations. The control-chart-type quantile-quantile (Q-Q) graphical display of multivariate data combines the effect of a formal test procedure and an informal graphical display into one powerful multiple outlier identification procedure. The scatter plot of the robustified square root leverage distances vs the residuals identifies all regression outliers and distinguishes between significant and insignificant leverage points. The procedures discussed here unmask multiple anomalies and provide reliable estimates of the population parameters in several areas of interest, including linear regression models, discriminant and principal component analyses, and variogram modeling in geostatistical applications. The U.S. EPA, through the Office of Research and Development (ORD), has research interests in optimizing its quality assurance program by developing statistical procedures that are insensitive to outliers (resistant) and the underlying assumptions (robust).

List of Symbols and Abbreviations

ANOVA	analysis of variance
CC	confidence coefficient
CI	confidence interval
CLP	Contract Laboratory Program
R^2	coefficient of determination
DF, ν	degrees of freedom
Huber	Huber's influence function
Biweight	Tukey's Biweight influence function
IRLS	iteratively reweighted least squares
LCL	lower confidence limit for the population mean
UCL	upper confidence limit for the population mean
LPL	lower limit for the prediction interval
UPL	upper limit for the prediction interval
LSL	lower limit for the simultaneous confidence interval
USL	upper limit for the simultaneous confidence interval
LMS	least median squares
M	median
M-estimator	generalized maximum likelihood estimator
MAD	median absolute deviation
$\hat{\sigma}_{MAD}$	estimate of σ based upon MAD
Max	maximum
MD	Mahalanobis distance
Max(MDs)	largest Mahalanobis distance
MLE	maximum likelihood estimation
MS	mean square
MVE	minimum variance ellipsoid
MVT	multivariate trimming
OLS	ordinary least squares
PCA	principal component analysis
PE	performance evaluation
PLS	partial least squares
PROP	proposed influence function
QA/QC	quality assurance/quality control
Q-Q	Quantile-Quantile
sd	standard deviation
SEDOP	statistical experimental design and optimization
sgn	the signum function
SIMCA	Soft Independent Modelling of Class Analogy
SS	sum of squares
TSP	three step procedure
n	sample size

p	dimension of the data set
μ	univariate population mean
σ	univariate population standard deviation
$\bar{x}$	sample mean
s	sample sd, and min(g–1,p)
$\overline{x^*}$	robust estimator of population mean, μ
s^*	robust estimator of population sd, σ
k	number of outliers, cutoff constant from the Gaussian distribution, and number of populations (groups)
x	p-dimensional random vector representing an observation
$f(x)$	the density function of the vector, x
$\sum$	summation sign
μ	p-dimensional population mean vector (location)
Σ	$p \times p$ population dispersion matrix (scale)
μ^*	robust estimator of population location
Σ^*	robust estimator of population scale
h	a spherically symmetric density in p-dimensional space
d_{i}^2	Mahalanobis distance for the i-th observation
d_{α}^{b}	α100% critical value of the test statistic Max(MDs)
$d_0^2, d_{\mathrm{ind}}^2$	α100% critical value from the distribution of d_{i}^2
$d_{\mathrm{m},\alpha}^2$	α100% critical value from the distribution of the Max(d_{i}^2)
$\psi(d_{\mathrm{i}})$	the PROP influence function
$w(d_{\mathrm{i}})$	the weight function
*wsum*1	sum of the weights, $w(d_{\mathrm{i}})$
*wsum*2	sum of the squared weights, $w^2(d_{\mathrm{i}})$
$\bar{x}_{\mathrm{bi}}^*$	Biweight estimator of μ
$\bar{x}_{\mathrm{H}}^*$	Huber estimator of μ
$\psi_{\mathrm{bi}}(u)$	Biweight influence function
$\psi_{\mathrm{H}}(u)$	Huber influence function
u_{i}	i-th standardized observation
c	tuning constant
α	level of significance
t_v	Student's t-value with v DF
t_{bi}	t-value associated with the Biweight function
$t_{0.7(\mathrm{n}-1)}$	Student's t-value with $0.7(n-1)$ DF
$t_{\alpha/2,v}$	$(\alpha/2)$100% Student's t-value with v DF
t_{c}	classical critical value of Student's t distribution
t_{r}	robust critical value of Student's t distribution
$B(a,b)$	beta distribution with parameters a and b
Γ	gamma function
χ^2	chi-square distribution
Q_1	Dixon's statistic for finding a single upper outlier
Q_2	Dixon's statistic for finding two upper outliers
$R_{\mathrm{p,n}}$	the correlation coefficient

y	observed response variable
e_i	normally distributed error term in regression model
$\hat{\sigma}$	estimate of the sd of the error term
r_i^2	the residual sum of squares
β	vector of regression coefficients
$\hat{\beta}_{OLS}$	ordinary least squares estimate of β
$\hat{\beta}_R$	robust estimator of β
$\hat{\beta}_{LMS}$	least median squares estimate of β
$\hat{\beta}_{PROP}$	PROP estimate of β
Ld_i^2	MDs using the x-explanatory variables
Ld_α^b	$\alpha 100\%$ critical value from the distribution of Ld_i^2
$w(x_i, d_i)$	weight function used in robust regression
p_i	eigenvector corresponding to the i-th eigenvalue
$q_{(k)}$	normal quantile
g	number of distinct populations (groups)
π_i	the i-th population
μ_i	mean vector of the i-th population
Σ_i	dispersion matrix of the i-th population
$\hat{B}^*$	between-groups matrix
W^*	within-groups matrix
S^*_{pooled}	pooled estimate of the common dispersion matrix, Σ
$\hat{\lambda}_i$	an eigenvalue of ${W^*}^{-1}\hat{B}^*$
l_i	normalized eigenvector corresponding to $\hat{\lambda}_i$
y_i	i-th discriminant function

Introduction

Outliers are almost inevitable in most applied and scientific disciplines. In a manufacturing process, outliers typically represent some mechanical disorder of the system, unexpected experimental conditions and results, raw material of an inferior quality, and misrecorded values. In biological dose-response applications, outlying observations may indicate an entirely different type of reaction (an unusual response) to a newly developed drug. In this case, outliers may be more informative than the rest of the data. In environmental and ecological applications, outliers could be indicative of highly contaminated areas, sections of a forest in poor or degraded states, inconsistent analytical results in a typical quality assurance and quality control (QA/QC) program, or gross typing errors. Several univariate classical, as well as robust, outlier identification procedures exist in the literature.

In environmental chemistry, since multiple compounds are analyzed simultaneously, use of multivariate outlier detection procedures is recommended. The estimation of population parameters and the identification of anomalous observations are closely related problems. The successful identification of outliers depends upon the accurate estimates of the population parameters of concern. The classical MLEs and the OLS estimates are distorted by the presence of outlying observations. The use of robust and resistant estimation procedures has been recommended in the literature to identify outliers. Here we present robust statistical theory for the identification of outliers and the estimation of parameters of p-dimensional multivariate populations. Univariate procedures can be derived as special cases by choosing p equal to unity. Some univariate examples have been discussed for the sake of completeness and for interested readers.

Grubbs [1] and Dixon-type [2] classical univariate test statistics are fairly popular among analytical chemists [3, 4]. For higher dimensional data sets, Mahalanobis generalized distances (MDs) and their functions, such as Max(MDs) [5] and Mardia's multivariate kurtosis [6] are used to identify outliers. Stapanian et al. [7, 8] used a sequential procedure based upon these two test statistics to identify multivariate anomalies in samples from multinormal populations. These statistics depend upon the classical MLEs [9] of population location (mean vector) and scale (dispersion matrix). These classical estimates have a "zero" breakdown point and are vulnerable to severe masking effects in the presence of multiple outliers. Masking means that the outliers are hidden and the procedure used cannot find them. For example, when outliers arise in clusters, the regression model gets attracted toward these outliers resulting in misleading normal-looking residuals, leading to masking (Example 5, below). Even the sequential use of classical outlier identification procedures cannot help unmask these multiple outliers.

"Zero" breakdown point means that the presence of even a single discordant value (large or small) can completely distort the classical estimates obtained using the OLS and MLE procedures. This means that the resulting estimates may not be reliable. Thus, all other related classical statistics, including interval estimates of the population means, variances, discriminant functions, principal components, and regression parameters, may also be grossly distorted by these anomalies.

Moreover, due to masking, sometimes large MDs do not necessarily correspond to the outlying observations. This can happen for example: a) when the data set has multiple outliers, or b) when one is dealing with a mixture sample from two or more component populations, or c) when one is dealing with many dimensional data sets of small or moderate sizes. The use of approximate distributions of the MDs, such as chi-square [10–12] or normal [13], can also lead to the incorrect ordering of the MDs (or equivalently can lead to the misidentification of discordant observations). The use of the exact c^* beta distribution (c being the probability density constant) of the MDs and an initial robust start (e.g., median, MAD pair) in the iterative procedure of deriving robust M-estimates of location and scale overcomes this problem effectively.

The M-estimators are robust generalizations of the maximum likelihood estimators [14]. It is desirable for these robust estimates – of location and scale and of regression parameters in linear models – and the MDs, *with or without* the outliers, to be in close agreement with the corresponding classical MLEs, OLS estimates, and the MDs when no outlying observations are present (or after all of the outlying observations have been correctly identified and removed). Estimates with this property will relieve the user from the burden of performing the data analysis twice: once using the whole data set and once without the discordant observations after their correct identification.

The breakdown point of an estimator [15, 16] is the smallest possible fraction of observations that have to be replaced to distort the estimator over all bounds. Robust statistics deal with developing statistical procedures that are insensitive to violations of the assumptions under which they were developed. A resistant measure of any aspect (e.g., mean, sd) of a distribution is relatively unaffected by changes in the numerical values of some of the observations, no matter how large these changes are. Robust procedures with high breakdown points (the maximum possible is 50%) are desirable.

Several researchers, including Rousseeuw and van Zomeren [11, 12], Campbell [13], Huber [14], Hampel et al. [16], Tukey [17], Andrews [18], Maronna [19], Hawkins et al. [20], Rousseeuw and Leroy [21], Singh [22], and Singh and Nocerino [23, 24] have researched developing robust/resistant procedures for the estimation of population parameters and the identification of multiple outliers and leverage points in univariate and multivariate populations. These efforts resulted in several robust procedures and influence functions, including: Biweight [17, 25], Huber [14, 26], Hampel [15, 16], univariate and Multivariate Trimming (MVT) [10], Minimum Volume Ellipsoid (MVE) [11], and the recently "proposed" PROP influence function [22].

Here, robust procedures based upon the PROP function are used to locate outliers in several applications, including interval estimation, regression models, principal component and discriminant analyses. The procedure based upon the PROP influence function works quite effectively at identifying multiple outliers in univariate as well as multivariate data sets of all sizes. Small sample correction factors are not required to provide appropriate coverage as suggested by Rousseeuw and van Zomeren [11, 12]. Also, no tuning constants [16, 25], except for an appropriate choice of an α-value, are needed in the definition of the influence function (often the choice of different tuning constants leads to significantly different results). Most practitioners are familiar with choosing an α-value in their statistical applications.

In layman's terminology, outliers are observations that are "well separated" from the main stream of data points. These observations could inflate the variances and covariances inappropriately or might not comply with the correlation structure imposed by the majority of data. It can be challenging to locate these anomalies in data sets of dimensionality larger than two. In higher dimensions it is difficult to visualize this main stream of data points and the points which lie far away from the main stream.

We provide a systematic statistical definition of this separation. A control-chart-type graphical display of the multivariate data combines the effect of a formal test procedure and an intuitive graphical display into one multiple outlier identification procedure. This procedure provides a natural operational tool for unmasking multiple anomalies in several areas of interest, including quality control, multiple linear regression models, discriminant and principal component analyses, and variogram modeling in geostatistical applications. The robust regression procedure based upon the PROP influence function identifies all regression outliers and effectively distinguishes between significant and insignificant leverage points.

In an effort to keep the environment clean, government agencies such as the U.S. EPA need to assess the extent of contamination at polluted sites. This site characterization is then sometimes used to determine remediation and monitoring activities and also to set up cleanup standards. The presence of discordant observations in samples obtained from these environmental applications may distort the entire estimation process. Thus the use of robust procedures is preferable in the estimation phase. For example, the use of robust variogram models is desirable to obtain more precise and accurate kriging estimates. Suppose a sample of n observations (soil samples) is available from a polluted site. In this case, outliers may indicate the presence of a highly contaminated area. The sample may also represent a mixture of several populations with a varying degree of contamination requiring different levels of remediation activities. The robust procedure presented by Singh et al. [27] works effectively in breaking down a univariate mixture sample into component populations (e.g., separating the background from the polluted areas).

Site characterization is, in part, based upon the chemical analyses of environmental samples taken from the site. Those samples are routinely analyzed by commercial and research laboratories who participate in the various programs (e.g., the Superfund Contract Laboratory Program (CLP)) of the U.S. EPA. The performance of those laboratories is typically monitored through QA/QC techniques. The performance of a population of laboratories may be evaluated through the estimates of population parameters, such as the mean, standard deviation, coefficient of variation, and the range. However, laboratories giving QA/QC results discordant with the dominant population may greatly influence the estimation process and may result in inappropriate statistical regions within which the majority of the laboratories are expected to report their analytical results. The robust procedure presented by Singh and Nocerino [23] assigns reduced weights to the outlying observations, resulting in more precise and accurate statistical intervals.

The robust statistical methods are becoming more popular among practicing scientists with the increasing availability of personal computers and computer software packages. Most robust procedures are iterative and require substantial computations and several passes through the same data set. This of course would be impossible (too tedious and time consuming) to do without available statistical software and current computing power. Some statistical software packages, such as PROGRESS [21] and Scout [8, 24, 28, 29], are available to perform robust statistical analysis. All computations presented here are performed using the ROBUST module of the Scout software package. The robust regression routine, REGRESS [24], based on the PROP influence function, has been used in all regression examples.

The structure of this discussion is as follows. We start with the simple one-step univariate robust (median, MAD) estimators of population mean, μ, and standard deviation (sd), σ. Then we briefly present the multivariate mathematical formulation for more sophisticated robust generalized maximum likelihood estimators called the M-estimators. Following this we summarize the PROP procedure. Various robust confidence intervals useful in environmental applications are then described, followed by robust multiple linear regression models. Robust procedures for multivariate chemometric applications such as principal component and discriminant analyses are then discussed and the scope of robust statistical methods in environmental and chemometric applications are summarized, giving some general recommendations. Several well known examples from the literature have been used.

Outliers in Univariate and Multivariate Populations

Experimentalists, especially environmental scientists, generate and analyze large amounts of data. Most of these practitioners, therefore, are familiar with the situations when some of their experimental results look suspicious or significantly different from the rest of the data. These observations are known as discordant observations (outliers, anomalies, extremes). In data sets of large dimensionality, it becomes tedious to identify these anomalies. Appropriate multivariate procedures should be used to identify multivariate anomalies. Some of these procedures are discussed later. Before proceeding any further, we first discuss the simple univariate cases.

In statistical terminology, the univariate experimental data can be represented as follows. Let $x_1, x_2, \ldots, x_n$ represent a sample of size n from a population with mean, μ, and standard deviation (sd), σ. The sample mean and sd are

$\bar{x} = \Sigma x_i/n$, and $s = \sqrt{(\Sigma x_i^2 - n\bar{x}^2)/(n-1)}$, respectively. The use of Grubbs [1], Dixon-type [2] test-statistics, and Rosner's [30] test for finding "Many-Univariate-Outliers" are fairly common among practicing scientists [3, 31]. Grubbs test-statistic depends upon the sample mean and sd. The sample mean and sd have a zero breakdown point. This means that even the presence of a single discordant value can completely distort these sample statistics. Dixon [2] knew that the correct identification of outliers can suffer from masking. He recommended the use of multiple hypotheses testing to identify upper and lower outliers separately. Several Dixon-type test-statistics for finding upper and lower outliers are listed by Barnett and Lewis [32]. In practice, the user does not know the number of discordant values that might be present in a data set. It can become quite tedious to test multiple hypotheses, H_k: k outliers are present, k being a finite positive integer. Moreover, separate critical values are typically required for each test. For example, selected critical values of Rosner's [30] test for finding up to ten outliers are listed by Gilbert [31]. The robust procedure based upon the PROP function can identify these multiple anomalies in a single execution.

Alternatively, simple one-step robust statistics [4], such as the sample median (M) and $\hat{\sigma}_{MAD}$, are used to estimate μ and σ, respectively. The computations of M and $\hat{\sigma}_{MAD}$ are described as follows. First arrange the data in ascending order, $x_{(1)} \leq x_{(2)} \leq \ldots \leq x_{(n)}$. Find the median value, M, and compute the absolute deviations,

$$|(x_{(i)} - M)|; \; i = 1, 2, \ldots, n .$$

The median absolute deviation from the median (MAD) is the median of these deviations. For data sets from Gaussian populations, $\hat{\sigma}_{MAD} = \text{MAD}/0.6745$ is an unbiased estimator of the population sd, σ. Both of these estimates suffer from masking in the presence of multiple outliers. Moreover, these two estimates are based on the ranks of the sample observations and do not utilize most of the information contained in a given data set. The use of M and $\hat{\sigma}_{MAD}$ as the initial start estimators in the iterative process of obtaining more precise robust M-estimators of location and scale have been recommended in the literature [10, 17, 22]. The following example explains these ideas.

Example 1. Consider a simulated small data set with seven observations generated from a normal population, $N(0, 1)$. Next, three discordant observations from a different normal $N(5, 1)$ population are included in this data set. The generated data are: 1.0927, 0.9309, –0.4867, 0.1471, 1.3154, –0.8546, and –0.3176 from the $N(0, 1)$, and 5.3566, 4.4920, 5.08 from the $N(5, 1)$. The classical mean and sd based upon the first seven observations (without the outliers) are 0.26 and 0.857, respectively. The classical mean and sd using all ten observations get distorted and are given by 1.68 and 2.39. In order to obtain the simple robust estimates, arrange the data in ascending order: –0.8546, –0.4867, –0.3176, 0.1471, 0.9309, 1.0927, 1.3154, 4.4920, 5.08, 5.3566. The median, M, the mean of the two middle values, is 1.012. The absolute deviations from M are: 1.87, 1.50, 1.33, 0.86, 0.08,

0.08, 0.303, 3.48, 4.068, 4.34. The MAD is 1.415, obtained by taking the mean of the middle two values 1.33 and 1.50. Thus an estimate of population spread is given by $\hat{\sigma}_{MAD} = MAD/0.6745 = 2.096$. Notice that both M and $\hat{\sigma}_{MAD}$ are influenced by the presence of the three discordant values.

Due to masking, Dixon-type tests for upper outliers [32] also failed to identify these discordant values. Dixon's test-statistics to identify a single upper outlier and two upper outliers are given by $Q_1 = (x_{(n)} - x_{(n-1)})/((x_{(n)} - x_{(1)})$ and $Q_2 = (x_{(n)} - x_{(n-2)})/((x_{(n)} - x_{(1)})$, respectively. For the above data set, these test-statistics are $Q_1 = (5.3566 - 5.08)/(5.35 + 0.8546) = 0.0445$, and $Q_2 = (5.3566 - 4.4920)/(5.35 + 0.8546) = 0.139$, respectively. The critical values of Q_1 and Q_2 for various values of n and α are listed by Barnett and Lewis [32]. For an $\alpha = 0.05$ level of significance, and $n = 10$, the critical values for Q_1 and Q_2 are 0.412 and 0.531, respectively. Since the test-statistics are smaller than the corresponding critical values, we are led to the wrong conclusion that all of the observations come from a single population and no discordant values are present. This is a simple example of univariate data masking. The problem gets quite complicated for data sets of higher dimensions. We will continue with this example after formally defining the generalized maximum likelihood M-estimators.

Robust M-Estimators of Location and Scale

Wilks [5] introduced classical procedures to test for $k\,(k \geq 1)$ outliers in samples from multinormal populations. For $k = 1$, this test is equivalent to the well-documented classical test [7, 8, 33, 34] based upon the largest Mahalanobis distance, Max(MDs), for the identification of a single outlier. Observations with MDs greater than the $\alpha 100\%$ critical value of the Max(MDs) are potential outliers. For the univariate case, this test is equivalent to the Grubbs test [1]. Extensive simulation studies have been performed [7, 34] to obtain the critical values of the test-statistics Max(MDs) and multivariate kurtosis [6]. The simulated critical values of Max(MDs) and multivariate kurtosis have been incorporated in the Scout [29] software package for sequential identification of outliers based on classical test-statistics Max(MDs) and multivariate kurtosis [6, 7]. Under multinormality, Schwager and Margolin [35] discussed some optimal properties of classical multivariate kurtosis. Their study proved that the test based upon kurtosis is less susceptible to masking by multiple outliers than the test based upon the Max(MDs). It should be noted that, just like the statistic Max(MDs), sample kurtosis, being a function of sample mean vector and dispersion matrix, can get distorted by the presence of anomalies.

Singh [22] pointed out that the critical values for the distribution of Max(MDs) can be computed directly using the Bonferroni inequality. Using Wilks' procedure as a model for the identification of outliers, one should be using the critical values of the distribution of Max(MDs) rather than the exact or approximate critical values from the distribution of individual MDs, d_i^2. Several authors [10–13, 22, 36] have used robustified MDs based upon the robust estimators of location and scale to identify multiple multivariate outliers. The robust procedures based upon Multivariate Trimming (MVT), the Huber, and the PROP influence functions have been incorporated in the ROBUST module of the Scout software package.

It is a well known fact that the individual classical MDs, d_i^2, under multinormality, follow a $(n-1)^2\,\beta(p/2,(n-p-1)/2)/n$ distribution [33]. However, in order to obtain robust estimators of location and scale, either a normal approximation [13] or a chi-square, χ^2, approximation [10–12] is routinely used to obtain the critical values for the distribution of the individual distances, d_i^2, which are needed to define the influence functions and minimum volume ellipsoids (MVEs). The MVEs [11] use a chi-square correction factor to provide appropriate coverage for the sample observations, especially for small data sets. Singh [22] noted that the calculations based upon the approximate and the exact distributions of the MDs differ significantly. For example, for a 15-dimensional sample of size 30, the 5% critical value of the exact distribution of the individual MDs is 20.33 (based upon the beta distribution), whereas the approximate χ^2_{15} value is 25.0, which is already larger than the 5% critical value, 23.837, of the test-statistic Max(MDs).

In the presence of multiple outliers, the generalized distances get distorted (even the robust MDs) to such an extent that the cases with large MDs may not correspond to the outlying observations. This multivariate data masking can completely distort the estimates of the population parameters and the correct ordering of the MDs and often leads to the misidentification of outliers. A robust start, such as that of the M and $\hat{\sigma}_{MAD}$ pair in the iterative process of obtaining the M-estimators, helps to overcome this problem by producing estimates which are resistant to masking effects. This is especially true for the M-estimators obtained using the PROP influence function. The mathematical derivation of the M-estimators based upon the PROP influence function is briefly given as follows.

The PROP Robust Procedure

Let $x_1, x_2, \ldots, x_n$ represent a random sample from a p-variate population having a density function, $f(x) = |\Sigma|^{-1/2} h[(x-\mu)'\Sigma^{-1}(x-\mu)]$, which is elliptically symmetric about the mean vector, μ. Here, Σ is the variance-covariance (dispersion) matrix, and h represents a spherically symmetric density in p-dimensional space [19]. The MDs are given by $d_i^2 = (x_i - \mu^*)'{\Sigma^*}^{-1}(x_i - \mu^*)$; $i = 1, 2, \ldots, n$, where μ^* and Σ^* represent the M-estimators (classical or robust) of μ and Σ, respectively. The robust M-estimators of location and scale are obtained by solving the following system of equations iteratively:

$$\mu^* = \sum_{i=1}^{n} w_1(d_i)x_i \Big/ \sum_{i=1}^{n} w_1(d_i) \tag{1}$$

$$\sum^* = \sum_{i=1}^{n} w_2(d_i)(x_i - \mu^*)(x_i - \mu^*)' \Big/ \left[\sum_{i=1}^{n} w_2(d_i) - 1\right] . \tag{2}$$

The PROP Redescending Influence Function

The PROP redescending influence function is given by

$$\begin{aligned} \psi(d_i) &= d_i \quad &; d_i \leq d_0 \\ &= d_0 \exp[-(d_i - d_0)]; &d_i > d_0 . \end{aligned} \tag{3}$$

Here, d_0^2 is the $\alpha100\%$ critical value obtained from the distribution, $(n-1)^2\, \beta(p/2, (n-p-1)/2)/n$, of the individual MDs, d_i^2. The weight functions are given by $w_1(d_i) = \psi(d_i)/d_i$ and $w_2(d_i) = w_1^2(d_i)$. The weights in Eqs. (1) and (2) are obtained using these weight functions.

The PROP M-estimators and the MDs, with or without outliers, and the corresponding classical MLEs and the MDs for multinormal populations free of discordant observations (after removal of discordant values) are usually in close or complete agreement.

In this iterative process of obtaining μ^* and Σ^*, all observations are assigned some weights, which can be used to characterize the extremeness of the various observations. Extreme observations too far from the center of the data are assigned reduced-to-negligible weights, whereas the observations coming from the central part of the data (with similar PROP and classical MDs) receive full weight. Singh [22], and Singh and Nocerino [23,24] used these weights to obtain more accurate estimates of degrees of freedom associated with the various robust statistics. Robust distances corresponding to these inlying observations roughly follow a $(wsum2 - 1)^2\ \beta(p/2, (wsum2 - p - 1)/2)/wsum2$ distribution, where $wsum2 = \sum w_2(d_i)$.

Construction of the Q-Q Plot of the MDs

The control-chart-type quantile-quantile (Q-Q) plot of the MDs given as follows provides a single graphical display of the underlying multivariate data set, which identifies these well-separated observations effectively. This graph also provides a useful tool for assessing the multinormality of the underlying population. The

Q-Q plot of the MDs can be constructed as follows.

(a) Compute the MDs, $d_i^2 = (x_i - \mu^*)'\Sigma^{*^{-1}}(x_i - \mu^*)$ for $i = 1, 2, \ldots, n$, where μ^* and Σ^* are the M-estimators obtained using an appropriate classical or robust procedure.
(b) Order $d_i^2 : d_{(1)}^2 \leq d_{(2)}^2 \leq \ldots \leq d_{(n)}^2$.
(c) Compute the expected beta quantiles, $b_{(i)}$, using the following equation:

$$\int_0^{b_{(i)}} \frac{\Gamma(\alpha+\beta)}{\Gamma(\alpha)\Gamma(\beta)} x^{\alpha-1}(1-x)^{\beta-1}\,dx = (i-\alpha)/(n-\alpha-\beta+1) \tag{4}$$

where $\alpha = (a-1)/2a$, $\beta = (b-1)/2b$, $a = p/2$, and $b = (n-p-1)/2$. This choice of parameters gives fairly good estimates of the expected beta quantiles. Compute the theoretical quantiles, $c_{(i)}$, for the distribution of the MDs using the equation $c_{(i)} = (n-1)^2\, b_{(i)}/n$.
(d) Finally, plot the pairs, $(c_{(i)}, d_{(i)}^2)$: $i = 1, 2, \ldots, n$.

For multinormal data, this plot resembles a straight line. Systematic curved patterns suggest a lack of normality. On this graphical display of multivariate data, points well-separated from the bulk of the data represent potential outliers. A systematic procedure defining this separation statistically is given in the following section. It should be noticed that the presence of outliers also destroys the linearity of the Q-Q plot of the MDs. This graphical display of the multivariate data is much more enlightening in identifying the cause for the rejection of the hypotheses (multinormality of the underlying population, or no outliers are present) than any other formal or informal test. A formal test statistic, $R_{p,n}$, and its critical values for assessing multinormality are given by Singh [22]. Multiple outliers, when present, typically represent observations from a different population(s), which should be treated appropriately. For example, $R_{p,n}$ should be obtained using an appropriate robust regression procedure.

Operational Guideline for the Identification of Multiple Outliers

The three-step procedure (TSP) to identify anomalous observations is described as follows.

(i) Construct the Q-Q plot of the MDs as described above. The initial robust start (e.g., M, $\hat{\sigma}_{MAD}$) is recommended, especially for small data sets of large dimensionality.
(ii) Draw a horizontal line at the $\alpha 100\%$ critical value (or at the $(1-\alpha)100\%$ confidence coefficient (CC) level), d_α^b, of the distribution of the Max (MDs) satisfying the simultaneous probability statement, $P(d_i^2 \leq d_\alpha^b, i: = 1, 2, \ldots, n) = 1 - \alpha$. Points lying above this horizontal line are potential outliers. The critical values, d_α^b, are computed using the Bonferroni inequality as described by Singh [22].

(iii) Draw a horizontal line at the $\alpha 100\%$ critical value, d^2_{ind}, obtained from the distribution, $(n-1)^2 \beta(p/2,(n-p-1)/2)/n$, of the individual d^2_{i}. These distances satisfy the probability statement, $P(d^2_{\mathrm{i}} \leq d^2_{\mathrm{ind}}) = 1 - \alpha$; $i: = 1, 2, \ldots, n$. Note that the critical values in steps (ii) and (iii) come from two different distributions.

It is important to emphasize here that in practice, probably due to computational ease, many authors [10–12, 36] use a chi-square, or even a normal [13] approximation to obtain the critical value, d^2_{ind}. Individual MDs, d^2_{i}, are then compared to this value. The probability statement in (iii), above, represents the confidence ellipsoid for individual distances, d^2_{i}. This ellipsoid is expected to cover the bulk of the data simultaneously. A more appropriate comparison is obtained by using the simultaneous confidence ellipsoid given in (ii), above. This statement has the built-in outlier detection criterion and provides appropriate simultaneous coverage for all of the sample observations. This also eliminates the need to assign small sample correction factors to provide the desired coverage (e.g., 90%, 95%) for the sample MDs, as suggested in the literature [12]. Moreover, the use of the simultaneous confidence ellipsoid provides a well-defined cutoff point for potential anomalies. This of course is further enhanced by the formal control-chart-type Q-Q plot using the robustified MDs. Formally, points above the horizontal line in (ii) (i.e., for which $d^2_{\mathrm{i}} > d^{\mathrm{b}}_{\alpha}$) are obvious outliers. These points are well-separated from the main stream of data. The points falling between the horizontal lines in (ii) and (iii) (i.e., for which $d^2_{\mathrm{ind}} < d^2_{\mathrm{i}} < d^{\mathrm{b}}_{\alpha}$) could be discordant and need further investigation (border-line cases). All other observations for which $d^2_{\mathrm{i}} \leq d^2_{\mathrm{ind}}$ form the main stream of inlying observations.

Example 1 (**continued**). We continue with the small data set of Example 1 to further explain how the classical procedures suffer from masking. The ROBUST module of the Scout software has been used in all subsequent calculations. The classical as well as the robust M-estimators (PROP) of μ and σ using the seven observations from $N(0, 1)$ are 0.26 and 0.857, respectively. Recall that the classical estimates of μ and σ obtained using all of the ten observations are 1.68 and 2.39, respectively. The PROP estimates (with $\alpha = 0.1$) based upon all the observations are 0.27 and 0.857. These are in close agreement with the classical/PROP estimates without the three added discordant values. The PROP robust estimates are not influenced by the outlying observations. The classical Q-Q plot of the MDs is given in Fig. 1a. Since all of the distances are smaller than the 5% critical value, 5.24, of the test-statistic, Max(MDs), it can be wrongly concluded that no discordant observations are present. The sequential classical test (Fig. 1a) based upon the Max(MDs) failed to identify the three discordant values present. Figure 1a suggests that all the observations come from a single population and no outliers are present, which is not the case. The robust Q-Q plot (Fig. 1b) of the PROP MDs identified the three discordant values in a single execution. No sequential outlier testing process or multiple hypotheses testing are needed here.

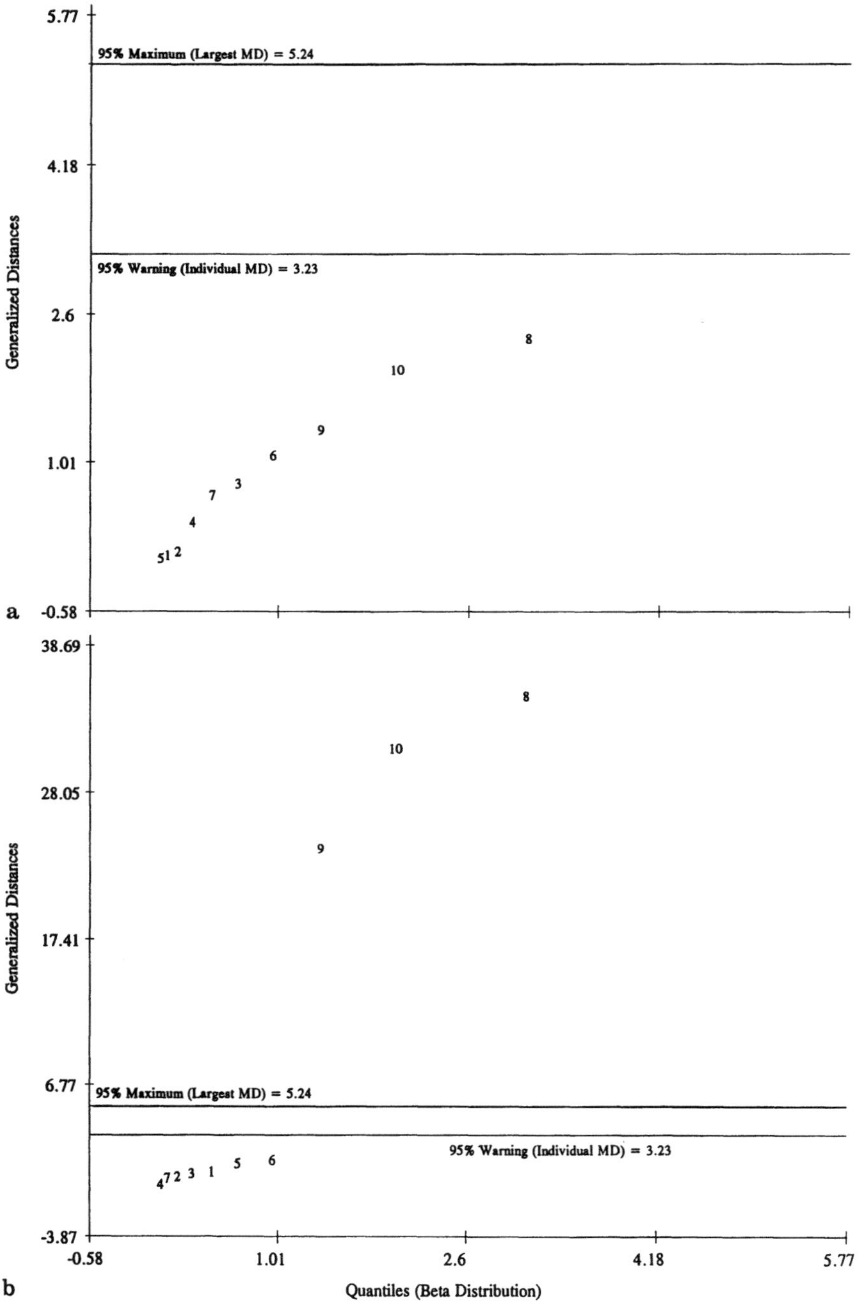

Fig. 1. **a** Classical Q-Q plot of MDs. **b** Robust (PROP, alpha=0.1) Q-Q plot of MDs

Contour Ellipse Plots

The contour probability plots of the MDs based upon the classical and robust estimators of location and scale can be used to enhance further the identification of outlying observations in bivariate data sets. Contour plots of the MDs have been displayed at the same two levels as the horizontal lines in the Q-Q plot of the MDs. The warning-point corresponds to a confidence ellipsoid given by the probability statement, $p(d_i^2 \leq d_{\text{ind}}^2) = (1 - \alpha)$; $i = 1, 2, \ldots, n$, and the maximum-point corresponds to an ellipsoid given by the probability statement, $p(d_i^2 \leq d_\alpha^b, i = 1, 2, \ldots, n) = 1 - \alpha$. The plots based upon the classical MDs accommodate outliers as part of the same population and hence may not present the true picture of all of the discordant observations present in the data set. The outlying observations are more prominent in the contour plots obtained using the robustified estimates. Observations falling outside the outer contour (maximum-point) are anomalous whereas the observations lying between the inner (warning-point) and the outer contours may be discordant. On this graph, the elliptical scatter suggests bivariate normality.

Example 2. This bivariate data set represents the ages and salaries of electrical engineers in the United Kingdom and is selected from Barnett and Lewis [32]. Using the standard classical procedures, they pointed out that observation 20 is an outlier and that the status of observation 19 is questionable. Figure 2a, b shows the Q-Q plots for the classical and the PROP MDs, respectively. Using the classical outlier test based on the Max(MDs) given in Fig. 2a, one concurs with Barnett and Lewis. From this Fig it is also noticed that observations 8, 26, and 37 are separated from the rest of the data set. The robust Q-Q plot of the PROP MDs, shown in Fig. 2b, identified observations 8, 19, 20, and 37 as outliers. The 95% classical and the robust contour plots are given in Fig. 2c, d, respectively. In Fig. 2d, the four outliers, observations 8, 19, 20, and 37, lie outside the outer ellipse. The outer and the inner ellipses are drawn at the 5% critical levels, 12.52, and 5.77, of the Max(MDs) and d_i^2, respectively. Observation 54 is right on the outer ellipse (compare it with Fig. 2b) and observations 4, 45, and 52 fall between the inner and the outer ellipses. Also notice that observation 26 is not discordant as implied from the classical presentations given in Fig. 2a, c. This example illustrates how the classical MDs and their correct ordering can get distorted by discordant observations.

Outliers in Interval Estimation

Several types of interval estimates exist in the literature [23, 37]. Here, we review three univariate interval estimates, namely: a) the confidence interval (LCL, UCL) for the population mean, μ, where LCL and UCL represent the lower and the upper confidence limits; b) the simultaneous confidence interval (LSL, USL) to

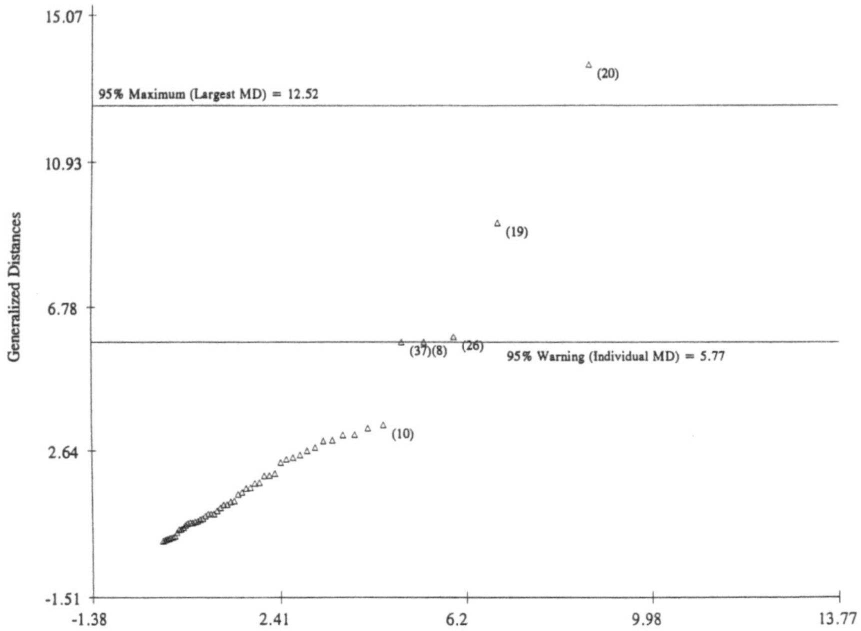

a

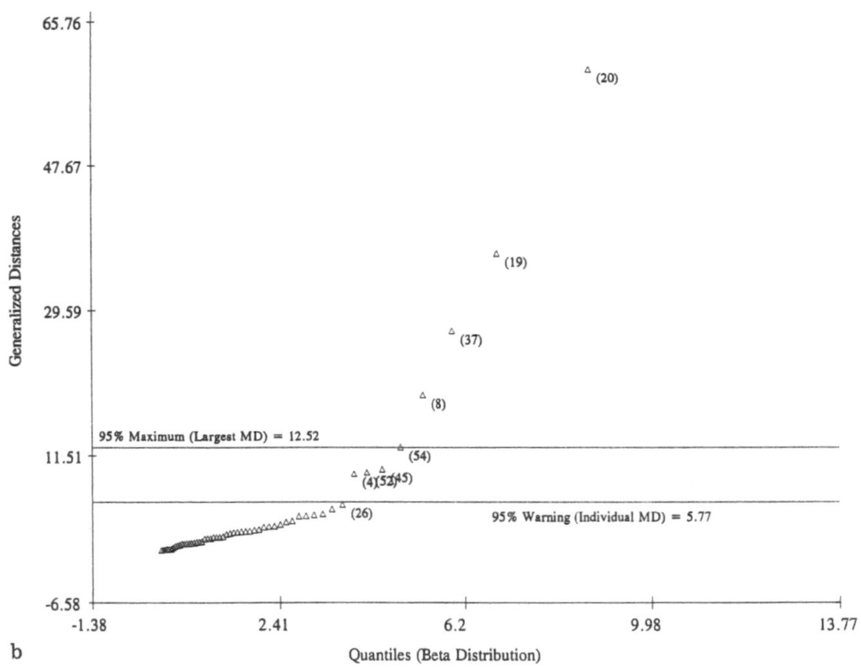

b

Fig. 2. a Classical Q-Q plot of MDs. **b** Robust (PROP, alpha=0.05) Q-Q plot of MDs. **c** Classical contour plot of age vs salary data. **d** Robust contour plot of age vs salary data

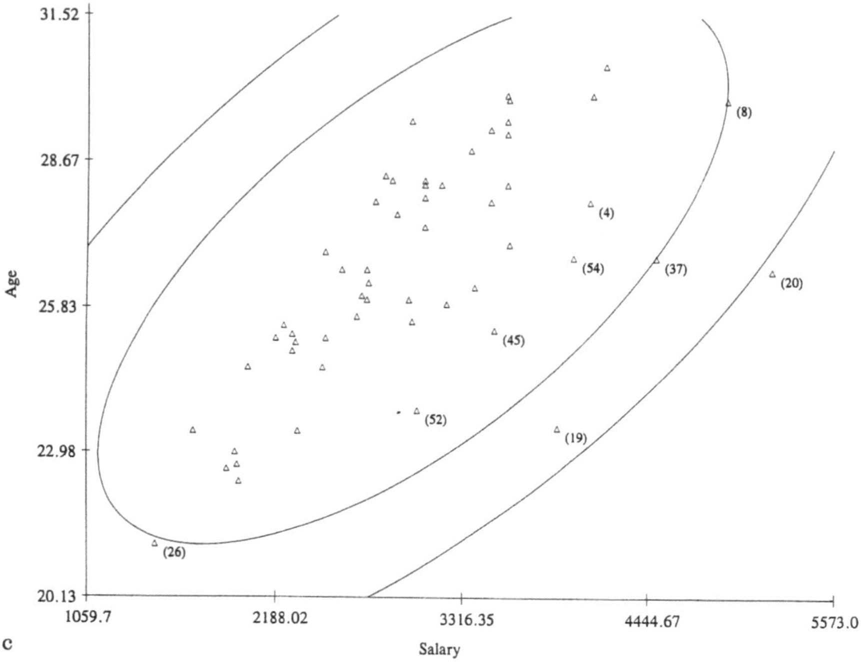

c

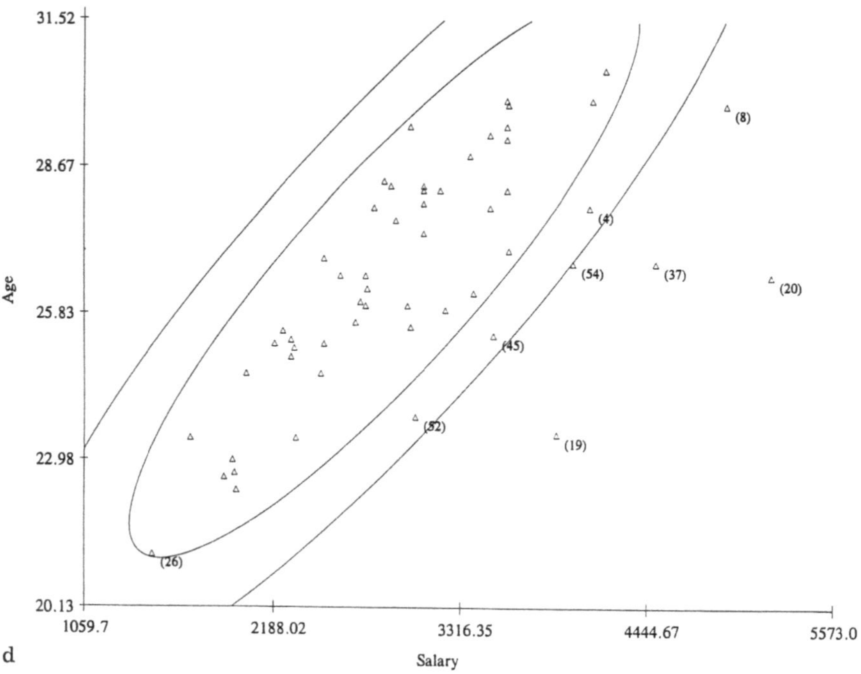

d

Fig. 2. *(Continued)*

contain all of the sample observations, x_i, $i := 1, 2, \ldots, n$, simultaneously, where LSL and USL are the lower and the upper simultaneous limits, respectively; and c) the prediction interval (LPL, UPL) for a future observation, x_0, from the underlying population, where LPL and UPL represent the lower and the upper prediction limits, respectively.

These intervals are significantly different from each other and care must be exercised to use them appropriately. For example, at a polluted site it is important to obtain a threshold value estimating the background level contamination at that site prior to any industrial activity that might have polluted the site. Here, the USL given by (b), and not the upper limit UCL for the mean (a), should be used. It is inappropriate to compare (as is sometimes done in practice) individual observations, x_i, with the UCL for the population mean, μ, and expect adequate coverage for x_i. A prediction interval given by (c) should be used when the coverage for a future and/or delayed observation, x_0, is needed. For simultaneous coverage of the results reported by the participants in a typical performance evaluation (PE) study of the U.S. EPA, the intervals given by (b) are more appropriate, by definition. These differences are illustrated later in Example 4.

The presence of a few anomalous observations can contaminate the underlying population and result in distorted classical estimates, $\bar{x}$ and s^2, of the population mean, μ, and variance, σ^2. This in turn leads to inflated and imprecise estimates of the above-mentioned intervals. Therefore, robust estimators, which assign reduced weights to these outliers, should be used. Robust interval estimation of the population mean has been studied by several authors [23, 25, 38, 39]. Student's t-like statistics with over-estimated numbers of degrees of freedom (DF) $(n-1)$ have been used to obtain the robust confidence interval (CI) for μ [25, 39]. The classical CI for μ is $\bar{x} \pm t_c\, s/\sqrt{n}$, where t_c is the critical value of Student's t distribution with $(n-1)$ DF. Using this as a model, the robust CI for μ has been defined [25, 39] as $\bar{x}^* \pm t_r s^*/\sqrt{n}$, where $\bar{x}^*$ and s^* are a pair of robust estimators of location, μ, and scale, σ, t_r is a critical t-value appropriately obtained.

The number of DF to be associated with t_r is not well established in the literature [26]. Kafadar [25], using a Monte Carlo simulation, tried to estimate a relationship between the classical Student's t-values, t_c, and the robustified t-values, t_{bi}, based upon the Biweight influence function. Iglewicz [40] suggested using an approximate relationship given by $t_{bi} = t_{0.7(n-1)}$ to obtain the critical t-values to be used with the Biweight function. However, the Biweight function does not perform well enough in small samples ($n \leq 20$) when multiple outliers are present. The robust interval estimates based upon the PROP influence, on the other hand, seem to perform very well for samples of all sizes. Each observation is assigned some weight according to its extremeness. These weights have been used effectively to obtain more accurate estimates of the degrees of freedom associated with the underlying test-statistics. This results in more precise interval estimates. This has been demonstrated by using simulated and real data sets [23].

Mathematical Formulation of Robust Interval Estimates

Robust M-estimators of μ and σ using the PROP function (Eq. 3) are given by

$$\overline{x^*} = \sum w_1(d_i)x_i \Big/ \sum w_1(d_i); \quad s^{*2} = \sum w_2(d_i)(x_i - \overline{x^*})^2/v \qquad (5)$$

where $w_1(d_i) = \psi(d_i)/d_i$, $w_2(d_i) = [w_1(d_i)]^2$; $i := 1, 2, \ldots, n$, and $wsum1 = \sum w_1(d_i)$, $wsum2 = \sum w_2(d_i)$. The DF are given by $v = wsum2 - 1$. The univariate robustified distances, $d_i^{*2} = (x_i - \overline{x^*})^2/s^{*2}$, follow a $v^2\beta\langle 1/2, (wsum2-2)/2\rangle/wsum2$ distribution. Tukey's Biweight estimator, $\bar{x}^*_{bi}$, or Huber's estimator, $\bar{x}^*_H$, of location, μ, are solutions of the equation, $\sum \psi(u_i) = 0$, where $\psi(u_i)$ is an influence function chosen accordingly. The Biweight and the Huber influence functions are given as follows:

$$\psi_{bi}(u) = \begin{cases} u(1-u^2)^2 & |u| \le 1 \\ 0 & |u| > 1 \end{cases}$$

$$\psi_H(u) = \begin{cases} u & |u| \le k \\ k \; sgn\, u & |u| > k \end{cases}$$

where $u_i = [x_i - \bar{x}^*]/cs^*$, c is a tuning constant chosen appropriately, k is a cutoff constant obtained from the Gaussian distribution, and *sgn* stands for the signum function. The details of the computational process to obtain the simultaneous M-estimators of location and scale are given by Kafadar [25] and Huber [26].

Robust estimates based on the PROP function for the above-mentioned intervals are given by the following probability statements.

(a) $(1-\alpha)100\%$ confidence interval (LCL, UCL) for μ:

$$P(\overline{x^*} - t_{v,\alpha/2}s^*/\sqrt{wsum2} \le \mu \le \overline{x^*} + t_{v,\alpha/2}s^*/\sqrt{wsum2}) = 1 - \alpha \qquad (6)$$

where $t_{v,\alpha/2}$ represents an appropriate critical value obtained from the Student's t-distribution.

(b) $(1-\alpha)100\%$ simultaneous confidence interval (LSL, USL) for all x_i; $i = 1, 2, \ldots, n$. Let $d^2_{m,\alpha}$ represent the $\alpha(100\%)$ critical value for the distribution of the Max (d_i^2). The simultaneous interval is given by $P(\text{Max}\,(d_i^2) \le d^2_{m,\alpha}) = 1 - \alpha$, which is equivalent to

$$P(\overline{x^*} - s^* d_{m,\alpha} \le x_i \le \overline{x^*} + s^* d_{m,\alpha};\ i = 1, 2, \ldots, n) = 1 - \alpha\,. \qquad (7)$$

This simultaneous confidence interval has the built-in outlier detection test. An observation outside this interval is a potential outlier. The critical values, $d_{m,\alpha}$, are obtained using the Bonferroni inequality and are listed in Table 1 for selected values of n and α. Finally, the prediction interval is given by:

(c) $(1-\alpha)100\%$ prediction interval (LPL, UPL) for a future observation, x_0:

$$P(\overline{x^*} - t_{v,\alpha/2}s^*\sqrt{[1/wsum2 + 1]} \le x_o \le \overline{x^*} + t_{v,\alpha/2}s^*\sqrt{[1/wsum2 + 1]}) = (1 - \alpha)\,. \qquad (8)$$

Table 1. Critical values of d_0^2 and $d_{m,\alpha}^2$ for selected values of sample size n and significance level α

		d_o^2			$d_{m,\alpha}^2$	
n\\α	0.1	0.05	0.01	0.1	0.05	0.01
4	1.822	2.031	2.205	2.139	2.194	2.239
5	2.076	2.469	2.941	2.794	2.941	3.111
6	2.216	2.743	3.505	3.320	3.561	3.892
7	2.305	2.928	3.933	3.756	4.080	4.576
8	2.366	3.059	4.264	4.128	4.523	5.173
9	2.410	3.158	4.525	4.450	4.906	5.697
10	2.445	3.234	4.735	4.735	5.244	6.161
11	2.471	3.295	4.908	4.990	5.545	6.575
12	2.493	3.345	5.053	5.221	5.816	6.947
13	2.511	3.387	5.175	5.431	6.062	7.284
14	2.527	3.422	5.280	5.625	6.287	7.592
15	2.540	3.452	5.371	5.803	6.494	7.874
16	2.551	3.478	5.451	5.970	6.686	8.134
17	2.561	3.501	5.521	6.125	6.864	8.375
18	2.569	3.521	5.584	6.270	7.031	8.599
19	2.577	3.539	5.639	6.407	7.187	8.809
20	2.584	3.555	5.690	6.536	7.335	9.005
25	2.609	3.614	5.880	7.091	7.962	9.830
30	2.626	3.653	6.007	7.536	8.459	10.472
40	2.646	3.702	6.165	8.223	9.218	11.429
50	2.658	3.730	6.259	8.744	9.786	12.128
60	2.666	3.749	6.322	9.162	10.238	12.673
70	2.672	3.762	6.367	9.511	10.612	13.117
80	2.676	3.772	6.400	9.809	10.930	13.490
90	2.680	3.780	6.427	10.070	11.207	13.811
100	2.682	3.786	6.447	10.301	11.452	14.093
120	2.686	3.796	6.479	10.697	11.869	14.567
140	2.689	3.802	6.501	11.028	12.216	14.956
160	2.691	3.807	6.518	11.312	12.512	15.286
180	2.693	3.811	6.531	11.561	12.771	15.571
200	2.694	3.814	6.541	11.781	13.000	15.822
250	2.696	3.820	6.560	12.245	13.479	16.342
300	2.698	3.823	6.573	12.620	13.865	16.757

Example 3. This example uses an historical data set consisting of 66 experimental results measured in millionths of a second for the speed of light (y), collected by Simon Newcomb in 1882. The transformed values, x, given by the equation, $y = 10^{-3}x + 24.8$, are listed by Stigler [41]. The data consists of two gross outliers. The classical and robust normal probability plots are given in Fig. 3a, b, respectively. The horizontal lines on these graphs are obtained using the simultaneous probability statement given by Eq. (7) above. The classical, Huber ($\alpha = 0.05$), and PROP ($\alpha = 0.05$) interval estimates for the population mean are summarized in Table 2. A systematic bias apparently existed in Newcomb's measurements. The currently assumed "true value" of the speed of light has been shown to be 33.02 decameters per millionth of a second. The discrepancy between the estimates presented here and the assumed "true value" of 33.02 relates to the calibration and the design of the measurement instrument rather than the estimation procedure employed. This should not obscure the fact that the classical

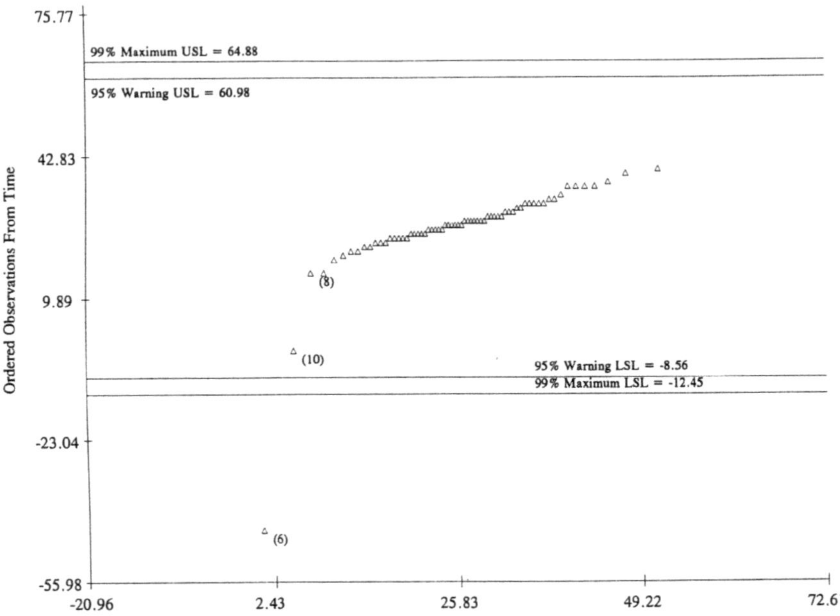

a

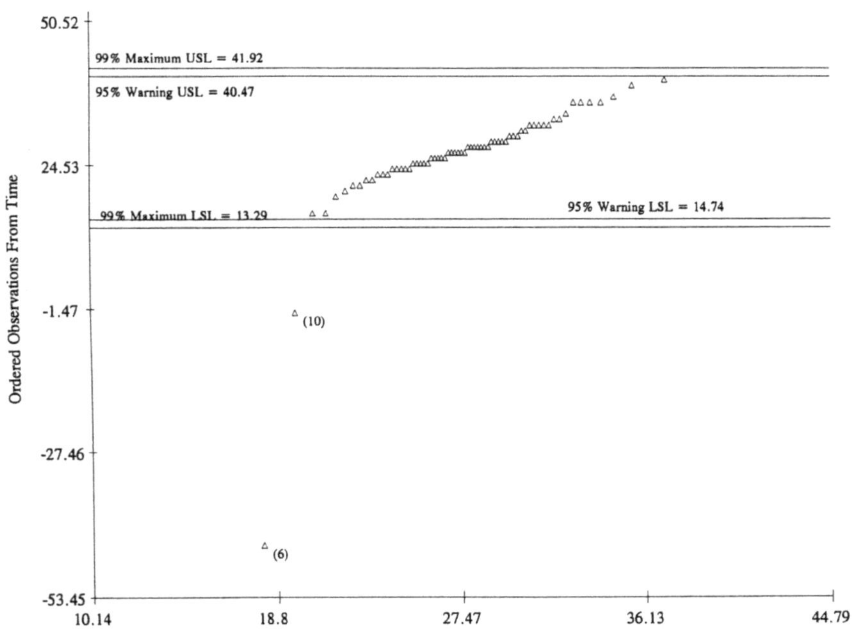

b

Fig. 3. **a** Classical probability plot of Newcomb's data. **b** Robust probability plot of Newcomb's data

Table 2. Simon Newcomb's speed of light data (n=66), used in Example 3

	Classical	Huber, $\alpha = 0.05$	PROP, $\alpha = 0.05$
mean	26.21	27.41	27.61
sd	10.745	5.23	4.04
LCL	23.57	26.11	26.57
UCL	28.85	28.71	28.64

estimates are distorted by the presence of discordant values. As can be seen, the PROP procedure resulted in the narrowest and most precise 95% confidence interval for the population mean for Newcomb's data.

Example 1 (**continued**). Using the small data set of Example 1, the robust interval estimates for μ using the four different methods are summarized in Table 3. Robust interval estimates for μ with CC = 0.95 based upon the PROP influence function alone are in close agreement for the contaminated and the uncontaminated populations. Due to the masking effect, the classical, the Huber, and the Biweight procedures could not identify the three discordant values. This resulted in inflated and unreliable estimates of the population mean, μ, and the sd, σ. The estimate of σ (using these three approaches) was distorted to such an extent that the resulting interval estimates are wider even though no adjustment has been made in the estimate of DF. Interval estimates obtained using the Huber, the Biweight, and the classical approaches are based on the same $(n - 1) = 9$ DF. The PROP procedure, on the other hand, uses the adjusted number of DF given by $wsum2 - 1 = 6$.

Table 3. Interval estimates with CC = 0.95 for the simulated data set of Example 1

	$N(0.1) - (n = 7)$				$7 \sim N(0,1)$ & $3 \sim N(5, 1) - (n = 10)$			
	Classical	Huber	Biweight([a])	PROP	Classical	Huber	Biweight([a])	PROP
mean	.26	.26	.26	.26	1.68	1.68	1.53	.27
sd	.857	.857	.97	.857	2.39	2.39	2.81	.857
LCL	−.531	−.531	−.635	−.531	−.035	−.035	−.48	−.517
UCL	1.053	1.053	1.155	1.053	3.387	3.387	3.54	1.064

[a]No adjustment in the degrees of freedom have been made while computing these statistics which in turn will only inflate these intervals

Example 4. Analytical laboratories participating in some programs of the U.S. EPA requiring QA/QC monitoring receive performance evaluation (PE) samples periodically. These samples contain known amounts of various organic or inorganic compounds. Laboratories are expected to achieve analytical results that are relatively close to the known values. However, in practice, the recoveries reported by the laboratories are lower (especially for volatile and semi-volatile compounds) than the known spiked amount. One of the objectives of the PE studies is to obtain

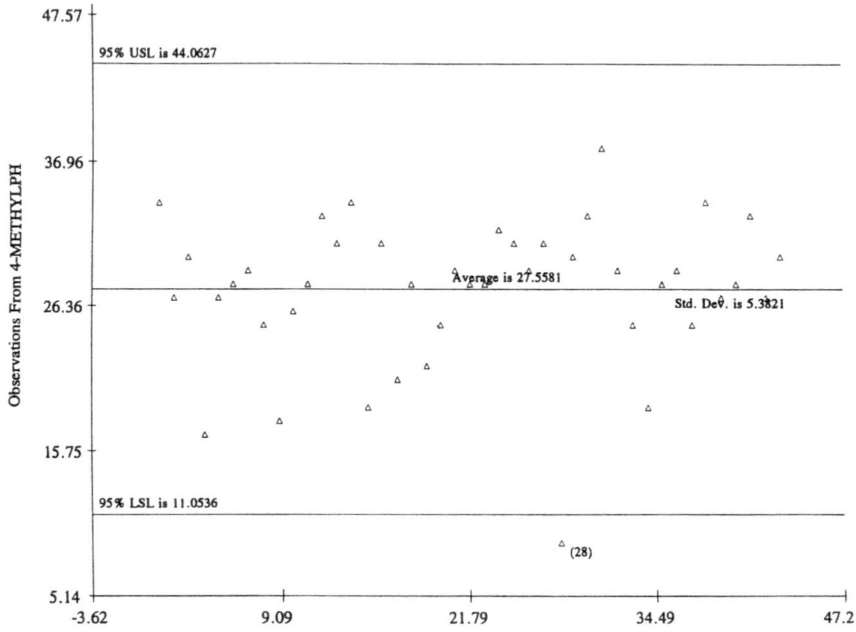

a

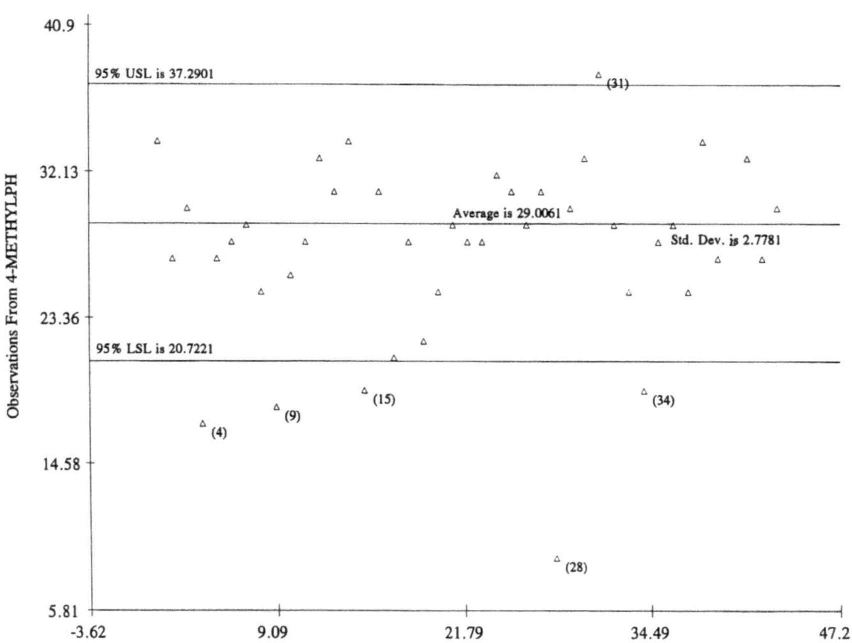

b

Fig. 4. a 95% classical simultaneous confidence intervals for 4-methylphenol. **b** 95% PROP simultaneous confidence intervals for 4-methylphenol. **c** 95% classical prediction intervals for 4-methylphenol. **d** 95% PROP prediction intervals for 4-methylphenol

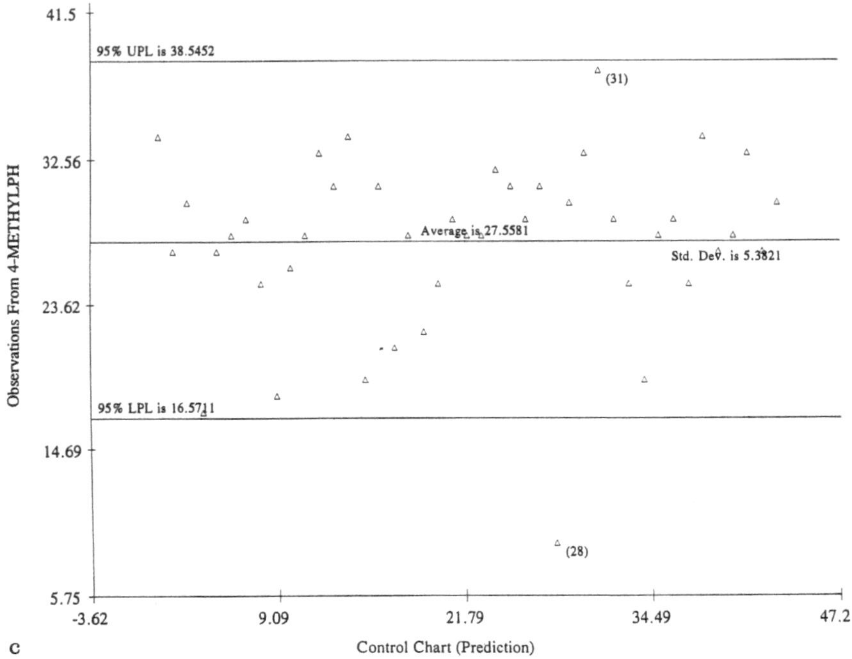

c

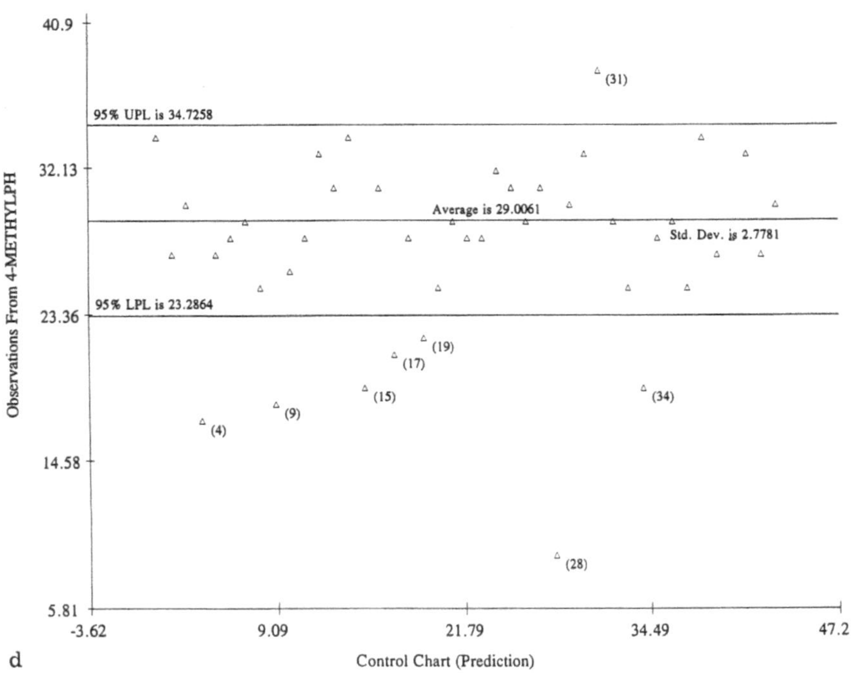

d

Fig. 4. *(Continued)*

Table 4. Interval estimates with CC = 0.95 using the performance evaluation data discussed in Example 4

	ν	mean	sd	LCL	UCL	LPL	UPL	LSL	USL
MLE	42.00	27.56	5.38	26.18	28.94	18.40	36.72	11.97	43.15
Huber	40.49	27.83	4.62	26.63	29.02	19.95	35.70	14.5	41.15
PROP	34.39	29.01	2.78	28.23	29.79	24.25	33.77	21.18	36.83
Bi-wt	42.00	28.40	4.21	27.32	29.47	21.00	40.00	N/A	N/A

N/A = not available

rigorous statistical regions within which most of the participants are expected to report their recoveries simultaneously. Horn et al. [38] used the Biweight function to obtain a prediction interval for a single future observation from a possibly noisy sample. The robust (PROP) simultaneous confidence intervals given by Eq. (7) provide precise and accurate estimates of those acceptance regions within which the participants are expected to perform. Computations for the various intervals described here are summarized in Table 4 using the analytical results reported by 43 laboratories for the semi-volatile chemical, 4-methylphenol, in one such PE study. Note the smaller and more precise estimate of the DF, $\nu = 34.4$, for the PROP method; this is to be expected because of the reduced weights assigned to the outlying observations. Using Iglewicz's [40] recommendation, one might use a substantially smaller number of DF, $(0.7)(42) \approx 29$. Also notice that the PROP sd is much smaller, again due to the negligible contribution of the outliers; this is a more precise estimate of the *sd* of the dominant population of this data set. The graphical displays of the classical and the PROP simultaneous confidence intervals are given in Fig. 4a, b, respectively. Figure 4c, d presents the graphs of the corresponding prediction intervals. These graphs suggest that it is inappropriate to use prediction intervals when the simultaneous coverage for all of the participants is desired.

Outliers in Linear Regression Models

Robust regression has been studied by several researchers, including: Rousseeuw and van Zomeren [11], Andrews [18], Hawkins et al. [20], Rousseeuw and Leroy [21], Singh and Nocerino [24], Ruppert and Carroll [42], Carroll and Ruppert [43], and Jongh et al. [44].

The multiple linear regression model with p explanatory variables is given by

$$y_i = x_{i1}\beta_1 + x_{i2}\beta_2 + \ldots.. + x_{ip}\beta_p + e_i \qquad (9)$$

where e_i is assumed to be normally distributed as $N(0, \sigma^2)$; $i = 1, 2, \ldots, n$. Let $x_i' = (x_{i1}, x_{i2}, \ldots, x_{ip})$, $\beta' = (\beta_1, \beta_2, \ldots, \beta_p)$. The objective here is to obtain a robust and resistant estimate, $\hat{\beta}$, of β using the data set, (y_i, x_i'); $i = 1, 2, \ldots, n$. The ordinary least squares (OLS) estimate, $\hat{\beta}_{\text{OLS}}$ [45], of $\hat{\beta}$ is obtained by minimizing the residual sum of squares, namely, $\sum_{i=1}^{n} r_i^2$, where $r_i = y_i - x_i' \hat{\beta}_{\text{OLS}}$. Like the sample classical mean, $\bar{x}$, the estimator, $\hat{\beta}_{\text{OLS}}$, of β has a "zero" breakdown point. This means that the estimator, $\hat{\beta}_{\text{OLS}}$, can take an arbitrarily aberrant value even through the presence of a single outlier or leverage point, leading to a distorted regression model (9). The use of robust procedures that eliminate and/or dampen the influence of discordant observations on the estimates of regression parameters is desirable.

In regression applications, anomalies arising in the p-dimensional space of the design variables, $x' = (x_1, x_2, \ldots, x_p)$ (e.g., due to unexpected experimental conditions), are called leverage points. Outliers in the response variable, y (e.g., due to unexpected outcomes such as unusual reactions to a drug), are called regression or vertical outliers. The leverage points are typically divided into two categories: significant leverages ("bad" or inconsistent) and insignificant leverage ("good" and consistent) points. Insignificant leverage points either a) do not distort the underlying regression model when present in a typical regression data set, and/or b) extend the domain of the explanatory variables with the corresponding y-values to be consistent with the regression model displayed by the bulk of the data.

These consistent leverage points given above by (b) generally increase the coefficient of determination, R^2 [45], and improve the precision by decreasing the standard deviations of the estimates of the regression parameters, thus enhancing the underlying regression model. The leverage points described in (a) may arise when a subset of $q (1 \leq q \leq p)$ explanatory variables is insignificantly correlated with the response variable y, but the position of these q-variables in the x-space causes the corresponding p-dimensional vector, x, to be discordant. Significant leverage points and regression outliers, on the other hand, do notably distort the regression model. Thus, the significant leverages are outliers in the x-space with the corresponding y-values being inconsistent with the regression model displayed by the bulk of the data. These anomalies generally cause a decrease in the value of R^2, as well as distort and deflate the resulting residuals. All of these regression outliers and multiple leverage points can be easily identified by using the TSP outlined earlier.

In robust regression, the objective is twofold: 1) identification of vertical (y) outliers and distinction between significant and insignificant leverage points, and 2) estimation of regression parameters that are not influenced by the presence of these anomalies. Moreover, these estimates should also be in close agreement with the corresponding classical OLS estimates when no outlying observations are present. Singh and Nocerino [24] presented a formal graphical approach based upon the square root distances vs standardized residuals scatter plot for the identification of regression outliers and significant leverages. This graphical approach identifies all regression outliers and distinguishes between significant and insignificant leverages effectively. This is briefly described in the sequel.

Hawkins et al. [20] used the notion of elemental sets for locating multiple outliers in multiple linear regression models. Rousseeuw and van Zomeren

[11, 12] used the least median square (LMS) estimator, $\hat{\beta}_{LMS}$, of β, obtained by minimizing the median of the residual sum of squares (r_i^2); $i := 1, 2, \ldots, n$. They mentioned that the estimates based on the LMS procedure are very robust and resistant with respect to regression outliers as well as leverage points. Both of these techniques require a massive amount of computation to obtain the robust estimator, $\hat{\beta}_R$, of β. It has been noticed that the robust estimator, $\hat{\beta}_{LMS}$, usually does not agree with its classical counterpart after the removal of discordant observations (see Example 5).

Mathematical Formulation of Robust Regression

The robust estimator, $\hat{\beta}_{PROP}$, of β, is obtained by using the following PROP influence function:

$$\begin{aligned}\psi(y_i - x_i'\beta) &= d_i && \text{if } d_i \le d_0 \\ &= d_o \exp[-(d_i - d_o)] && \text{if } d_i > d_o\end{aligned} \tag{10}$$

where $d_i = |(y_i - x_i'\beta)/\hat{\sigma}| = |r_i/\hat{\sigma}|$, with $\hat{\sigma}$ being a scale estimate appropriately obtained, and d_0^2 is the $100 * \alpha\%$ critical value obtained from the distribution of $(n-1)^2\ \beta(1/2, (n-2)/2)/n$. The corresponding regression weight function, $w(d_i) = \psi(d_i)/d_i$, is given by

$$\begin{aligned}w(y_i - x_i'\beta) &= 1 && \text{if } d_i \le d_0 \\ &= d_0 \exp[-(d_i - d_0)]/d_i && \text{if } d_i > d_0\,.\end{aligned} \tag{11}$$

This influence function works well in identifying multiple regression outliers.

The TSP given earlier identifies all leverage points ("good" and "bad") using the robustified distances, Ld_i^2, corresponding to the p-explanatory variables. All distances, Ld_i^2, exceeding Ld_α^b are leverage points. These leverage points can be conveniently accommodated in a robust regression procedure by using the following function of the leverage distances, Ld_i^2, at the initial iteration:

$$\begin{aligned}\psi(x_i, y_i - x_i'\beta) &= d_i && \text{if } d_i \le d_0,\ Ld_i^2 \le Ld_\alpha^b \\ &= d_0 \exp[-(d_i - d_0)] && \text{if } d_i > d_0 \\ &= d_i \exp\left[-(Ld_i - \sqrt{Ld_\alpha^b})\right] && \text{if } d_i \le d_0, Ld_i^2 > Ld_\alpha^b\end{aligned} \tag{12}$$

where Ld_α^b is the $100 * \alpha\%$ critical value from the distribution of $\mathrm{Max}(Ld_i^2)$. The corresponding weight function is given by $w(x_i, d_i) = \psi(x_i, d_i)/d_i$.

Here, the function at Eq. (12) and the associated weights are used only in the first iteration to accommodate all pre-identified leverage points in the regression model. The influence function, Eq. (10), should be used in all subsequent iterations. The function obtained using the combination of Eq. (12) at the initial iteration and Eq. (10) in all subsequent iterations is referred to as the "modified PROP influence function". This procedure starts with reduced weights assigned to all pre-identified leverage points. The iteratively re-weighted least squares (IRLS) [26] procedure using this modified PROP function works effectively in separating significant leverage points from insignificant leverage points. All significant leverages are also identified as regression outliers. Convergence is generally achieved in less than ten iterations. In all of the examples discussed here, twenty iterations have been used. The graphical procedure is briefly described in the following section.

Identification of Leverage Points and Regression Outliers

The graphical display of robust leverage distances vs the robust standardized residuals can be used to identify the regression outliers and to distinguish between significant and insignificant leverage points. In this graphical display, the square root of the leverage distances, Ld_i, corresponding to the p-explanatory x variables are plotted along the horizontal axis, and the standardized residuals, $r_i/\hat{\sigma}$, are plotted along the vertical axis. The two-dimensional graphical display is obtained by plotting the pairs, $(Ld_i, r_i/\hat{\sigma}), i: = 1, 2, \ldots, n$. The square root distances, Ld_i, and the residuals, r_i, are obtained using an appropriate regression procedure (classical or robust). The critical values for Ld_i^2 and the Max (Ld_i^2) are obtained using the procedure presented in the section entitled "The PROP Robust Procedure" above. The $(1 - \alpha)100\%$ confidence limits for the standardized residuals, $r_i/\hat{\sigma}$, are obtained using the simultaneous confidence interval as described under "Outliers in Interval Estimation".

On this graphical display, points with square root distances, Ld_i, greater than the critical value, $\sqrt{(Ld_\alpha^b)}$, represent leverage points. Observations with standardized residuals outside the $(1-\alpha)100\%$ simultaneous confidence band are the regression outliers. Leverage points with residuals outside this confidence band are significant leverages and all other leverages are insignificant leverage points.

The robust multiple regression procedure based upon the PROP influence function has been incorporated into a computer program called REGRESS. A copy of the executable version of this program can be obtained by writing to the author(s). Some well known data sets from the literature are discussed next in Examples 5 and 6. In these examples, the estimators based upon the PROP function (Eq. 10) or the modified PROP function (combination of Eqs. 10 and 12) produce robust regression models consistent with the literature.

Example 5. This calibration example from astronomy is taken from Rousseeuw and Leroy [21]. The data consist of 47 stars in the direction of Cygnus (Hertzsprung-Russell Diagram of Star Cluster CYG OBI). This data set is known to consist of 43 stars following the main sequence and 4 giant stars from a different population. The objective is to obtain a calibration model that describes the main sequence of stars. For this data set the M-estimators are obtained using the modified PROP function. Figure 5a has the OLS (solid line) and the robust (dashed line) fits. Due to masking, the classical OLS procedure resulted in normal looking residuals as seen in Fig. 5b, giving the impression that all observations came from the same sequence of stars. Figure 5c, d shows the scatter plots of the OLS and the robust leverage square root distances vs the standardized residuals, respectively. Here, also, due to masking, Fig. 5c could not identify the four discordant observations, leading to the wrong conclusion that all of the stars might have come from a single population. The discordant observations, 11, 20, 30, and 34, are evident in Fig. 5d. From this figure, it can be seen that observations 7 and 9 are also somewhat different from the rest of the data set. Figure 5e has the classical OLS and the robust fits after the removal of the four giant stars and Fig. 5f has these fits (both fits are very similar; in fact, they overlap in the figure) after the removal of two more discordant observations, 7 and 9. In all of these graphs the robust fit of Fig. 5a is in close agreement with all the other PROP fits of Fig. 5e, f, and also with the classical fit of Fig. 5f.

Next, the robust PROP fit is compared with the corresponding LMS robust fit of Rousseeuw and Leroy [21]. The LMS fit is given by $\hat{y} = 3.898x - 12.298$, which is quite different from the robust PROP fit (Fig. 5d-f) and the classical fit of Fig. 5f.

Example 6. This example uses Brownlee's famous stack-loss [46] data set. Several authors, Rousseeuw and van Zomeren [11], Andrews [18], Hawkins et al. [20], and Ruppert and Carroll [42], have applied robust regression techniques on this data set. Here we present the results using the PROP function described above. Figure 6a, b represents the classical and the robust Q-Q plot of the MDs using all four variables. From Fig. 6b, it is obvious that observations 1–4, 21, and possibly 13 are outlying observations, including leverage points and regression outliers. Figure 6c, d represents plots of, respectively, the OLS and the robust leverage distances vs the standardized residuals. Figure 6d identified all regression and leverage points correctly as cited in the literature. Using the graphical procedure described earlier, we would conclude that observation 2 is an insignificant leverage point, observation 4 is a regression outlier, and observations 1, 3, and 21 are significant leverages. Observation 13 is also a mild regression outlier. The robust regression model and the relevant ANOVA statistics are given in the following.

PROP Estimates at the 20th Iteration

$$\hat{\sigma} = 1.029$$

$$R^2 = 0.985\,.$$

Regression coefficients are given by $\hat{\beta}_{\mathrm{PROP}} = [-36.497, 0.841, 0.456, -0.079]$, and the corresponding beta sds are $= [4.012, 0.058, 0.144, 0.052]$.

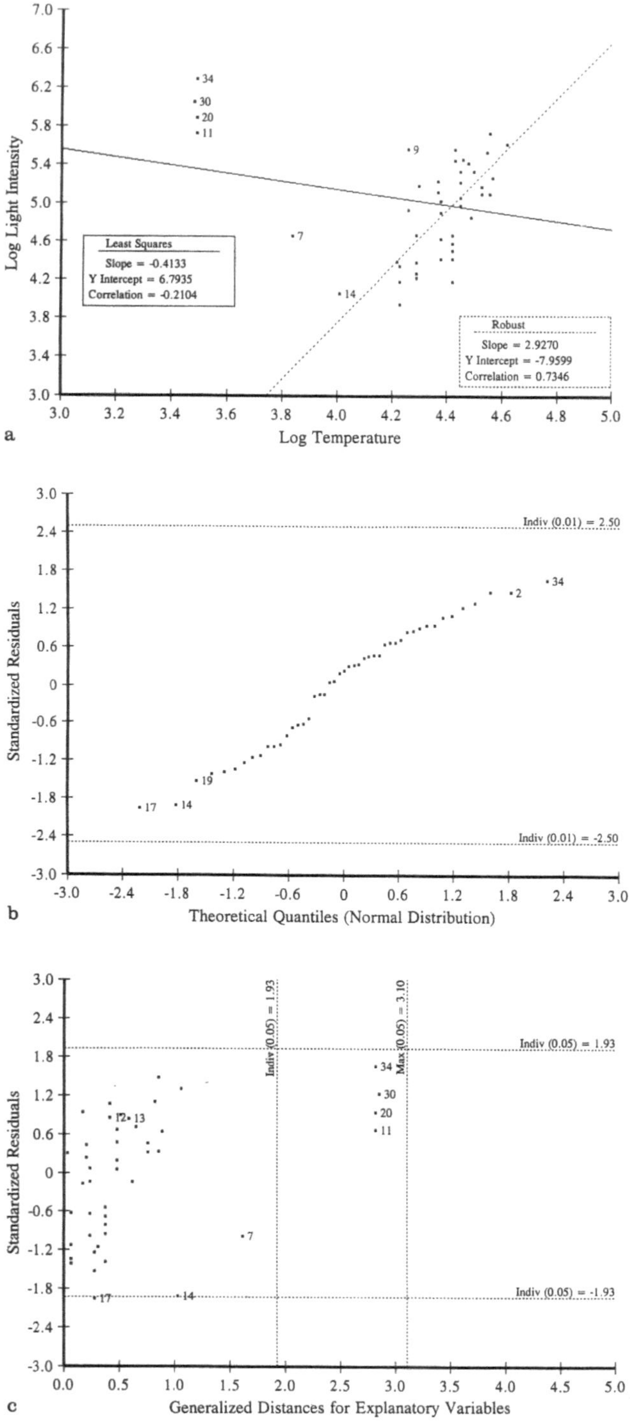

Fig. 5. a Star data (alpha = 0.05). **b** Star data - classical residuals. **c** Star data - Classical OLS plot. **d** Star data - PROP (alpha = 0.05) plot. **e** Star data - without four giants. **f** Star data - without 7, 9 and 4 giants

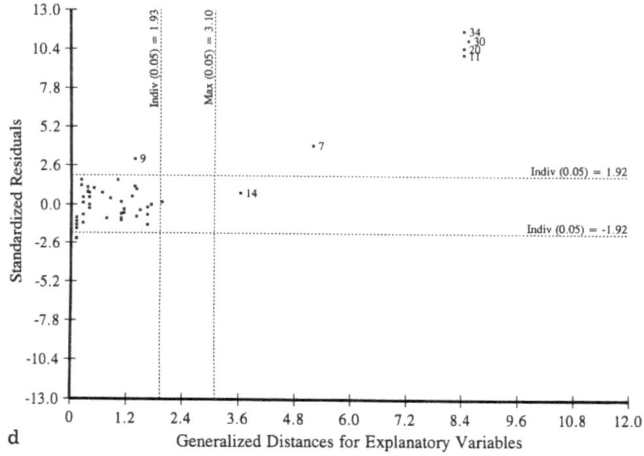

d

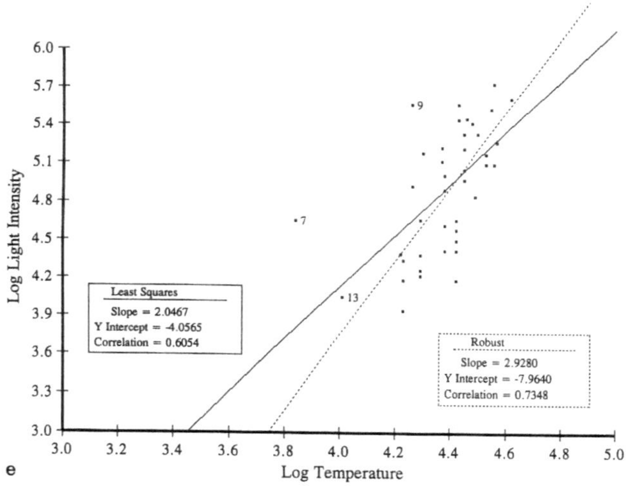

e

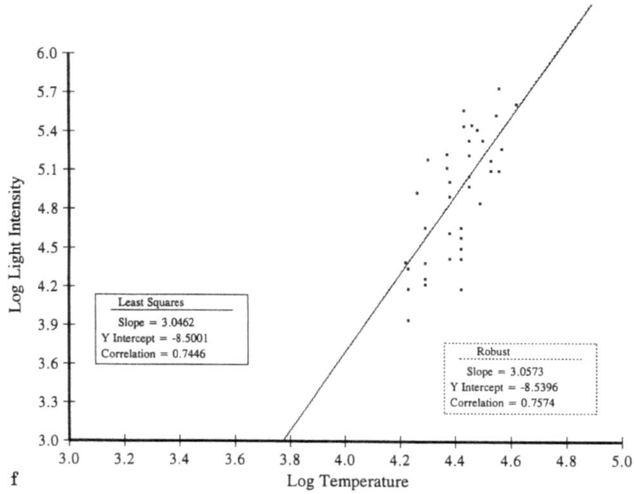

f

Fig. 5. *(Continued)*

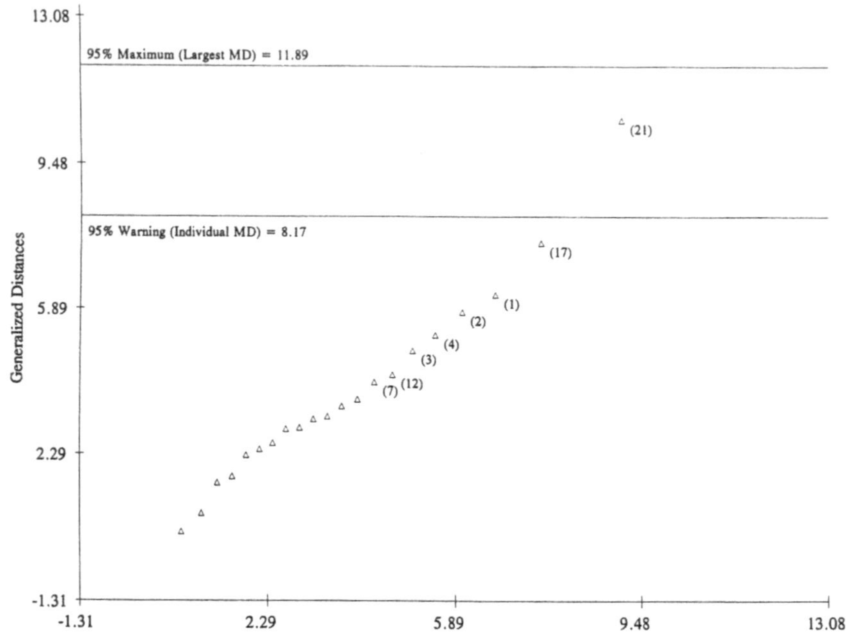

a

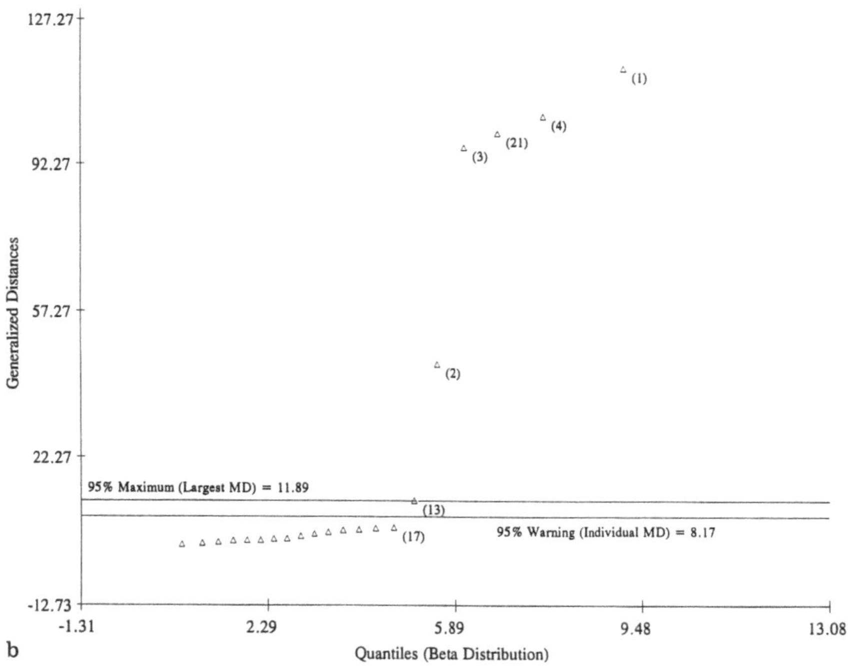

b

Fig. 6. **a** Classical Q-Q plot of MDs (four variables). **b** Robust (PROP, alpha = 0.15) Q-Q plot of MDs (four variables). **c** Stack - loss data, OLS plot. **d** Stack - loss data, PROP (alpha = 0.15) plot

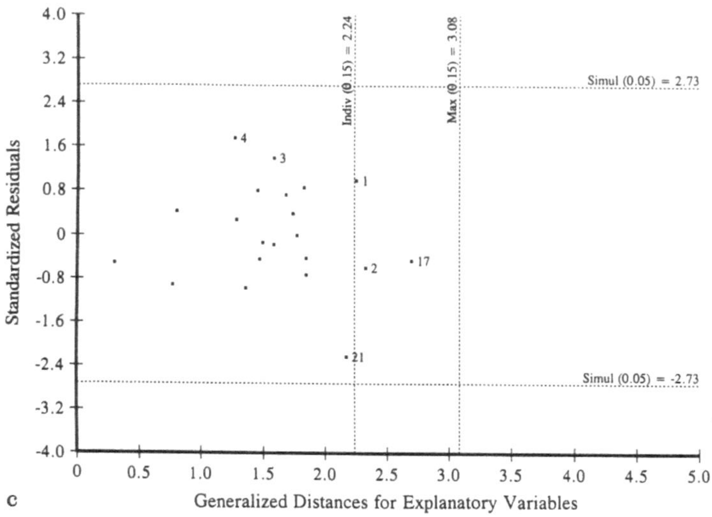

c

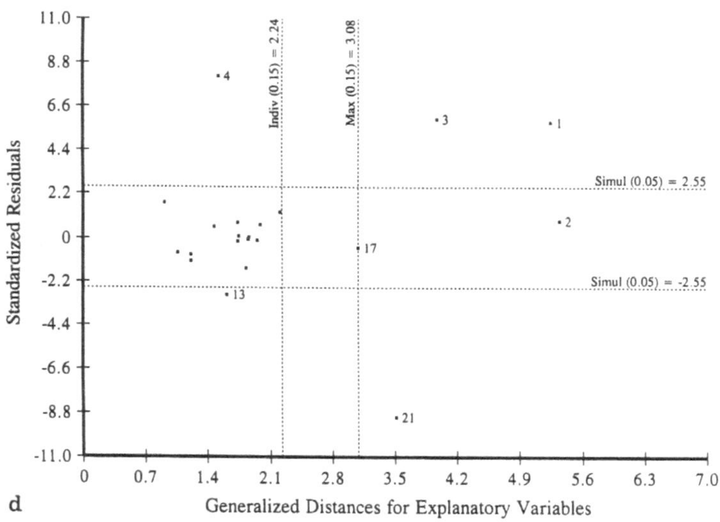

d

Fig. 6. *(Continued)*

ANOVA summary statistics (PROP) using IRLS

Source	Weighted SS	DF	MS	F-Statistic
Regression	795.018	3	265.006	
Residual	12.227	11.545	1.059	250.218
Total	807.245	14.54574		

Outliers in Other Chemometric Applications

Statistics play an important role in several chemometric, environmental, and ecological applications. Canonical correlation [47], discriminant and classification analyses [48, 49], principal component analysis (PCA) [50–54], regression, and calibration [55] are some of the statistical techniques adopted by researchers in these areas. Several researchers are developing new methods, or modifying existing techniques, to address the various mathematical and statistical issues which arise in these applications.

Chemometricians have found many useful applications of PCA. Some of the well known chemometric techniques based upon PCA are supervised and unsupervised pattern recognition [50, 51, 54], partial least-squares (PLS) [52, 56], PCA regression, and Soft Independent Modeling of Class Analogy (SIMCA) [50, 52]. The use of statistical experimental design and optimization (SEDOP) is fairly common among practicing scientists and researchers. Metallurgists [57–59] and research chemists [60–62] in their quest and need for new improved materials, alloys, and chemicals use SEDOP and response surface methodology to obtain optimal experimental regions. Discordant observations can arise naturally in data sets originating from these applications. These anomalies should be treated appropriately in all subsequent statistical analyses. Several researchers have applied robust methods [4, 63, 64] in these applications.

It is often overwhelming and tedious to deal with the raw data generated from these applications. PCA is one of the well-recognized data reduction techniques in the chemometric literature. It has been well-established that, while the first few high-variance principal components (PCs) represent most of the variation in the data, the last few low-variance PCs provide useful information about the noise present in the experimental results. The classical statistical procedures mentioned above may not perform well enough in the presence of anomalous observations. In this section, we briefly describe the robust estimation of PCs and discriminant functions.

Graphical displays of the first few PCs are routinely used as pattern recognition and classification techniques (supervised and unsupervised). Gnanadesikan and Kettenring [65] encouraged the use of normal probability plots and the scatter plots of PCs for the detection of multivariate outliers. The robust PCs give more precise estimates of the variation in the data by assigning reduced weights to the outlying observations. We will compare the classical PCs with the robustified PCs based upon the robust estimate of the dispersion matrix given by Eq. (2).

Outliers in Principal Component Analysis

Q-Q plots of the principal components can reveal any suspect observations as well as provide checks on the assumption of normality. Scatter plots of the first few high-variance PCs detect outliers which may inappropriately inflate variances and covariances. Plots of the last few low-variance PCs may be able to pinpoint observations that violate the correlation structure imposed by the bulk of the data but that are not necessarily discordant with respect to the individual variables.

Formally, let $\boldsymbol{P} = (\boldsymbol{p}_1, \boldsymbol{p}_2, \ldots, \boldsymbol{p}_p)$ be the matrix of eigenvectors of the sample dispersion (correlation) matrix, Σ^* (classical or robust), where the eigenvector, $\boldsymbol{p}_1$, corresponds to the largest eigenvalue and the vector, $\boldsymbol{p}_p$, corresponds to the smallest eigenvalue of Σ^*. The equation, $y = P'x$, represents the p-principal components with $y_i = \boldsymbol{p}_i'\boldsymbol{x}$ representing the i-th PC ; $i = 1, 2, \ldots, p$. A brief description of the construction of the probability plots for the PCs is presented in the next section.

Construction of the Q-Q Plot of the PCs

(i) Order the PC scores:

$$y_{(1)i} \leq y_{(2)i} \leq y_{(n)i}; \quad i = 1, 2, \ldots, p$$

(ii) Generate the normal quantiles, $q_{(k)}$, by solving:

$$P[z \leq q_{(k)}] = \int_{-\infty}^{q_{(k)}} \frac{1}{\sqrt{2\pi}} \exp\left(-z^2/2\right) dz = \frac{k - 3/8}{n + 1/4}; \quad k = 1, 2, \ldots, k.$$

(iii) Plot the pairs, $(q_{(k)}, y_{(k)i})$; $k = 1, 2, \ldots, n$, for the i-th PC. If the data are from a normal population, then these pairs will be approximately linearly related. Outlying observations are well-separated from the bulk of the data.

Contour ellipses of constant MDs on the scatter plots of the PCs provide another useful graphical representation of multidimensional data in two dimensions. The scatter diagrams for pairs of the first few PCs and the last few PCs also identify the discordant observations effectively. This is especially true for the PCs based upon the PROP estimate of the dispersion matrix. Next, we present an example illustrating these ideas.

Example 7. The ten-dimensional data set of size 25, used in this example, is taken from Jennings and Young [34]. We use this data set to illustrate that PCA represents a good data reduction technique. However, since the outlying observations can distort the PCs, robust estimation of the PCs is recommended. Several graphical displays are presented here. Figure 7a, b shows the classical and robust (PROP) Q-Q plots of the MDs, respectively. Figure 7b alone identifies all discordant observations. The discordant observations arranged in decreasing order of extremeness are 12, 21, 14, 22, 5, 8, and 6. One can also use the scatter plots of the PCs to identify the anomalies. The scatter plots of the first two and

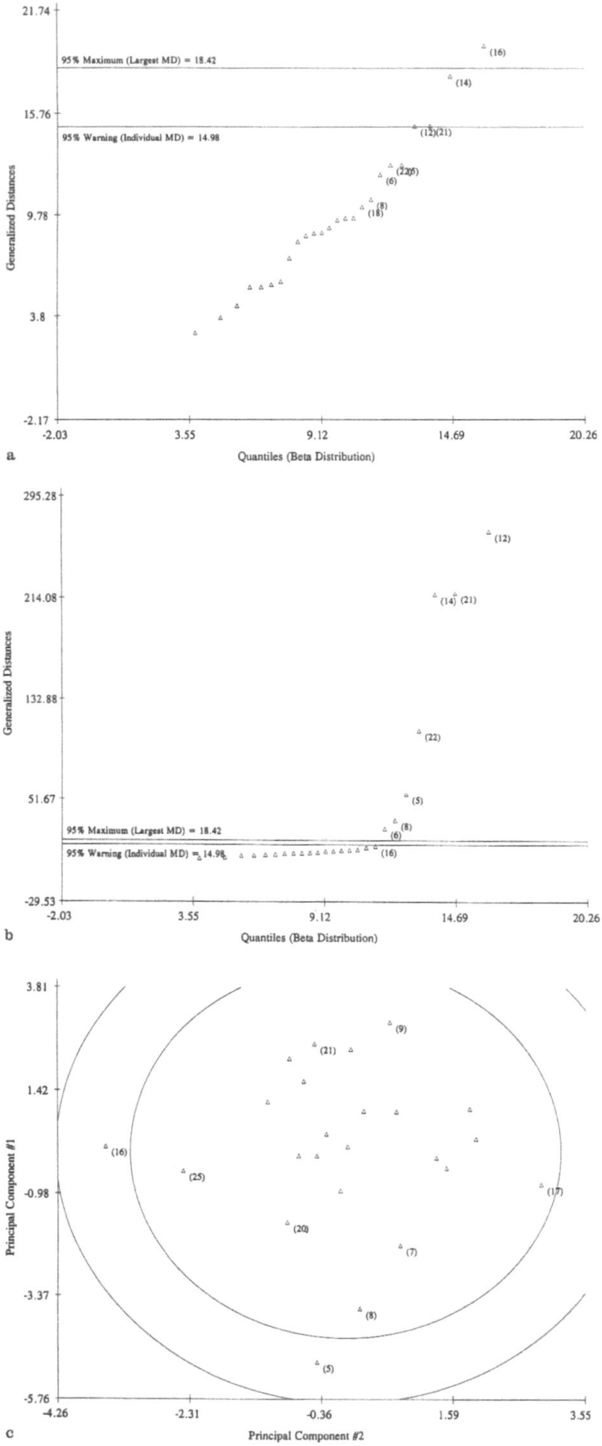

Fig. 7. **a** Classical Q-Q plot of MDs. **b** Robust (PROP, alpha = 0.05) Q-Q plot of MDs. **c** Classical scatter plot of PCs. **d** Classical scatter plot of PCs. **e** Robust (PROP, alpha = 0.05) scatter plot of PCs. **f** Robust (PROP, alpha=0.05) scatter plot of PCs

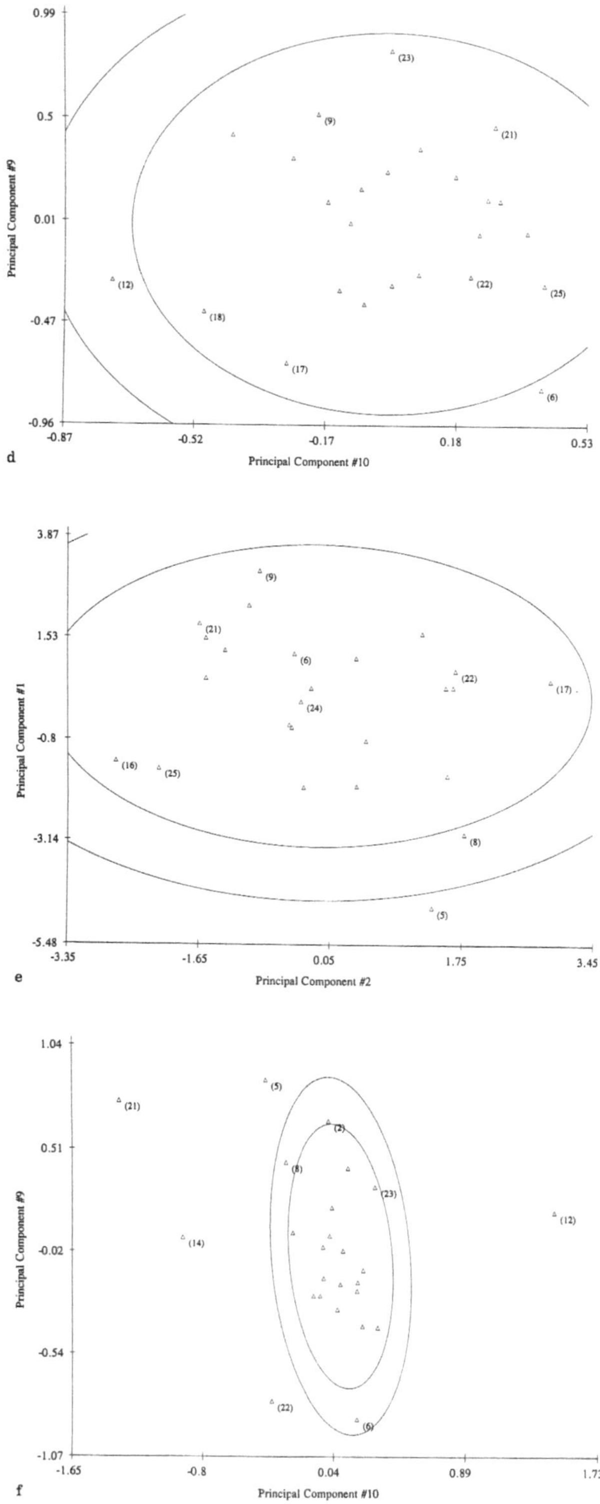

Fig. 7. *(Continued)*

the last two classical PCs are given in Fig. 7c, d respectively. Figure 7e, f shows the corresponding scatter plots of the robust (PROP) PCs. The 95% confidence ellipses on these graphs are obtained using the bivariate probability statements given by (ii) and (iii) in the section "Contour Ellipse Plots". The extent of distortion of the classical PCs by the discordant observations can be seen by comparing Fig. 7c, e, and Fig. 7d, f. One can also use the plots of the original variables taken two at a time, as is sometimes done in practice. For this data set alone, one would have to look at 45 scatter plots of the original 10 variables! The two scatter plots of Fig. 7e, f of the robustified PCs alone provide a good description of the anomalies present in this data set.

The point we are making here is that, although PCA provides a useful data reduction technique, the PCs may get distorted by discordant observations. Therefore, it is advisable to use a robust estimator (MVT, Huber, PROP) as well as the usual classical MLE of scale to obtain the PCs. If the results obtained using the classical and the robust procedures differ significantly, then the PCs based on a robust procedure should be used in all subsequent analyses, such as pattern recognition, PCA regression, and partial least-squares.

Outliers in Discriminant and Classification Analysis

Discriminant and classification analyses are multivariate techniques concerned with separating distinct groups (discriminant analysis) of observations and with allocating new observations (classification analysis) of previously defined groups (populations).

The separatory procedure is rather exploratory. In practice, the investigator has some knowledge about the nature and the number of groups. The study might be about k known groups, for example k geographic regions, k treatments, k analytical methods, k species, or k laboratories. In these cases the investigator knows the origin of each of the objects in a sample of size n obtained from these k populations. However, some of these k groups may be similar in nature and can be merged together. The objective here is to establish $g \leq k$ significantly different groups. Let $s = \min(g-1, p)$, then s discriminant functions can be computed for these g p-dimensional groups [9]. These functions are then used in all subsequent classifications. However, if the investigators have no prior information about the observations and their origin, then they have to search for natural groupings of observations (unsupervised classification). This grouping can be done on the basis of similarities or distance measures obtained from the observed variables or characteristics (analytes, defects, etc.).

Principal component analysis, or cluster analysis techniques, such as complete linkage, single linkage, average linkage, and Wards minimum distance, are used to separate observations into various groups [66, 67]. Several clustering techniques should be applied on the same data set. If the outcomes of these clustering techniques are roughly consistent with one another, then probably there are some well-separated groups. This separation process is often performed only once, preferably on training sets with known group membership to investigate the differences among the various groups. Discriminant functions are then obtained using these separated groups.

Classification procedures are less exploratory. Discriminant functions obtained in the separatory process are used to assign current and new observations into previously defined groups. The correct classification of the current observations with known group membership is the basis for the validity of the discriminant functions. This information is typically summarized in an error or "confusion" matrix.

However, discordant observations, when present, can distort the discriminant functions and the corresponding discriminant scores significantly. This can result in serious misclassification results. For example, in environmental applications, it is possible that a distorted discriminant function can classify a reasonably clean sample as coming from the contaminated population and a contaminated sample as coming from the clean population (the background).

In this section, we discuss the Fisher's robust classification procedure [9, 68]. The robust procedure, based upon the PROP function (Eq. 3) accommodates outlying observations by down-weighting them appropriately. This results in resistant discriminant functions which are not distorted by the anomalies. In turn, this leads to more accurate classifications of all future observations. This procedure has been tested on several well known examples available in the literature [50]. Fisher's discriminant analysis is briefly described in the following section.

Fisher's Robust Method for Discriminating Among *k* Populations

Let π_i represent the i-th population with mean vector, μ_i, and dispersion matrix, $\Sigma_i, i := 1, 2, \ldots, g$. One of the key objectives of Fisher's method of discriminant analysis is to separate these g populations. This is done by extracting $s = \min(g - 1, p)$ discriminant functions. These functions are derived to separate the populations as much as possible. Fisher's method also provides a very convenient and effective way of graphical separation of the p-dimensional data in terms of a few discriminant functions ($\leq s$).

The graphical displays of the first few Fisher's discriminant functions reveal possible groupings and clustering of the g populations. It should be pointed out that the derivation of Fisher's discriminants does not require multinormality of the distribution of the underlying g populations. Under normality and equal covariance matrices, Fisher's discriminant functions reduce to the linear discriminant functions [9]. The discriminants are extracted by maximizing the between-groups variability relative to the within-groups variability, Σ.

The population parameter, μ_i, and the common covariance matrix, Σ, need to be estimated based on training samples of size n_i from population, π_i,

$i := 1, 2, \ldots, g$. These estimates can be obtained using an appropriate classical or robust procedure. In the following, these estimates have been derived using the PROP influence function individually for each of the g populations. The relevant statistics are given as follows.

The group mean vectors are given by $\bar{\boldsymbol{x}}_i^* = \sum_{l=1}^{n_i} w_{il}\, \boldsymbol{x}_{il} \,/\, wsum1_i$, and the dispersion matrices are given by $\boldsymbol{S}_i^* = \sum_{l=1}^{n_i} (\boldsymbol{x}_{il} - \bar{\boldsymbol{x}}_i^*)(\boldsymbol{x}_{il} - \bar{\boldsymbol{x}}_i^*)'(w_{il})^2 \,/\, [wsum2_i - 1]$; where $wsum1_i = \sum w_{il}$, $wsum2_i = \sum w_{il}^2$, $i := 1, 2, \ldots, g$, and w_{il} is the weight associated with the l-th observation, $l := 1, 2, \ldots, n_i$, of the sample from population π_i.

The grand mean vector is given by $\bar{\boldsymbol{x}}^* = \sum_{i=1}^{g} wsum1_i \bar{\boldsymbol{x}}_i^* / \sum_{i=1}^{g} wsum1_i$ and the between-groups matrix is given by $\hat{\boldsymbol{B}}^* = \sum wsum1_i (\bar{\boldsymbol{x}}_i^* - \bar{\boldsymbol{x}}^*)(\bar{\boldsymbol{x}}_i^* - \bar{\boldsymbol{x}}^*)'$.

The within-groups matrix is $\boldsymbol{W}^* = \sum (wsum2_i - 1)\, \boldsymbol{S}_i^*$, and the pooled dispersion matrix is $\boldsymbol{S}_{\text{pooled}}^* = \boldsymbol{W}^* / [\sum wsum2_i - g]$.

Let $\hat{\lambda}_1 \geq \hat{\lambda}_2 \geq \ldots \geq \hat{\lambda}_s > 0$ denote the $s = \min(g - 1, p)$ non-zero eigenvalues of $\boldsymbol{W}^{*-1} \hat{\boldsymbol{B}}^*$. Let $\boldsymbol{e}_i$, $i = 1, 2, \ldots, s$, be the corresponding eigenvectors, scaled such that $\boldsymbol{e}'_i \boldsymbol{S}_{\text{pooled}}^* \boldsymbol{e}_i = 1$, $i = 1, 2, \ldots, s$. This can be done by choosing $\boldsymbol{l}_i = \boldsymbol{e}_i / \sqrt{\boldsymbol{e}'_i \boldsymbol{s}_{\text{pooled}}^* \boldsymbol{e}_i}$. These vectors, $\boldsymbol{l}'_i$, satisfy the equality $\boldsymbol{l}'_i \boldsymbol{S}_{\text{pooled}}^* \boldsymbol{l}_i = 1$, $i = 1, 2, \ldots, g$.

The linear combinations, $y_i = \boldsymbol{l}'_i \boldsymbol{x}, i = 1, 2, \ldots, s$, are called Fisher's discriminant functions. Scatter plots of the pairs, $(y_i, y_j), i \neq j = 1, 2, \ldots, s$, represent valuable graphical displays of between-group separation. The constant-distance ellipses can also be drawn individually for each of the g groups on the scatter plots of the discriminant scores. These plots provide a formal visual separation among the various groups. The Fisher's classification rule is: assign an observation $\boldsymbol{x}_0$ to π_h, $h := 1, 2, \ldots, g$, if

$$\sum_{i=1}^{s} [\boldsymbol{l}'_i(\boldsymbol{x}_0 - \bar{\boldsymbol{x}}_h^*)]^2 = minimum[\sum_{i=1}^{s} [\boldsymbol{l}'_i(\boldsymbol{x}_0 - \bar{\boldsymbol{x}}_l^*)]^2; l = 1, 2, \ldots, g]. \tag{13}$$

Graphical displays of the discriminant functions coupled with the contour ellipses reveal the group separation (or overlap) very effectively. Moreover, the scatter plots of the discriminants vs the original variables can also be used to achieve additional insight for graphically identifying those variables that are the most significant in discriminating among the g populations under consideration. The robust discriminants based upon the PROP function seem to enhance this grouping further. Discordant observations apparently have a negligible influence on the PROP discriminants. Confusion or error matrices [9] can be computed using these discriminant functions.

Finally, this chapter would not be complete without a discussion of Fisher's four dimensional taxonomic data set given in the following example.

Example 8. Fisher's four-dimensional data set consists of a training set of size 150, with 50 observations from each of the three species of iris. This data set was obtained from the depository of reference data sets established by the Editors of Chemometrics and Intelligent Laboratory Systems [50]. The four variables are sepal length, sepal width, petal length, and petal width. The data are fairly well behaved in the sense that three populations are roughly multinormal with equal covariance matrices. There are no significant differences in the classification results obtained using Fisher's classical and robust procedures. These are summarized in Table 5. Figure 8a,b shows the classical and robust scatter plots of Fisher's first two discriminant scores. These two graphs are in close agreement. The 95% confidence ellipses for the first two discriminants on these graphs are obtained separately for each of the three populations using the first two discriminant scores and the bivariate probability statement given earlier.

Next, in order to show the effectiveness of the classification procedure based upon the PROP function, 16 discordant observations (given in Table 5) were included in the data set. The differences between the classical and the robust classification results and the corresponding discriminants are significant in the presence of outliers, as can be seen from the results summarized in Table 5. However, robust estimates of the eigenvalues and the discriminant function coefficients given by vector l_i', with or without the outliers, are in close agreement (Table 5). Moreover, in the presence of discordant observations, the classical procedure misclassified some of the original observations; but for the original 150

Table 5. Classification results for Fisher's iris data (Example 8)

16 discordant observations added to the iris data set:

Obs No	Grp Id	Sp Ln	Sp Wd	Pt Ln	Pt Wd	Obs No	Grp Id	Sp Ln	Sp Wd	Pt Ln	Pt Wd
151	3	9.0	5.0	2.3	4.5	159	2	9.0	5.8	6.0	3.4
152	3	9.3	5.5	1.4	5.5	160	1	3.2	2.2	0.1	2.5
153	1	7.0	2.3	4.8	3.2	161	1	3.0	2.0	0.2	3.0
154	1	7.0	3.3	4.2	4.2	162	3	5.1	3.6	1.5	0.3
155	1	6.7	3.6	3.9	3.9	163	3	5.6	4.5	1.5	0.5
156	2	9.4	5.0	6.1	0.2	164	1	7.0	3.3	5.7	2.2
157	2	7.8	5.2	5.9	0.1	165	1	7.2	3.4	5.6	2.1
158	2	9.2	5.4	6.3	4.0	166	1	9.0	4.0	7.0	4.0

Results using the original 150 observations:

Classical

Nonzero Eigenvalues		Eigenvectors				
1	32.1920	1	−0.8294	−1.5345	2.2012	2.8105
2	0.2854	2	0.0241	2.1645	−0.9319	2.8392

Robust (PROP)

Nonzero Eigenvalues		Eigenvectors				
1	38.8440	1	−0.7252	−2.1498	2.2875	3.1364
2	0.2885	2	−0.0746	2.2523	−0.8806	2.8155

Confusion Matrix (Classical/Robust)

Actual	Predicted Pop1	Pop2	Pop3
Pop1	50	0	0
Pop2	0	48	2
Pop3	0	1	49

Misclassifications (Classical/Robust)

Obs.	Actual	Predict
71	2	3
84	2	3
134	2	2

Results using the original 150 observations and 16 outliers:

Classical

Nonzero	Eigenvalues		Eigenvectors			
1	2.1591	1	−0.1137	−0.3067	0.9025	0.2483
2	0.0824	2	−1.3282	1.8052	0.2556	1.2180

Robust (PROP)

Nonzero	Eigenvalues		Eigenvectors			
1	38.8334	1	−0.7249	−2.1483	2.2863	3.1359
2	0.2887	2	−0.0749	2.2536	−0.8791	2.8133

Confusion Matrices

	Classical Predicted				Robust Predicted		
Actual	Pop1	Pop2	Pop3	Actual	Pop1	Pop2	Pop3
Pop1	52	0	6	Pop1	50	2	6
Pop2	0	45	9	Pop2	2	48	4
Pop3	4	5	45	Pop3	2	1	51

Misclassifications

Classical

Obs. #	57	67	71	84	85	86	120	124	130	134	147	151
Actual	2	2	2	2	2	2	3	3	3	3	3	3
Predicted	3	3	3	3	3	3	2	2	2	2	2	1
Obs. #	152	153	154	155	157	158	159	162	163	164	165	166
Actual	3	1	1	1	2	2	2	3	3	1	1	1
Predicted	1	3	3	3	3	3	3	1	1	3	3	3

Robust

Obs.#	71	84	134	153	154	155	156	157	158	159	160	161	162	163	164	165	166
Actual	2	2	3	1	1	1	2	2	2	2	1	1	3	3	1	1	1
Predicted	3	3	2	3	3	3	1	1	3	3	2	2	1	1	3	3	3

observations, the robust classification results, with or without the 16 outlying observations, are in complete agreement, as can be seen in Table 5. Thus, the discordant observations have a negligible influence on the discriminant functions obtained using the PROP function. When the 16 discordant observations are included in the data set, the classical results are no longer reliable.

Figure 8c is the classical and Fig. 8d is the robust (PROP) scatter plot of the first two discriminant scores for the contaminated data set. Also, Fig. 8e displays the three populations for the robust (PROP) scatter plot of the first two discriminant scores for the contaminated data set. The robust procedure offers more accurate and refined group separation and classification. Thus, when the data are well-behaved, there are no statistically significant differences between the classical and robust classification estimates. However, in the presence of discordant observations, the classical estimates get distorted, whereas the discordant observations have little or no influence on the corresponding robust estimates (Fig. 8d).

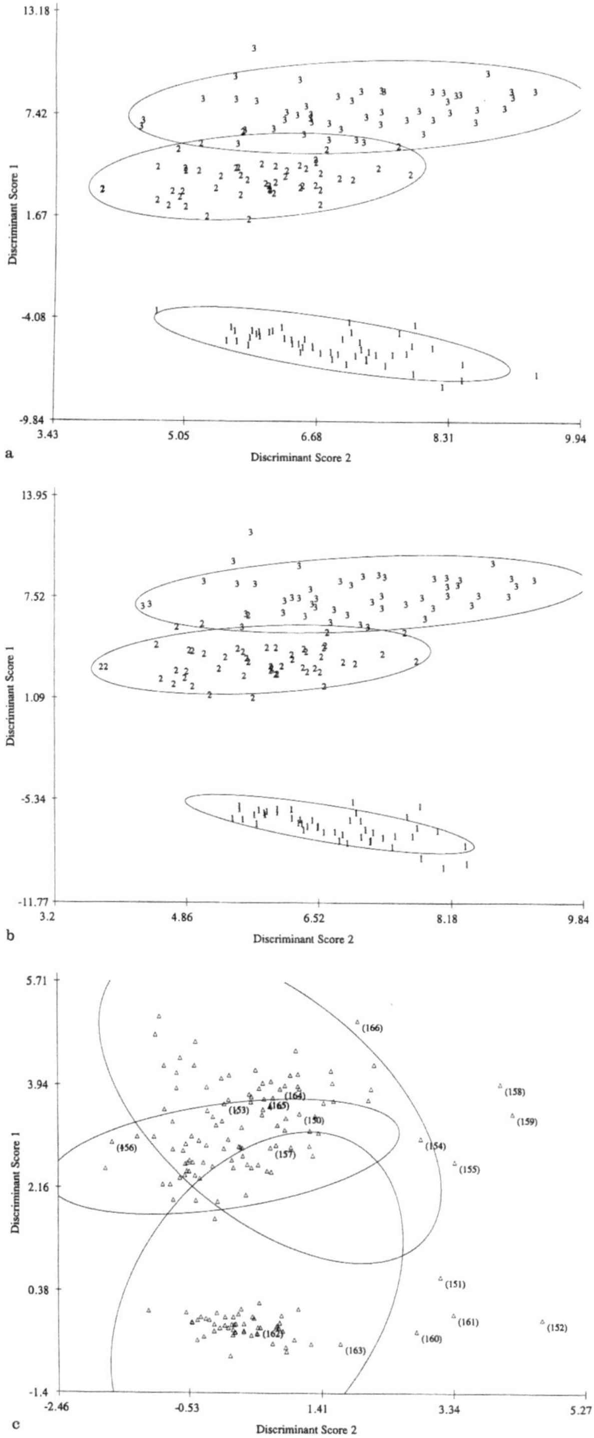

Fig. 8. **a** Classical discriminant scores. **b** Robust (PROP) discriminant scores. **c** Classical discriminant scores. **d** Robust (PROP) discriminant scores. **e** Robust (PROP) discriminant scores

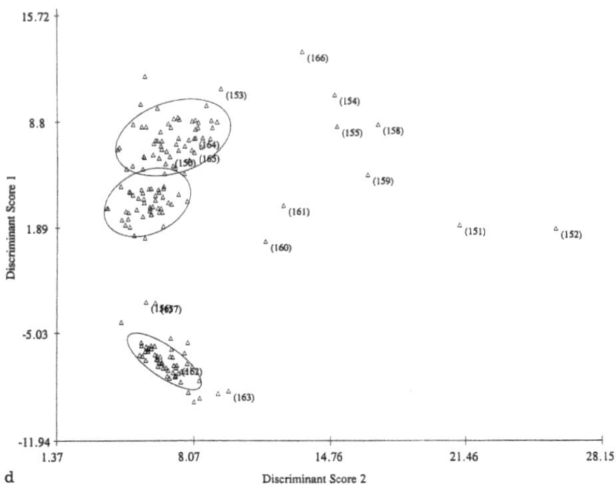

d

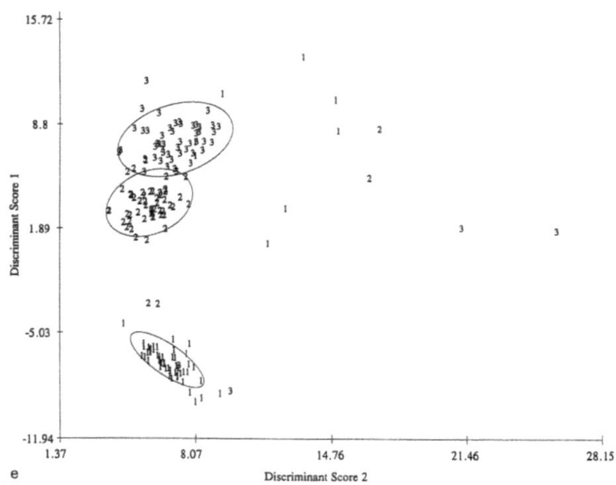

e

Fig. 8. *(Continued)*

Conclusions and Recommendations

Identification of discordant observations in samples from univariate and multivariate populations have received considerable attention over the past few decades. Identification of these anomalies is especially important in populations consisting of pollutants which may have potentially adverse effects on human health and the environment.

Robustified MDs are widely used for the identification of multiple outliers in multivariate populations. Mahalanobis distances exceeding the $\alpha 100\%$ critical value from the distribution of the Max(MDs) correspond to potential outliers.

The presence of discordant observations distorts these MDs to such an extent that the cases with the largest MDs may not necessarily correspond to the outlying observations. It is therefore necessary to obtain robust estimators that are resistant to multiple outliers with a high break-down point.

The robust estimators, based on the PROP influence function with an initial robust start and a suitable value of α, are shown to achieve a high (50%) break-down point in simulated data sets. Moreover, the PROP estimates and the MDs, with or without the discordant values, are in close agreement with the corresponding classical MLEs and the MDs when no outliers are present (or after their removal). The robust initial start in the iterative estimation process yields estimates that are resistant to masking effects caused by the presence of groups of outliers. This results in the correct ordering of the MDs leading to adequate identification of potential outliers.

The graphical control-chart-type Q-Q plot of the PROP MDs is an indispensable outlier identification tool. All kinds of outliers and leverage points can be adequately identified by using the formal graphical procedures presented earlier. Simultaneous confidence ellipsoids used here provide appropriate coverage to the bulk of the data points. No small sample correction factors or approximations are needed to provide appropriate coverage for the sample MDs. In addition, the PROP procedure answers the often raised question as to which cut-off values are to be used to determine which large MDs may correspond to potential outliers [12].

The robust regression procedure, using the PROP (or modified PROP) influence function and the associated graphical displays, identifies all of the regression outliers and distinguishes between significant and insignificant leverage points effectively. The procedure was tried on several simulated and well known data sets from the literature, including multiple linear regression models, and resulted in the correct identification of outliers and regression fit each time. We have found that the use of robust regression in relevant environmental calibration applications and in other chemometric applications enhances the understanding of the inherent complex chemical structure existing among various characteristics.

We are currently investigating statistical procedures to obtain robust and resistant variogram models used in kriging. Kriging has become a useful tool in estimation of various contaminants at polluted sites. However, kriging parameters are typically estimated using a suitable variogram model that best fits the data. In practice these estimates can be distorted by the presence of a few discordant observations.

The graphical displays presented here are the combination of informal exploratory techniques and formal statistical procedures making them appealing to inexperienced occasional users as well as to statistically-oriented users demanding statistical rigor. Robust procedures described in this article support John Tukey's message "THE CLASSICAL METHODS – means, variances, covariances and related statistics, and least-squares – ARE UNSAFE"[69]. The routine use of robust statistical methods together with the classical procedures for the identification of outliers and the estimation of population parameters of concern in various ecological, chemometric, environmental monitoring and QA applications is recommended as it provides more accurate and precise estimates of the population parameters.

Notice. The U.S. Environmental Protection Agency (EPA), through its Office of Research and Development (ORD), funded and collaborated in the research described here. It has been subjected to the Agency's peer review and has been approved as an EPA publication. The U.S. Government has a non-exclusive, royalty-free license in and to any copyright covering this article.

References

1. Grubbs FE (1950) Ann Math Statist 21:27
2. Dixon WJ (1953) Biometrics 9:74
3. Miller JC, Miller JN (1990) Statistics for analytical chemistry, 2nd edn Ellis Horwood, Chichester
4. Miller JN (1993) Analyst 118:455
5. Wilks SS (1963) Sankhya 25:407
6. Mardia KV (1972) Biometrika 57:519
7. Stapanian MA, Garner FC, Fitzgerald KE, Flatman GT, Englund EJ (1990) Communication in Statistics-Simulation. 20:667
8. Stapanian MA, Garner FC, Fitzgerald KE, Flatman GT, Nocerino JM (1993) J of Chemometrics 7:165
9. Anderson TW (1984) An Introduction to Multivariate Statistical Analysis. John Wiley, New York
10. Devlin SJ, Gnanadesikan R, Kettenring JR (1981) J Amer Statist Assoc 76:354
11. Rousseeuw PJ, van Zomeren C (1990) J Amer Statist Assoc 85:633
12. Rousseeuw PJ, van Zomeren C (1991) In: Stahel S, Weisberg S (ed) Direction in Robust Statistics and Diagnostics. Springer- Verlag, part II, Vol 34, New York, p 195
13. Campbell NA (1980) Applied Statistics 29(3):231
14. Huber PJ (1964) Ann Math Statist 35:73
15. Hampel FR (1974) J Amer Statist Assoc 69:383
16. Hampel FR, Rousseeuw PJ, Ronchetti R (1981) J Amer Statist Assoc 76:643
17. Tukey JW (1977) Exploratory Data Analysis. Reading Ma: Addison Wesley
18. Andrews DF (1974) Technometrics 16(4):523
19. Maronna RA (1976) Annals of Statist 4:51
20. Hawkins DM, Bradu D, Kass GV (1984) Technometrics 26(13):197
21. Rousseeuw PJ, Leroy AM (1987) Robust Regression and Outlier Detection. John Wiley, New York
22. Singh A (1993) In: Patil GP, Rao CR (ed) Multivariate Environmental Statistics. Elsevier Science Publishers, Amsterdam, 445
23. Singh A, Nocerino JM. Proceedings of the Ninth International conference on Systems Engineering. July 14–16, 1993, Las Vegas, NV, 370
24. Singh A, Nocerino JM under review
25. Kafadar K (1982) J of the Amer Statist Assoc 77(378):416
26. Huber PJ (1981) Robust Statistics. John Wiley, New York
27. Singh A, Singh AK, Flatman GT (1994) Int J Math Geology 26(3):361
28. Lavine BK (1992) J of Chemometrics 6:357
29. Scout: A Data Analysis Program, Technology Support Project, U.S. EPA, EMSL-LV, Las Vegas, NV 89193–3478
30. Rosner B (1975) Technometrics 17:221
31. Gilbert RO (1987) Statistical Methods for Environmental Pollution Monitoring. Van Nostrand, Reinhold Company, New York
32. Barnett V, Lewis T (1984) Outliers in Statistical Data. John Wiley, New York
33. Gnanadesikan R (1977) Methods for Statistical Data Analysis of Multivariate Observations. John Wiley, New York

34. Jennings LW, Young DM (1988) Communications in Statistics-Simulation 17(4):1359
35. Schwager SJ, Margolin BH (1982) Ann Statist 10:943
36. Neykov MN, Neytchev PN (1991) In: Stahel W and Weisberg S (ed) Direction in Robust Statistics and Diagnostics, part II, Vol 34, Springer-Verlag, New York, p 115
37. Hahn GJ, Meeker WQ (1991) Statistical Intervals. John Wiley, New York
38. Horn PS, Britton PW, Lewis DF (1988) The Statistician 37:165
39. Gross AM (1976) J Amer Statist Assoc 71(356):409
40. Iglewicz B, In: Hoaglin DC, Mosteller F, Tukey JW (ed) Understanding Robust and Exploratory Data Analysis. John Wiley, New York, p 404
41. Stigler SM (1977) The Annals of Statistics 5(6):1055
42. Ruppert D, Carroll RJ (1980) J of Amer Statist Assoc 75:828
43. Carroll RJ, Ruppert D (1985) Technometrics 27:1
44. Jongh PJ, De Wet T, Welsh AH (1988) J of Amer Statist Assoc 83:806
45. Draper NR, Smith H (1981) Applied Regression Analysis, 2nd ed. John Wiley, New York
46. Brownlee KA (1965) Statistical Theory and Methodology in Science and Engineering, 2nd ed. John Wiley, New York
47. Gittins R (1985) Canonical Analysis, A review with applications in Ecology, Springer-Verlag, Berlin, Heidelberg
48. Coomans D, Jonckheer M, Massart DL, Broeckaert I, Blockx P (1978) Anal Chim Acta 103:409
49. Coomans D, Massart DL, Broeckaert I, Tassin A (1981) Anal Chim Acta 133:215
50. Hopke PK, Massart DL (1993) Chemometrics and Intelligent Laboratory Systems 19:35
51. Sharaf MA, Illman DL, Kowalski BR (1986) Chemometrics, John Wiley, New York
52. Derde MP, Coomans D, Massart DL (1984) J of the Assoc of Official Analytical Chemists 67:721
53. Swain D, Dunn III WJ, Talaat RE (1993) Anal Chim Acta 277:305
54. Lavine BK, Stine A, Mayfield HT (1993) Anal Chim Acta 277:357
55. Massart DL, Kaufman L, Rousseeuw PJ, Leroy A (1986) Anal Chim Acta 187:171
56. Wold S, Johnson J, Sjostrom M, Sandberg M, Rannar S (1993) Anal Chim Acta 277:239
57. Scherer A, Inal OT, Singh AJ (1983) Solar Energy Materials 9:139
58. Patel S, Inal OT, Singh AJ (1985) Solar Energy Materials 11:381
59. Jiang H, Lee K, Singh Anita, Singh AK, Torma AE (1988) in Torma AE, Gundiler IH (ed) Precious and Rare Metal Technologies. Elsevier Science Publishers, Amsterdam, The Netherlands, p 547
60. Deming SN, Morgan SL (1973) Anal Chem 45(3):278A
61. Deming SN, Morgan SL (1983) Anal Chim Acta 150:183
62. Deming SN (1985) J of Research of the National Bur of Stds 90(6):479
63. Shoemaker AC, Kwok-Leung Tsui, Jeffwu CF (1991) Technometrics 33(4):415
64. Thompson M, Mertens B, Kessler M, Fearn T (1993) Analyst 118:235
65. Gnanadesikan R, Kettenring JR (1972) Biometrics 28:81
66. Anderberg MR (1973) Cluster Analysis for Applications. Academic Press, New York
67. Hartigan JA (1975) Clustering Algorithms. John Wiley, New York
68. Fisher RA (1936) Ann Eugenics 7:179
69. Tukey JW (1979) In: Launer RL, Wilkinson GN (ed) Robustness in Statistics. Academic Press, p 103

Pattern Analysis and Classification

Danny H. Coomans[1] *and Olivier Y. de Vel*[2]

[1] Department of Mathematics and Statistics, James Cook University, Townsville QLD 4811, Australia.
[2] Department of Computer Science, James Cook University, Townsville QLD 4811, Australia.

Summary

Various pattern classification techniques used in the context of environmental chemistry are presented. Both parametric and non-parametric Bayesian classifiers are outlined, as well as artificial intelligence-based techniques such as artificial neural networks and classification trees. A comparative analysis of the performance of some of the pattern classification methods is also given and several environmental applications are used to illustrate the various classifiers. A brief discussion on the software packages that are available is also included.

List of Symbols and Abbreviations

$\Re^Q$	Q dimensional real space
Q	number of variables or features
$\boldsymbol{x}$	Q dimensional pattern vector
N	total number of (training) data vectors in data set
f	function
$\boldsymbol{x}_j$	the jth pattern vector
χ	input feature space
$\mathcal{Y}$	output response space
y_i	response variable value for associated input data vector $\boldsymbol{x}_i$
h_i	smooth function of a single variable
B_m	multivariate spline basis function
K	number of training classes
I	indicator function
D_f	discriminant function
DS_i	discriminant score for test pattern vector $\boldsymbol{x}_i$
w_0	constant
$\boldsymbol{w}$	weight vector
^{T}	transpose
$\boldsymbol{W}$ or $\hat{\boldsymbol{Q}}_{\mathbf{pooled}}$	total sample within-class sum of squares and cross-products matrix
$\hat{\boldsymbol{Q}}_k$	estimator of the within ω_k sums of squares and cross products matrix
$\boldsymbol{x}_{\cdot 1}$	sample mean (centroid) vector of class ω_l
$\boldsymbol{S}$	pooled sample covariance matrix
$\boldsymbol{B}$	between-group sum of squares and cross-products matrix
ω_k	training class k, with k = 1,...,K
F	F-distribution
χ^2	χ^2-distribution
d_{ik}	distance between test data vector (object) i and class ω_k
$d_{k(\alpha)}$	critical distance for class ω_k at a α th tolerance level
$\boldsymbol{\Sigma}_k$	population covariance matrix for class ω_k
$\hat{\boldsymbol{\Sigma}}_k$	estimator of the population covariance matrix
$\boldsymbol{z}_{jk}$	z-standardized data vector from training class ω_k; $j = 1,\ldots,N_k$
N_k	number of data vectors (objects) in class ω_k
$\boldsymbol{B}_k$	matrix of principal components weight coefficients in the SIMCA method
A_k	dimensionality of the SIMCA method for class ω_k
$\boldsymbol{e}_{ik}$	a residual (error) vector for the SIMCA method of class ω_k
$P(\omega_k \vert \boldsymbol{x}_i)$	posterior probability of ω_k given test vector $\boldsymbol{x}_i$
$P(\omega_k)$	prior probability of class ω_k
L_{kl}	loss incurred if an object is classified as belonging to ω_k when in reality it belongs to class ω_l
$R(\omega_k \vert \boldsymbol{x}_i)$	risk taken in that $\boldsymbol{x}_i$ belongs to ω_k

$g_k(\boldsymbol{x}_i)$	classification (function) score of test data vector $\boldsymbol{x}_i$ for class ω_k
$\boldsymbol{\mu}_k$	population mean vector of class ω_k
$\hat{\boldsymbol{\mu}}_k$	estimator of the population mean vector μ_k of class ω_k
$\hat{\boldsymbol{\Sigma}}_k(\lambda, \gamma)$	regularized covariance matrix for class ω_k with two regularisation parameters λ and γ
$\phi(\boldsymbol{x}_i, \boldsymbol{x}_{jk}; \boldsymbol{u})$	kernel function for test data vector $\boldsymbol{x}_i$ and training data vector $\boldsymbol{x}_{jk}$ from class ω_k with smoothing vector $\boldsymbol{u}$
o_i	active output of a neural unit i in ANN
w_{in}	connection weight in ANN
θ_i	threshold of neural unit i in ANN
t_i	expected output of neural unit i in ANN
η	learning rate (ANN)
δ_j	error estimate for each output unit j in ANN
τ	iteration step
E	energy function in ANN
W	number of weights in ANN
$i(t)$	entropy function at tree node t
T	tree being set of branch and nodes
$\|T_t\|$	number of terminal nodes in a tree
VNIR	visible near infrared reflectance
SWIR	short wave infrared reflectance
PIMA	portable infrared mineral analysis
FLDA	Fisher's linear discriminant analysis
DLDA	Descriptive linear discriminant analysis
AIMS	Australian Institute of Marine Science
PDA	penalized discriminant analysis
ACE	alternating conditional expectation
MARS	multivariate adaptive regression splines
UNEQ	Mahalonolis distance-based class modeling using separate (unequal) class covariance matrices
SIMCA	soft independent modeling of class analogy
DASCO	discriminant analysis with shrunken covariances
LDA	linear discriminant analysis (as a Bayesian classifier)
QDA	quadratic discriminant analysis
RDA	regularized discriminant analysis
KNN	k-nearest neighbour method
ANN	artificial neural network
MLP	multi-layer perceptron
BP	back-propagation in ANN
CART	classification and regression trees
PARIS	pattern analysis and recognition interactive system

Patterns and Pattern Formation

Humans can distinguish very effectively between patterns that abound in nature (such as eye-like patterns on butterfly wings, etc.) despite the large variety of shapes, sizes, colours and so on. In all of these patterns we can identify sub-patterns which, in turn, are made up of even smaller sub-patterns (which, at the lowest level, we call primitives). In natural systems, the low-level semi-autonomous primitive components are subject to simple interactions with one another. More complex patterns will develop over time from all of the local interactions of the primitive components. These emerging complex patterns will also evolve in time because they themselves play a role in organising the behaviour of the lower-level components and providing a context in which these components interact. Patterns can therefore be static as well as dynamic. The nature of the pattern components is irrelevant, but it is rather the order of the components, their interaction and their structural interrelationship with other components which are important.

It is obvious that each pattern has a special meaning, depending on the context in which it is used. In environmental chemistry, visual patterns may also arise from imaging techniques [1]. However, more often the pattern will be a collection of measurements (i.e. a data profile) on a given physical sample. As a matter of fact, modern analytical chemical and physical measuring methods provide an ever increasing amount of data and information. Also, spectra obtained from spectroscopic techniques can be considered as patterns. An example is the use of visible-near infrared (VNIR) to short wave infrared (SWIR) reflectance spectra, as measured remotely and by hand-held field spectrometers, in the study of weathered soils. Weathering processes are interwoven into the inorganic aspects of soil production. In the savannah-style tropics, for instance, a range of unstable clay minerals are formed under the episotic wetting regimes of this climate type. These clays are liable to be expansive and dispersive. Recognition of these properties from spectral features is thus of great importance in the delineation of areas liable to suffer land degradation. This provides the ability to determine which area should be targeted for clearing or re-vegetation. Figure 1 illustrates the difference in digitised reflectance spectra (i.e. patterns) between weathered and un-weathered clay minerals in a semi-arid tropical environment. The first and third spectra (Kaolonite KGA-la and Dickite API-15) are examples of un-weathered clay minerals spectra. Both spectra are the result of an active, high temperature hydrothermal environment capable of producing the precise conditions of temperature, pressure and host rock composition for the formation of such minerals. At high temperatures, highly ordered clay mineral forms are formed [2]. The second spectrum (Kaolin KGa-2) is an example of a weathered clay mineral resulting from the effects of low temperature surface weathering processes. Such low temperature processes commonly produce poorly ordered clay minerals. As can be seen, there is a reasonable degree of variability in the spectra between the weathered and non-weathered spectra.

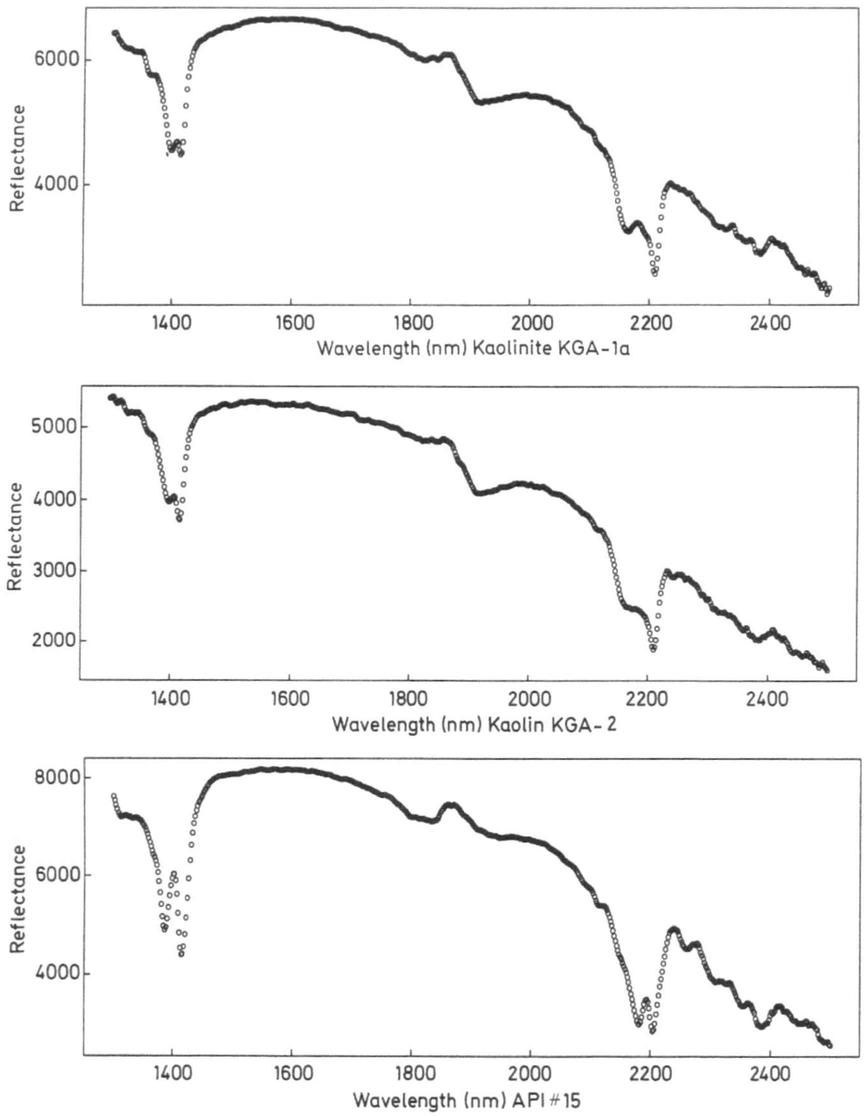

Fig. 1. Example PIMA reflectance spectra for weathered (first and third spectra) and un-weathered (second spectrum) clay minerals

More formally, we can define a pattern as a vector-like data structure of *features* (also referred to as *measurements, variables* or *attributes*) that includes information on the name of the feature, the values of the feature, and perhaps a measure of certainty or belief in the feature together with any information on the relationship between the given feature and any other feature. The number of features defines the *dimensionality* of the feature space-a large number of features gives rise to a high-dimensional pattern.

For example, the physico-chemical properties of an inter-tidal estuarine sample can be defined using various sediment features: the depth of the H_2S layer, the interstitial salinity, the median particle diameter, percentage organics, etc...

$$\begin{aligned} \textit{estuarine-sample} \equiv (& (\textit{depth of the } H_2S \textit{ layer}) \\ & (\textit{interstitial salinity}) \\ & (\textit{median particle diameter}) \\ & (\ldots)) \end{aligned}$$

For a given estuarine sample we measure the various physico-chemical properties and obtain, for example, the values as a set of *continuous* (real) numbers:

$$\textit{estuarine-sample} \equiv (0.3, 120.5, 434.0, \ldots)$$

One of the advantages of using a vector-like representation is that we can measure the *similarity* between any two given patterns by introducing the concept of *distance* and, therefore geometry. Example simple measures of distance include the Euclidean metric and Hamming distance.

Unfortunately, we do not always deal with continuous numeric features but sometimes with non-numeric (*symbolic*) features. For example, the "colour" feature may have various categories such as green, red, black etc. These are called *categorical* features. Some features may also be *discrete* and *ordered*, or *ordinal* features (as, for example, very low, low, high, and very high). To quantify symbolic feature, a method must be introduced to convert the feature into a numeric value. A discrete, ordered feature may be mapped onto the set of integers and then treated as a numerical variable.

One issue that does arise when we define a pattern and then set out to measure repeatedly all its values in an experiment is the relative number of features defining a pattern compared with the total number of patterns measured. For a high-dimensional problem (large number of features), the number of measurements that need to be undertaken is very much larger than the case for a low-dimensional problem. This is known as the "curse of dimensionality". If this is not observed a significant amount of bias is introduced into the recognition system i.e. the identification of two quite different patterns as being similar [3, 4, 5]. In some cases it may be necessary to choose (*feature selection*) and extract (*feature extraction*) the relevant features without discarding valuable information in order to achieve better recognition and be able to differentiate one class of objects from another. More formally, in feature extraction, we look for a function $f_{\mathrm{F}} : \Re^{Q} \rightarrow \Re^{Q'}$ (where $Q' \leq Q$) such that when f_{F} is applied to a pattern vector x, the new features are the image of x under f_{F} (say, $\mathrm{y} = f_{\mathrm{F}}(\mathrm{x})$). In feature selection we look to choose subsets of the original pattern features by taking f_{F} to be a projection onto some coordinate subspace of $\Re^{Q}$.

Pattern Analysis, Pattern Recognition and Pattern Classification

Pattern analysis attempts to understand the relationships and interactions between patterns. For example, we may be interested in identifying the biotic relationship between two biomass samples obtained from different ecological sites, or study the impact of oil spills on the community structure of sub-tidal macrobenthic communities as a function of time, or predict the amount of rainfall in Central Australia given the evolution in time of the pressure gradient between Darwin and Papeete etc. In each case we are interested in measuring the *similarity* between patterns.

Pattern recognition involves learning a set of decision-making rules given a set of *exemplars* or *training patterns* and their associated features. In some cases, the outputs of the decision rules for the training set are known and the decision rules must not only be able to reproduce the given training set-output associations, but they must also be able to accommodate any new (test) pattern. In general, pattern recognition is difficult because the patterns themselves are subject to noise distortion or there may be some variability among the patterns having the same output association. The minerals example in Fig. 1 illustrates the variability in weathered and un-weathered clay samples.

Of particular interest in pattern recognition is the case when the outputs of the decision-making rules correspond to a discrete label or *class*. In this case we have *pattern classification*. For example, we may wish to obtain a decision rule to determine if the impact of coral reef mining on the reef-fish species will be positive or negative, this being an example of a two-class problem. Pattern classification is of considerable importance in many application domains which require categories of decisions to be made, e.g. medical diagnosis, automatic image analysis, chemometrics, environmetrics etc. Many applications of pattern recognition and classification methods to environmental chemistry can be found in the literature. Examples include the identification of sources of contamination in environmental samples [6], heavy metals in suspended particulates of urban air [7], identification of oil spilled at sea [8], mass spectra of toxic compounds [9], trace organic pollutants in ambient air from mass spectra [10, 11], multi-species toxicity tests [12], metal contaminated soils [13], waste water pollution [14], neuropsychological effect of low lead exposure [15], etc.

There exist several paradigms to classify a pattern into one of several categories or classes. These include the *statistical* approach (also referred to as the *geometric* or *decision-theoretic* approach), the *syntactic* (or *structural*) approach, and the adaptive learning systems approach (e.g. *artificial neural networks, classification* or *induction trees*).

In the statistical approach each pattern is considered as a single entity (observational unit, object, pattern etc.) represented by an observation or data (feature) vector. The observational unit and the associated data vector are considered as an outcome of a random sampling process applied to a population of observations. The variability in the population is represented by a random variable vector for which a particular probability structure is assumed in the form of a specific probability distribution (parametric approach). Statistical methodology is used to estimate the parameters of the underlying theoretical population model from the set of measurements made on the patterns (i.e. sample of observational units and associated data vectors). More recently, distribution-free and non-parametric statistical models, which move away from the above parametic assumptions, have been developed. In general, however, class-conditioned probabilities and probability density functions are made use of in the statistical classification process. This approach is an integral part of multivariate statistics.

Rather than considering the underlying values of each single feature as being important information in the recognition process, the syntactic approach addresses the actual physical structural relationships that exist among the features. Structural relationships include the interrelationships or physical interconnections of features. Many patterns are best described by such relationships as, for example, musical patterns, hand-written characters, speech recognition, and fingerprint identification. Pattern representation is made by means of formal grammar rules and pattern recognition is undertaken by parsing a pattern's structure. The use of adaptive learning systems in pattern recognition is relatively new. Such systems do not generally make any assumptions on the functional form of the probability density distribution but are able to obtain relationships between the patterns and the required decision rules. Some of these techniques are still not widely understood, although some relationships between them and the statistical (especially non-parametric) paradigm have been established.

Learning

Environmental scientists are often confronted with the problem of having to make complex real-world decisions. To make such decisions, scientists rely almost invariably on previous experiences. For example, an environmental scientists must decide if a given ecosystem has exceeded a pollution threshold based on previous cases determined with attributes or variables such as concentration levels of lead, cadmium etc. The process of learning is the ability of a system to make effective and efficient decisions based on a history of previously accumulated experience.

One of the aims of a computational learning system (or machine learning system) is to develop a system that imitates the cognitive behaviour of human beings. There exist different kinds of learning as evidenced by the variety of different human learning paradigms. However, one that is particularly useful for decision-making is that of *inductive learning*-involving the extraction of general decision-making rules or procedures from a collection of samples of solved cases [16]. Inductive learning can be one of two types: *un-supervised learning* and *supervised learning*. In the former, the learning system looks for regularities or structures in the patterns or observations whereas, in the latter, learning is undertaken with the aid of a teacher who provides example input patterns and associated output "classes" [17]. Supervised learning is sometimes referred to as discriminant analysis by statisticians and un-supervised learning is known as clustering or numerical taxonomy. We will concentrate on supervised learning.

More formally, given a set of N input-output, data pairs $D \equiv \{(\boldsymbol{x}_i, y_i) \in \chi \times \mathcal{Y}\}$ for $i = 1, 2, \ldots N$, belonging to some input feature space χ (features are also referred to as *variables* or *attributes*) and output response space $\mathcal{Y}$ and assume that there exists a map

$$f : \chi \rightarrow \mathcal{Y} \tag{1}$$

with the property that

$$y_i = f(\boldsymbol{x}_i) \tag{2}$$

then learning is the process of estimating (or fitting) the function at points of its domain χ where data are not currently available. An example of such an estimation process involves the observation of the temporal evolution of coral growth on the Great Barrier Reef. It is known that the rate of coral growth is influenced by factors such as weather patterns, phosphate and nitrate run-offs from sugar cane fields etc. Here, we want to be able to predict the state of coral growth given the past samples. The predicted output has to be learned from the sequence of $L = Q$ input samples, each input time delay sample corresponding to a feature. The mapping corresponds to the function $f : \Re^L \rightarrow \Re$.

There are many different approaches for estimating f. Examples include a simple additive model of the form (for continuous $\mathcal{Y}$)

$$y = \sum_{i=1}^{n} h_i(\boldsymbol{x}_i) \quad \text{for } n \leq N \tag{3}$$

for some smooth arbitrary functions of a single variable, or by allowing interactions between the features using linear combinations (projections) [18]

$$y = \sum_{j=1}^{M} h_j \left(\sum_{i=1}^{n} \alpha_{ij} \boldsymbol{x}_i\right) \tag{4}$$

or, by allowing more explicit interactions [19]

$$y = \sum_{m=0}^{M} a_m B_m(\boldsymbol{x}_1, \boldsymbol{x}_2, \ldots, \boldsymbol{x}_N) \tag{5}$$

where each of the B_m is a multivariate spline basis function which is a product of univariate spline basis functions. Other estimation approaches include using kernel functions, nearest-neighbours, or feed-forward neural networks.

A particular case of this estimation procedure is when $\mathcal{Y}$ has finite range and assumes only a set C of K discrete values. Then we have

$$y_i = \sum_{j=1}^{K} j I_{[y_i = j]} \tag{6}$$

where I is the indicator function

$$I_{[y_i=j]} = \begin{cases} 1, & \text{if } y_i = j; \\ 0, & \text{otherwise.} \end{cases}$$

In case the output does not correspond exactly to one of the discrete values, we can choose the class k which is nearest to the observed value. This special case is the process of learning a *classifier*, where the output corresponds to K identifiable *classes*. For example, in the context of water quality monitoring, the input variables are the different laboratory tests used to evaluate the water quality. In this case, in the simplest form, it would be $K = 2$ output classes (i.e. acceptably good quality and unacceptable quality). The classifier learns by estimating the function f, $f : \Re \rightarrow C$ based on a set of measured water quality data samples.

Since data in chemistry are often subject to some amount of random variation the mapping can, in statistical terminology, be referred to as "regression" and the learning phase involves estimating the regression model function(s). Least-squares or another parameter estimation procedure can be used. It is clear that a supervised pattern recognition problem (discrimination) can be formulated as a special case of regression with the response variable being a categorical variable. The above, however, is only sound for the case of $K = 2$ or for $K > 2$ with ordered classes (e.g. $y_i = \{0, 1, 2, 3\}$ indicating classes "no", "low", "moderate" or "high" contamination). It is well known that, for two classes, linear regression gives linear discriminant analysis [20]. If we are dealing with the general case of K unordered classes then the multi-class problem can be regarded as a multivariate regression problem with K response (or output) variables y_i ($i = 1, 2, \ldots, K$):

$$y_i = \begin{cases} 1, & \text{if } y_i = i; \\ 0, & \text{otherwise.} \end{cases}$$

The implication is that any parametric and non-parametric regression method can be transformed into a classifier which may lead to a huge arsenal of models. In addition, artificial intelligence methods such as neural networks and adaptive classification techniques can be considered as adaptive regression methods. Recent papers that review and further develop statistical methodologies along these lines of thought include [21–23].

In the same context the well known multi-class extension of Fisher's linear discriminant analysis was an early historical example of the equivalence with canonical correlation analysis with the block of y variables defined as class indicator variables [24].

Other approaches of statistical learning relate merely to estimating class probability densities (e.g. Bayesian methods) and to develop independent class models (tolerance region methods).

Learning or training a classifier can be done in the context of two types of goals, which may succeed one another. First, in descriptive discriminant analysis, one is interested in the representation of the classifier itself, i.e. how a given set of features can be represented (usually spatially) in such a way that the K classes are clearly separated. This is done on the basis of the training samples. The spatial representation is called the "discriminant space". The graphical representation has an important "descriptive" value in analysing the data. An example is the canonical variate approach to linear discriminant analysis where the relevant information about the separation between classes is represented in a discriminant space of at most $K - 1$ dimensions. For small K, the data can be plotted in the reduced space, elucidating graphically the class discrimination [25].

Often, only two or three dimensions are needed, even for large K. This aspect of analysing data rather than purely classifying has been more the interest of statisticians. An environmental example of descriptive discriminant analysis is given by Pontasch et al. [12].

Second, the classifier is used to classify and estimate the accuracy of classification of future unknown observations (test samples). The second case is referred to as predictive classification.

In the predictive classification step, the focus is on correct classification. Some measure of correct classification is used to evaluate the performance of the classifier.

Statistical Classifiers

We present an overview of some of the classifiers that are proposed in the chemometrics literature and also some promising new algorithms. However, we have not made any attempt to be complete and we realise that a better classification of the methodologies is possible but may lead us too far. A very recent comprehensive review of the field of discriminant analysis and statistical pattern recognition is given in [26]. Simple introductory tutorials with chemical examples are available in textbooks such as [27–29]. We will distinguish the following areas.

1) Discriminant functions and regression methods.
2) Tolerance region or class modeling methods.
3) Parametric and non-parametric Bayesian classification.

Discriminant Functions and Regression Methods

Discrimination, strictly speaking, is based on discriminant functions which are combinations of the pattern variables with weight coefficients. Most often linear discriminant functions are used. They are usually of the form

$$D_f = \boldsymbol{w}^T \boldsymbol{x} + w_0 \tag{7}$$

where w_0 is a constant and $\boldsymbol{w}$ is the weight vector. A discriminant score DS_i is calculated by substituting the vector $\boldsymbol{x}$ with the pattern vector $\boldsymbol{x}_i$ of test pattern i in the above equation. That is

$$DS_i = \boldsymbol{w}^T \boldsymbol{x}_i + w_0 \, . \tag{8}$$

There are different ways of determining the weight coefficients. One of the best known and earliest methods in non-statistical pattern recognition is the linear learning machine which was extensively studied by Nilsson [30].

Fisher's Linear Discriminant Analysis (FLDA)

The statistical approach of the sample linear discriminant function for separating two classes (FLDA) was proposed by Fisher [31]. The weights are determined as follows:

$$\boldsymbol{w} = (\boldsymbol{x}_{\cdot 1} - \boldsymbol{x}_{\cdot 2})^T \boldsymbol{S}^{-1} \tag{9}$$

and

$$w_0 = -\frac{1}{2}(\boldsymbol{x}_{\cdot 1} - \boldsymbol{x}_{\cdot 2})^T \boldsymbol{S}^{-1}(\boldsymbol{x}_{\cdot 1} - \boldsymbol{x}_{\cdot 2}) \tag{10}$$

where $\boldsymbol{S}^{-1}$ is the inverse of $\boldsymbol{S}$, the pooled sample covariance matrix of the two training classes, and $\boldsymbol{x}_{\cdot 1}$ and $\boldsymbol{x}_{2}$ the sample mean vectors indicating the centroids of the two classes ω_1 and ω_2, respectively, in the pattern space. The vector $\boldsymbol{w}$ determines the direction of the separation plane between the two classes and w_0 indicates that the plane goes through the point located at half the distance between the two centroids $\boldsymbol{x}_{\cdot 1}$ and $\boldsymbol{x}_{2}$ of the classes. FLDA can be implemented as a least-squares multiple linear regression problem. Such an implementation was done in the chemometrics software ARTHUR [32]. A method related to the multiple regression approach of FLDA was the piece-wise least-squares multiple regression method also implemented in ARTHUR. It is in fact a combination of the k–nearest-neighbour classifier and FLDA.

Descriptive Linear Discriminant Analysis (DLDA)

The constraint of two classes used in FLDA can be relaxed to K classes. This is referred to as "descriptive linear discriminant analysis", also called canonical discriminant analysis. Here, the discriminant functions describe in a statistical optimal way the directions of class differentiation in the pattern space and are derived by maximizing the ratio of the between-class variation to within-classes variation. The weight vectors are the eigenvectors of the matrix

$$\boldsymbol{A} = \boldsymbol{W}^{-1}\boldsymbol{B} \tag{11}$$

where $\boldsymbol{W}$ is the "total within-group sum of squares and cross-products" matrix and $\boldsymbol{B}$ is the "between-groups sum of squares and cross-products" matrix. Descriptive LDA (DLDA) is often used for preprocessing in the context of feature extraction. Here, the original variables are linearly combined to generate a new set of variables spanning the discriminant space (discriminant functions, discriminant scores) – also known under the term canonical variables-which can further be used as input for another classification technique, usually a Bayesian classifier. The maximum number of significant discriminant functions is equal to the number of classes minus one ($K - 1$) or the number of original variables, Q, if $Q < K$. With a small number of discriminant axes the presentation of the separation between classes can be visualised graphically.

Water Quality Monitoring. As an illustration, consider the following example of the monitoring of dissolved nutrients on the Great Barrier Reef. In recent years there has been a growing concern among reef researchers that coral reefs worldwide appear to be degrading at an ever increasing rate. Among many different factors, pollution is one of the causes. In 1991 the Australian Institute of Marine Science (AIMS), in collaboration with the Great Barrier Reef Marine Park Authority, Australia, formally established a long term monitoring program of the Great Barrier Reef. Surveys of dissolved and particulate nutrients is one of their programmes. The objectives and methods for the above programme is fully described in the AIMS document [33].

The data is a subset of the water quality monitoring for which the following variables are measured: ammonia, nitrite, nitrate, dissolved inorganic nitrogen, dissolved organic nitrogen, dissolved inorganic phosphorus, dissolved organic phosphorus, and liquid extracted silicates. They are all logarithmic-transformed. The 116 observational units (data vectors) represent different monitoring stations. There are, however, different ways to classify the units. We will use a cross-reef shelf classification, i.e. inner, mid, and outer reef locations. The data analysis presented here is purely for illustrative purposes and is not intended to be a final analysis as other effects such as temporal and other differences in the data need to be taken into account. The two-dimensional DLDA plot shown in Fig. 2, obtained using the SPSS for MS-Windows software package [34], shows a substantial overlap between the inner, middle and outer reef locations.

Some separation is revealed between the outer and inner reef positions and the centroids on the plot indicate that the separation is in the direction of the first discriminant function. This function is the only one that is significant on the basis of a Wilk's Lambda test and accounts for 88% of the discrimination between the groups, whereas the second function consequently accounts for only 12%. The middle reef positions tend to overlap more with the outer reef ones (the centroid of the middle group is closer to the outer reef centroid) but there are many observations that also relate to the inner reef group. It is interesting to proceed further with only two training classes (inner and outer reef classes) and use the middle one as the test class. It means that we will investigate which (and how many) observational units of the middle reef samples have a nutrient profile relating to the inner reef or to the outer reef. This is the classification stage of linear discriminant analysis and will be dealt with in the Bayesian context later. In descriptive LDA the linear discriminant weight coefficients play an important

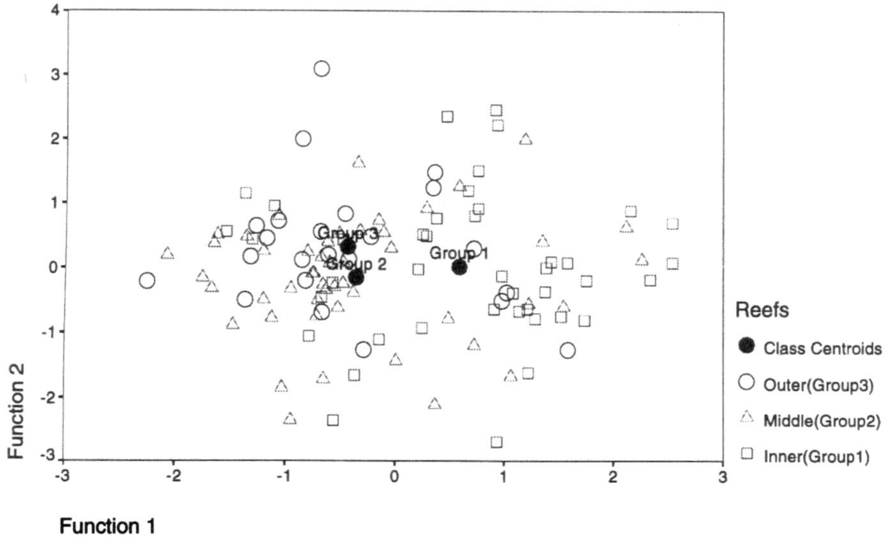

Fig. 2. DLDA plot of the nutrient data for the Great Barrier Reef

Table 1. Standardized discriminant function coefficients

Function 1	w_r
Dissolved inorganic nitrogen	−1.11772
Dissolved organic nitrogen	0.55272
Nitrite	−0.26815
Nitrate	0.07391
Dissolved inorganic phosphorus	−0.26627
Liquid extracted silicates	−0.43168
Ammonia	2.04435
Dissolved organic phosphorus	0.21750

role in interpreting the importance of the variables in the discrimination. We have repeated DLDA on the two training classes inner reef (Group 1) and outer reef positions (Group 3) and the discriminant coefficients for the single discriminant function are given in Table 1.

As the centroid of the inner reef class has a higher discriminant score (.37) on the discriminant axis than the outer reef class (−.71), it means that, from the weight coefficients table, it can be concluded that the discrimination can be primarily characterised by a difference in dissolved inorganic nitrogen and ammonia (high absolute values for the coefficients) and the inner reefs tend to have lower dissolved inorganic nitrogen (negative sign) and higher levels of ammonia (positive sign).

Regression Methods

It is well documented that discriminant functions can be approached from a regression point of view. Apart from this, another traditional example is "logistic regression" [35]. More recently, a theoretical framework for the transformation of modern (extending to nonlinear and non-parametric) regression methods into multi-class classifiers have been developed by statisticians [21–23, 36]. Because of its novelty, applications in chemistry are scarce, although there is great scope for it. Especially, techniques such as penalized discriminant analysis (PDA) [23] which is designed to be used in the context of situations with a large number (one hundred or more) highly correlated predictor variables, which is typical for digitised spectral data such as in the soil example shown in Fig. 1. The scope of penalized discriminant analysis is similar to that of regularised discriminant analysis (see later). On the other hand, linear discriminant analysis can be too rigid in situations where the class boundaries in pattern space are complex and nonlinear so that a linear discriminant mapping based on linear discriminant functions (with Descriptive LDA) does not use sufficient discriminant information in the data. Non-parametric regression techniques which can deal with noisy non-linear (environmental) data include for instance projection pursuit regression [18, 21], ACE [36, 37], MARS [19], flexible discriminant analysis [22] and generalised additive models [37]. The popular biased multivariate regression method in chemometrics, the two-block partial least-squares model and its non-linear extension [38], can also be used as a classifier. Also, it can be used to tackle the ill-posed problem created by many correlated variables in a small data sets.

Class Modeling Methods

So far, the pattern recognition techniques considered allow classification of observations only in predetermined classes, without the possibility of detecting a class not formerly included in the discrimination model (training set). This means that an aberrant pattern as such will not be detected by the discrimination model, but classified into one of the previously defined classes. It is sometimes important to have methods available that not only allow classification, but also the detection of aberrant patterns (or "outliers"). Moreover, there exist asymmetric classification problems where only one or a few classes are well-defined clusters in the pattern space and the rest of the data vectors are scattered without any structure. An example is the definition of a class of good water quality on the basis of a profile of analytical tests. The other class of bad water quality is much less defined and is a collection of all kind of aberrant test results. From a monitoring point of view it might be sufficient to model the good water quality class and classify new samples with respect to that class.

Tolerance region methods are founded on the idea that the territory of a training class can be represented by a hyper-volume in the original pattern space, confined by a multivariate confidence envelope. The multivariate tolerance region of a class ω_k is determined as the region in the pattern space enclosed by the class envelope determined on a statistical basis (F or χ^2 statistic) by a critical distance $d_{k(\alpha)}$ that is the largest distance allowed from an object belonging to that class at α level of significance. Object i belongs to class ω_k if

$$d_{ik} \leq d_{k(\alpha)} \tag{12}$$

otherwise it does not. A significance level of $\alpha = 0.05$ defines a 95% tolerance region. As in the case of the class modeling techniques, a distance measure d_{ik} with respect to each training class can be calculated, and they can also be employed at the level of discriminant analysis by assigning the test object i to the class closest to the object. The class modeling techniques can be distinguished by the way they define the distance measure d_{ik} and $d_{k(\alpha)}$. We now look at two of the techniques commonly used in chemometrics: UNEQ and SIMCA.

UNEQ

UNEQ is the Mahalonobis distance-based class modeling using separate (unequal) class covariance matrices. The squared metric d_{ik}^2 is the Mahalonobis distance to class ω_k and is given by

$$d_{ik}^2 = (\boldsymbol{x}_i - \hat{\boldsymbol{\mu}}_k)^T \hat{\boldsymbol{\Sigma}}_k^{-1} (\boldsymbol{x}_i - \hat{\boldsymbol{\mu}}_k) \tag{13}$$

where $\hat{\boldsymbol{\mu}}_k$ is the estimate of the population mean vector (i.e. the sample mean vector $\boldsymbol{x}_{.k}$, as given previously), and $\hat{\boldsymbol{\Sigma}}_k$ is the unbiased estimate of the covariance matrix $\boldsymbol{\Sigma}_k$ of class ω_k (i.e. the sample covariance matrix $\boldsymbol{S}$, also given previously). From a theoretical point of view a better (less biased) estimate of the Mahalonobis distance is given by the following equation [39, 40]

$$d_{ik}^2 = \frac{(N_k - Q - 2)}{(N_k - 1)} (\boldsymbol{x}_i - \boldsymbol{x}_{.k})^T \boldsymbol{S}_k^{-1} (\boldsymbol{x}_i - \boldsymbol{x}_{.k}) \tag{14}$$

where N_k is the number of samples in class ω_k. The Mahalonobis distance follows a χ^2 distribution with Q degrees of freedom assuming that the class data are multivariate normally distributed.

The UNEQ method is only reliable when the data have a multivariate normal distribution and there are sufficient observations to get stable estimates for the parameters of the covariance matrix and mean vector. Therefore, UNEQ is not useful in situations of low observation/variable ratios (i.e. in poorly or ill-posed problems) or when the variables are highly correlated. In those situations, the observations of a class lie in a subspace of the Q-dimensional pattern space and any direction orthogonal to that subspace has zero or near zero variance, making the covariance matrix singular or near-singular. In that case, the Mahalonobis distance is not well defined. A better method for high-dimensional settings and ill-posed problems is SIMCA.

Soft Independent Modeling of Class Analogy (SIMCA)

Instead of representing each class by a mean and covariance matrix, a disjoint principal components factor model is developed for each class. A class factor model translates the information relevant to the characterisation of a training class (also known as systematic variation or correlated information) in mathematical relationships and separates it from the non-class specific information (known as non-systematic or random variation or simply noise) which is filtered away. The factor model for each class ω_k is fitted to the data z_{ik} of the training set according to

$$z_{jk} = \boldsymbol{B}_k \boldsymbol{t}_{jk} + \boldsymbol{e}_{jk} \tag{15}$$

where z_j is the z-transformed data vector and $\boldsymbol{B}_k$ is the matrix of $Q \times A_k$ weight coefficients constituting the principal components model of class ω_k. The latter matrix collects A_k eigenvectors determined from the correlation matrix of class ω_k. The number of significant terms A_k determines the complexity (dimensionality) of the model and is traditionally found by cross-validation [41]. If structured information is present in the data of class ω_k one can expect that $0 < A_k \ll Q$. The $\boldsymbol{t}_{jk}$ is the orthogonal projection of the z_{jk} data vector on the principal components subspace. The residual vector $\boldsymbol{e}_{ik}$ quantifies the random part in z_{jk} and determines the deviation of the data vector with respect to the factor model. Note here that we use z-transformed data vectors as it is shown that better discriminating class models are obtained after class-specific z-transformation [42]. Based on the residual vector $\boldsymbol{e}_{ik}$, a distance with respect to the class factor model can be calculated. The squared distance of a test vector z_i to the factor model can be written in the following condensed way:

$$d_{ik}^2 = \frac{1}{(Q - A_k)}(z_i - \boldsymbol{B}_k \boldsymbol{B}_k^T z_i)^T (z_i - \boldsymbol{B}_k \boldsymbol{B}_k^T z_i)\,. \tag{16}$$

In addition, using Eq. (16) an average critical distance $d_{k(\alpha)}$ for the N_k training class data can be computed:

$$d_{k(\alpha)} = F_{crit}^{\frac{1}{2}} (N_k - A_k - 1)^{-\frac{1}{2}} \left(\sum_{j=1}^{N_k} d_{jk}^2 \right)^{\frac{1}{2}} \tag{17}$$

where F_{crit} is the critical F distribution value at a given significance level α (typical values of α are 0.05 and 0.01) with $(Q - A_k)$ and $(N_k - A_k - 1)$ degrees of freedom. Figure 3 illustrates the SIMCA method.

The training classes are described by two separate one-dimensional $A_k = 1$ principle components models. The tolerance region (here shown as a cylinder) around each principal components model is constructed on the basis of the class critical distance ($d_{k(\alpha)}$). Test patterns are assigned to a class if they fall inside the class cylinder.

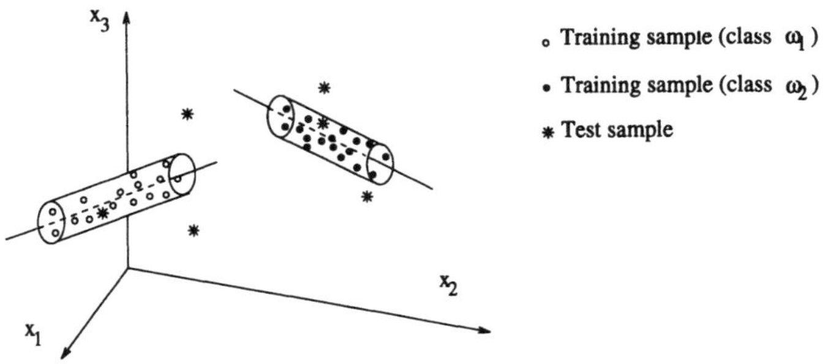

Fig. 3. Illustration of the SIMCA method (see text for further explanation)

SIMCA is a well established reference method in the chemometrics literature [43]. A further improvement to SIMCA is the DASCO method developed by Frank [44]. Examples of class modeling in environmental chemistry are, for example, in the use of SIMCA for the classification of mass spectra of toxic compounds [10] and waste water pollution modeling [14].

Bayesian Classification

In classification, input patterns may not always be classified correctly and consequently will give rise to errors. It is therefore an important characteristic of a classifier to be able to estimate the accuracy of classification of future unknown input samples. That is, we are interested in minimising the overall error rate for a given classifier. The problem can be formulated using the Bayesian approach.

Mathematically, let $\boldsymbol{x}_i$ be the input pattern i to be classified and let $\omega_k (k = 1, 2, \ldots, K)$ be the various possible classes of which $\boldsymbol{x}_i$ might be a member. The Bayes' decision rule for minimum classification error is given by

$$\begin{array}{lll} \text{if} & P(\omega_k|\boldsymbol{x}_i) > P(\omega_l|\boldsymbol{x}_i) & \text{then } \boldsymbol{x}_i \in \omega_k \quad \forall k \neq l \\ \text{else if} & P(\omega_l|\boldsymbol{x}_i) > P(\omega_k|\boldsymbol{x}_i) & \text{then } \boldsymbol{x}_i \in \omega_l \quad \forall k \neq l \end{array}$$

where $P(\omega_k|\boldsymbol{x}_i)$ is the conditional probability of a given feature vector $\boldsymbol{x}_i$ for a specific class ω_k or, alternatively, the posterior probability of ω_k given $\boldsymbol{x}_i$. Therefore, by assigning $\boldsymbol{x}_i$ to the class with maximum posterior probability, the error is minimised. In general, for any two neighbouring classes in feature space, classes ω_1 and ω_2 are separated by the following decision boundary;

$$P(\omega_1|\boldsymbol{x}_i) = P(\omega_2|\boldsymbol{x}_i) . \tag{18}$$

To calculate or estimate $P(\omega_k|\boldsymbol{x}_i)$ is, however, difficult because we require very large sample sizes to cover all possible values of $\boldsymbol{x}_i$. We can calculate

$P(\omega_k|\boldsymbol{x}_i)$ from the prior probability of class $\omega_k, P(\omega_k)$, and the conditional probability density of class ω_k, $p(\boldsymbol{x}_i|\omega_k)$, using Bayes theorem:

$$P(\omega_k|\boldsymbol{x}_i) = P(\boldsymbol{x}_i|\omega_k)P(\omega_k)/P(\boldsymbol{x}_i) \tag{19}$$

where $P\ (\boldsymbol{x}_i)$ is given by

$$P(\boldsymbol{x}_i) = \sum_{k=1}^{K} P(\boldsymbol{x}_i|\omega_k)P(\omega_k) \tag{20}$$

which is the probability density of $\boldsymbol{x}_i$. The $P(\boldsymbol{x}_i)$ is class-independent and thus can be omitted from the decision rule. We can therefore rewrite Bayes decision rule as

$$\begin{array}{lll} \text{if} & P(\boldsymbol{x}_i|\omega_k)P(\omega_k) > P(\boldsymbol{x}_i|\omega_l)P(\omega_l) & \text{then } \boldsymbol{x}_i \in \omega_k \quad \forall k \neq l \\ \text{else if} & P(\boldsymbol{x}_i|\omega_l)P(\omega_l) > P(\boldsymbol{x}_i|\omega_k)P(\omega_k) & \text{then } \boldsymbol{x}_i \in \omega_l \quad \forall k \neq l. \end{array}$$

We now need to compute distributions, $P(\boldsymbol{x}_i|\omega_k)$ and $P(\omega_k)$, to determine the minimum error rate decision rule. It is very difficult to obtain the optimal Bayes decision rule for minimum error classification because, in practice, we need large sample sizes to estimate the distributions. If we use random sampling we can simply estimate $P(\omega_k)$ to be the frequency of occurrence of each class in the sample. The estimation of the class conditional probability $P(\boldsymbol{x}_i|\omega_k)$ is more difficult, and is the subject of many Bayesian classification methods. Each method makes some assumptions with respect to a given class ω_k and then estimates the conditional probability such as by assuming a parametric form for $P(\boldsymbol{x}_i|\omega_k)$ or by estimating it locally around each vector $\boldsymbol{x}_i$. Methods that assume a parametric form for probability estimation are known as *parametric methods* while those that estimate the conditional probability locally without any assumptions about the underlying distribution are referred to as *non-parametric methods.* Both parametric and non-parametric methods make assumptions about the underlying model for estimating the distributions. If the sampled data set fits the model well, then the method will work well.

In some applications such environmental decision-making, we cannot assume that a decision has the same consequences when it is applied to all classes equally. For example, misclassifying a contaminated sample as uncontaminated has different consequences to misclassifying an uncontaminated sample as contaminated. Therefore, we have to introduce a "risk" or "cost" factor to each class membership as a linear combination of the posterior probabilities together with a *loss factor*, L_{kl}, where we define L_{kl} to be loss incurred if an object is classified as belonging to class ω_k when in reality it belongs to class ω_l. The risk taken in that x_i belongs to ω_k is given by

$$\begin{aligned} R(\omega_k|\boldsymbol{x}_i) &= \sum_{l} L_{kl}P(\omega_l|\boldsymbol{x}_i) \\ &= L_{kk}P(\omega_k|\boldsymbol{x}_i) + \sum_{k \neq l} L_{kl}P(\omega_l|\boldsymbol{x}_i) \end{aligned}$$

where L_{kk} are the losses connected with the correct classification (normally we assume no loss if we guess correctly, $L_{kk} = 0$). For a two-class problem ($k = 1, 2$) we get

$$R(\omega_1|\boldsymbol{x}_i) = L_{11}P(\omega_1|\boldsymbol{x}_i) + L_{12}P(\omega_2|\boldsymbol{x}_i) \tag{21}$$

and

$$R(\omega_2|\boldsymbol{x}_i) = L_{21}P(\omega_1|\boldsymbol{x}_i) + L_{22}P(\omega_2|\boldsymbol{x}_i)\,. \tag{22}$$

The decision rule now becomes

$$\text{if} \quad R(\omega_1|\boldsymbol{x}_i) < R(\omega_2|\boldsymbol{x}_i) \quad \text{then} \quad \boldsymbol{x}_i \in \omega_1$$

or

$$\text{if} \quad (L_{21} - L_{11})P(\omega_1|\boldsymbol{x}_i) > (L_{12} - L_{22})P(\omega_2|\boldsymbol{x}_i) \quad \text{then} \quad \boldsymbol{x}_i \in \omega_1$$

or, using Bayes Theorem, we get the *minimum risk decision rule*:

$$\text{if} \quad (L_{21} - L_{11})p(\boldsymbol{x}_i|\omega_1)P(\omega_1) > (L_{12} - L_{22})p(\boldsymbol{x}_i|\omega_2)P(\omega_2) \quad \text{then} \quad \boldsymbol{x}_i \in \omega_1\,.$$

A particular case occurs when $L_{11} = L_{22} = 0$ and $L_{12} = L_{21}$ (i.e. same loss either way we guess incorrectly) which is the same as the minimum error decision rule. Therefore, the choice of boundary that separates two classes is dependent on the purpose of the study and on how we define the loss factor L_{kl}.

Parametric Bayesian Classifiers

There exist various kinds of Bayesian classifiers. In particular, we consider Linear Discriminant Analysis, Quadratic Discriminant Analysis, and Regularised Discriminant Analysis.

Linear Discriminant Analysis (LDA)

Define a *classification (function) score* for class ω_k as

$$g_k(\boldsymbol{x}_i) = P(\omega_k|\boldsymbol{x}_i) \tag{23}$$

then, based on Bayes minimum error decision rule, classification is based on finding the largest classification (function) score for a given feature vector $\boldsymbol{x}_i$. We rewrite the above equation by taking the logarithm (any monotonically increasing function of $g_k(\boldsymbol{x}_i)$ is also a valid classification function):

$$G_k(\boldsymbol{x}_i) = \ln[P(\omega_k|\boldsymbol{x}_i)]\,. \tag{24}$$

Let us assume the underlying model for the conditional probability density, $P(\boldsymbol{x}_i|\omega_k)$, is normal (Gaussian):

$$P(\boldsymbol{x}_i|\omega_k) = (2\pi)^{-Q/2}|\boldsymbol{\Sigma}_k|^{-1/2}\exp\left[-\frac{1}{2}(\boldsymbol{x}_i - \mu_k)^T \boldsymbol{\Sigma}_k^{-1}(\boldsymbol{x}_i - \mu_k)\right] \quad (25)$$

where Q is the dimensionality, $\boldsymbol{\mu}_k$ and $\boldsymbol{\Sigma}_k$ are the mean vector and covariance matrix of class ω_k, respectively. In the context of a classification situation, the parameters are not generally known and have to be estimated from the N_k training samples for each class ω_k. Using the maximum likelihood estimate, we get [45]

$$\hat{\boldsymbol{\mu}}_k = \frac{1}{N_k}\sum_{i=1}^{N_k} \boldsymbol{x}_i$$

$$\hat{\boldsymbol{\Sigma}}_k = \frac{1}{N_k}\sum_{i=1}^{N_k} (\boldsymbol{x}_i - \hat{\boldsymbol{\mu}}_k)(\boldsymbol{x}_i - \hat{\boldsymbol{\mu}}_k)^T .$$

The classification function becomes (neglecting constants)

$$G_k((\boldsymbol{x}_i) = (\boldsymbol{x}_i - \hat{\boldsymbol{\mu}}_k)^T\hat{\boldsymbol{\Sigma}}_k^{-1}(\boldsymbol{x}_i - \hat{\boldsymbol{\mu}}_k) + \ln|\hat{\boldsymbol{\Sigma}}_k| - 2\ln(P(\omega_k)) . \quad (26)$$

Consider the particular case where all classes have identical class covariance matrices (alternatively, we could pool all the data from all classes and compute a pooled covariance matrix):

$$\boldsymbol{\Sigma}_1 = \boldsymbol{\Sigma}_2 = \ldots = \boldsymbol{\Sigma}_K = \boldsymbol{\Sigma} \quad (27)$$

then using Bayes minimum error decision rule, for all patterns on the boundary between a pair of classes ω_1 and ω_2 (where $G_1(\boldsymbol{x}_i) - G_2(\boldsymbol{x}_i) = 0$), we obtain

$$2\boldsymbol{x}_i^T\hat{\boldsymbol{\Sigma}}^{-1}(\hat{\boldsymbol{\mu}}_1 - \hat{\boldsymbol{\mu}}_2) - \hat{\boldsymbol{\mu}}_1^T\hat{\boldsymbol{\Sigma}}^{-1}\hat{\boldsymbol{\mu}}_1 + \hat{\boldsymbol{\mu}}_2^T\hat{\boldsymbol{\Sigma}}^{-1}\hat{\boldsymbol{\mu}}_2 - 2\ln\left(\frac{P(\omega_1)}{P(\omega_2)}\right) = 0 \quad (28)$$

which is a *linear* discriminant function of $\boldsymbol{x}_i$. That is, the resulting boundaries between the classes are linear $(Q-1)$-dimensional hyper-planes. Figure 4 illustrates the linear discrimination technique for the case of a three-dimensional pattern space with a training set comprising two training classes. The classes are separated by a two-dimensional hyper-plane (decision boundary). Test patterns (shown as asterisks) are to be classified. The test patterns are always classified in one of the two sub-spaces formed by the hyper-plane.

Since LDA only needs to evaluate a few parameters, it is less likely to over-fit the training data (i.e. over-fitting, or over-specification, occurs when the number of parameters to estimate is less than the training set size) and, consequently, it has often been the preferred method in high-dimensional settings. Over-fitting can give rise to unstable parameter estimates (and thus high variance). The decrease

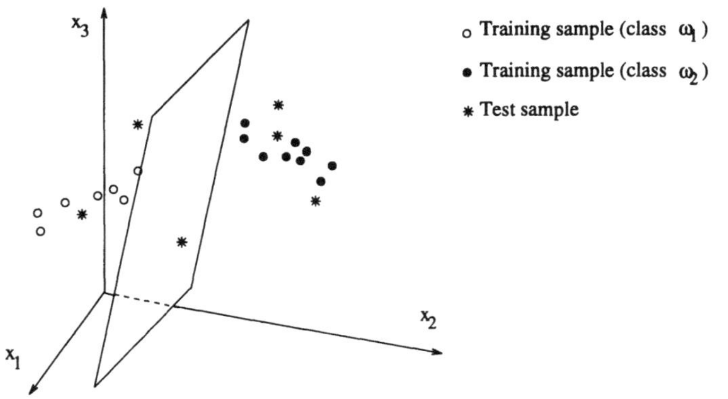

Fig. 4. Illustration of the linear discrimination technique

in variance accomplished by using the pooled covariance matrix is one of the reasons for the success of LDA. However, LDA has severe limitations in cases where the class covariance matrices differ significantly, or when the population means coincide, or when the total sample size is less than or comparable to the dimensionality [46].

Quadratic Discriminant Analysis (QDA)

In the case when all classes have different class covariance matrices, that is

$$\boldsymbol{\Sigma}_1 \neq \boldsymbol{\Sigma}_2 \neq \ldots \neq \boldsymbol{\Sigma}_K \tag{29}$$

the quadratic classification functions of Eq. (26) are used. In that case, the boundary between a pair of classes is no longer a hyper-plane, but a quadratic hypersurface. If only two features are involved, the boundary surface will be an ellipse, parabola or hyperbola. Figure 5b illustrates the boundary defined by a quadratic classification function and the corresponding univariate probability density functions (shown for variable x_1 only) with different class covariance matrices for the two classes (the example shown is for a bivariate case). A comparison is made with LDA (see Fig. 5a).

One disadvantage of QDA is that it can perform appreciably worse than LDA for small sample sizes. Since QDA requires many more parameters (i.e. reliable estimates of the class covariance matrices, $\boldsymbol{\Sigma}_k$) to be estimated than in the case of LDA, a very large number of training samples are required [47]. This is particularly so for situations involving a large number of features (i.e. high dimensions) [46]. Further, $\boldsymbol{\Sigma}_k$ cannot be inverted in ill-posed cases (i.e. the number of training samples N_k in each class ω_k is less than the dimensionality). QDA also performs poorly under conditions of non-normality [48].

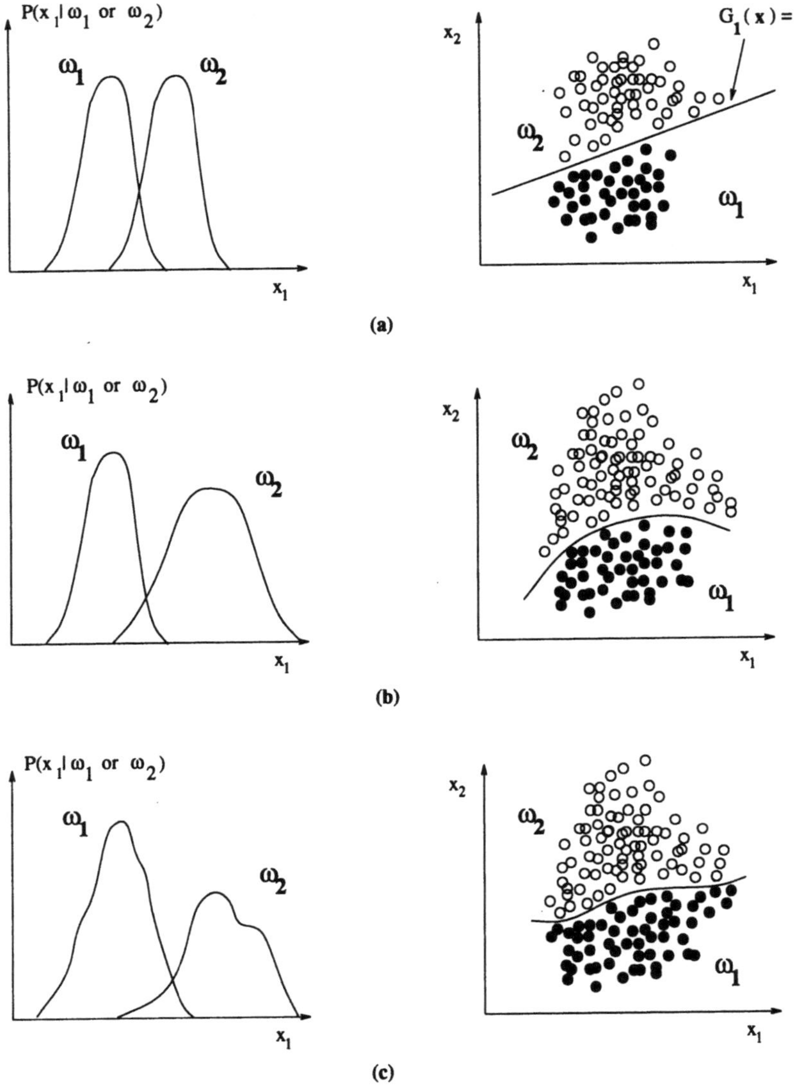

Fig. 5. Schematic representation of the univariate probability density functions (shown for x_1 only) for two classes ω_1 and ω_2 and the decision boundary in a bi-variate feature space (x_1 versus x_2) for the case of a) a linear classification function, b) a quadratic classification function, and c) the kernel density method

As an example, consider the reef classification problem discussed earlier. Based on two classification functions (one for inner reef and one for outer reef) and using the middle reef as the test class, we obtain (see Table 2) the following correct classification rates (one minus the mis-classification rate). Both LDA and QDA gave the same results. The reason is that the sample covariance matrices were not significantly different as tested by the Box's M test [34].

Table 2. Classification results for water quality monitoring

Actual Class	Number of Cases	Predicted Group Membership	
		Inner Reef	Outer Reef
Inner reef	44	32 (72.7%)	12 (27.3%)
Outer reef	23	5 (21.7%)	18 (78.3%)
Test cases (middle reef)	49	14 (28.6%)	35 (71.4%)

Percent of "training" cases correctly classified = 74.63%

Regularised Discriminant Analysis (RDA)

Regularised discriminant analysis [49] is closely related to QDA in that RDA also classifies a test object $\boldsymbol{x}_i$ into the class for which the classification score given by

$$G_k(\boldsymbol{x}_i) = (\boldsymbol{x}_i - \hat{\boldsymbol{\mu}}_k)^T \hat{\boldsymbol{\Sigma}}_k^{-1} (\boldsymbol{x}_i - \hat{\boldsymbol{\mu}}_k) + \ln \left|\hat{\boldsymbol{\Sigma}}_k\right| - 2\ln (P(\omega_k)) \qquad (30)$$

is minimised. The difference is that RDA makes use of a regularised covariance matrix $\hat{\boldsymbol{\Sigma}}_k(\lambda, \gamma)$ instead. $\hat{\boldsymbol{\Sigma}}_k(\lambda, \gamma)$ is the result of two successive regularisations involving the two new parameters λ and γ. The first regularisation reduces the number of parameters to be estimated by replacing $\hat{\boldsymbol{\Sigma}}_k$ with a linear combination $\hat{\boldsymbol{\Sigma}}_k(\lambda)$ of the class covariance matrix and the pooled sample covariance matrix:

$$\hat{\boldsymbol{\Sigma}}_k(\lambda) = \frac{(1-\lambda)\hat{\boldsymbol{Q}}_k + \lambda \hat{\boldsymbol{Q}}_{pooled}}{(1-\lambda)N_k + \lambda N} \qquad (31)$$

where

$$\hat{\boldsymbol{Q}}_{pooled} = \sum_{k=1}^{K} \hat{\boldsymbol{Q}}_k = \sum_{k=1}^{K} \sum_{i=1}^{N_k} (\boldsymbol{x}_i - \hat{\boldsymbol{\mu}}_k)(\boldsymbol{x}_i - \hat{\boldsymbol{\mu}}_k)^T . \qquad (32)$$

N is the total number of training objects and N_k is the number of training objects in class ω_k. The degree of regularisation is controlled by the parameter $\lambda\,(0 \leq \lambda \leq 1)$. For $\lambda = 0$, regularisation results in QDA and for $\lambda = 1$ it is LDA. Increased regularisation (increasing λ) results in decreased variance, which can lead to improved classification performance in small training sample size settings.

The second regularisation addresses the biasing that is inherent in sample-based estimation of eigenvalues of $\hat{\boldsymbol{\Sigma}}_k$. It is well known that, for limited training samples, the smallest eigenvalues are estimated too small and the largest eigenvalues are estimated too large. This biasing will result in a classification score $G_k(\boldsymbol{x}_i)$ that is seen to be weighted by the smallest eigenvalues. To reduce this bias effect, each class covariance matrix is "shrunk" towards the identity matrix multiplied by its average eigenvalue (trace $(\hat{\Sigma}_k/Q)$). That is,

$$\hat{\Sigma}_k(\lambda,\gamma) = (1-\gamma)\hat{\Sigma}_k(\lambda) + \frac{\gamma}{Q}\mathrm{trace}\left[\hat{\Sigma}_k(\lambda)\right]\boldsymbol{I}\,. \tag{33}$$

Q is the dimensionality (number of variables) and $\gamma\,(0 \le \gamma \le 1)$ is the regularisation parameter. Applying both regularisations and using $\hat{\Sigma}_k(\lambda,\gamma)$ instead of $\hat{\Sigma}_k$, the classification score to be minimised becomes

$$G_k(\boldsymbol{x}_i) = (\boldsymbol{x}_i - \hat{\boldsymbol{\mu}}_k)^T\hat{\Sigma}_k^{-1}(\lambda,\gamma)(\boldsymbol{x}_i - \hat{\boldsymbol{\mu}}_k) + \ln\left|\hat{\Sigma}_k(\lambda,\gamma)\right| - 2\ln\left(P(\omega_k)\right)\,. \tag{34}$$

To obtain values for the regularisation parameters λ and γ, an estimation procedure based on the training samples has to be undertaken. This is achieved by evaluating the resulting classifier for a number of λ and γ pairs and choosing the values that give the best classification performance [49, 50].

RDA is often superior to QDA and LDA in a high-dimensional setting with a limited number of training samples. LDA is only able to outperform RDA in the case of identical class covariance matrices and with many training samples [46]. One might also consider the option of initially performing dimensionality reduction by using feature extraction methods (such as Fisher's discriminant plane (i.e. Descriptive LDA with only two discriminant functions selected to give a two-dimensional plot), Fisher radius transform [51] etc.) and then applying LDA or QDA. However, [46] shows that, for a wide variety of problems, reducing the dimensionality of a particular problem leads to inferior classification results compared to those achieved by RDA in the full feature space.

Non-Parametric Classification

Non-parametric classification in the Bayesian context has concentrated on the estimation of class-conditional probability densities $P(\boldsymbol{x}_i|\omega_k)$ $(k = 1,2,\ldots,K)$ for the test vector $\boldsymbol{x}_i$ using the training objects surrounding i in the pattern space. This has given rise to methods such as k-nearest neighbours and kernel (also called potential) discrimination. Theory and examples in chemistry can be found in the monograph by Coomans and Broeckaert [52].

The probability density estimation performed by the kernel function methods is a direct procedure so that the complete density function $P(\boldsymbol{x}_i|\omega_k)$ is not estimated (in contrast to the parametric methods), but only the probability density in the position $\boldsymbol{x}_i$ of each test pattern vector to be classified. The probability density in $\boldsymbol{x}_i$ for a given training class ω_k containing N_k samples is obtained by creating individual kernels around each training sample point $\boldsymbol{x}_{jk}$ and averaging the N_k different kernel influences in the position $\boldsymbol{x}_i$. The influence of the kernel of $\boldsymbol{x}_{jk}$ on $\boldsymbol{x}_i$, i.e. the contribution of $\boldsymbol{x}_{jk}$ to the probability density $P(\boldsymbol{x}_i|\omega_k)$, is given by

$\phi(\boldsymbol{x}_i, \boldsymbol{x}_{jk}; \boldsymbol{u})$, the kernel function. It means that the class ω_k probability density is given by

$$P(\boldsymbol{x}_i|\omega_k) = \frac{1}{N_k \prod_{r=1}^{Q} u_r} \sum_{j=1}^{N_k} \phi(\boldsymbol{x}_i, \boldsymbol{x}_{jk}; \boldsymbol{u}) . \tag{35}$$

$\boldsymbol{u}$ is the vector of coefficients determining the smoothness of the kernels and consequently the smoothness of the probability density. The kernel function is a symmetric function with integral one, and usually non-negative, so it is a probability density function itself. Different types of kernels have been proposed among which the Gaussian is the most used. Figure 6 shows the resulting probability density $P(\boldsymbol{x}_i|\omega_k)$ for the case of a Gaussian kernel function.

All of the kernel functions depend on a smoothing coefficient $\boldsymbol{u}$ which determines implicitly to what distance a particular training object exerts its influence in the pattern space. To obtain good classification results the choice (optimisation) of the smoothing values seems to be more critical, followed by the actual functional form of the kernel. Kernel discrimination can lead to flexible decision boundaries between classes, more flexible than is the case with the Quadratic classifier. However, their performance seems to deteriorate quickly in high-dimensional settings (i.e. with many variables). Figure 5c illustrates the decision boundary defined by a quadratic classification function and the corresponding univariate probability density functions with different class covariance matrices for two classes.

A simple but powerful technique is that of the k-nearest neighbours classifier. In the 1-nearest neighbour method, a test vector (in a scaled pattern space), is classified in the class of the nearest training object using one or other distance metric. An extension to this is the k-nearest neighbour method (kNN), in which the test vector is assigned to a class according to the so-called majority vote, i.e. to the class which is most represented in the set of k nearest training points in the pattern space. The kNN method is considered as a reference method and used extensively in chemical classifications. Although the method performs well in many situations, it gives no data analytical information and little about the degree of uncertainty in the classification of a test vector. However, the kNN can be modified to represent a non-parametric density estimator for Bayesian classification [53].

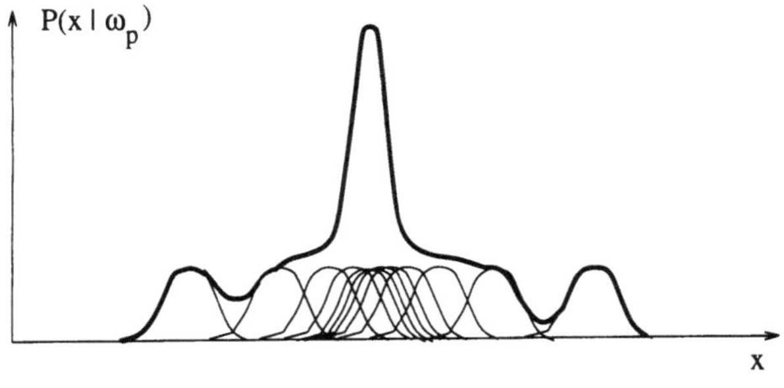

Fig. 6. An example probability density function generated from a sum of Gaussian kernel functions

Recently, attempts have been made to accommodate the modern non-parametric regression methods (mentioned before) including adaptive methods such as neural networks (see below) in an approximate Bayesian classification rule [21,22]. Another class of non-parametric methods that can be related to Bayesian heuristics are the classification trees machine learning methods (see below).

Machine Learning-Based Classification Methods

Much of the research work in traditional pattern recognition undertaken in the 1960s and 1970s concentrated on the mathematical or computer science aspects of pattern information processing. There was an emphasis on Bayesian classifiers together with a focus on the asymptotic properties (infinite number of training samples) of classifiers and on the problem of finding bounds on error rates for determining the adequacy of a given classifier.

More recently, practical classifier systems based on different paradigms have been developed in other scientific disciplines. Most of these paradigms fall into the class of non-parametric methods as they make no assumptions on the functional forms of the underlying probability density distribution. They are generally based on the idea of specifying a decision boundary, such as a hyper-plane with unknown coefficients, and then learning the coefficients of the boundary from the available training set as opposed to specifying or estimating the class-conditional density function.

Artificial Neural Networks

Artificial neural networks (ANN) are dense aggregates of interconnected simple computational elements. These networks are inspired by our current understanding of biological neural nets and are employed to undertake complex cognitive and computational tasks. Like biological neural nets, ANN perform distributed processing, are adaptive (can be trained to produce more desirable outputs), and are fault-tolerant (damage to a few computational elements does not significantly impair overall performance).

One of the most seminal developments in the early days of ANN (and, in fact, of pattern recognition) was the McCulloch and Pitts' simple model of a neuron as a binary threshold unit [54]. McCulloch and Pitts showed that such a neuron model was capable of realising any finite logical expression and, when connected up as a network, could perform any type of computation that a digital computer could (albeit not always so conveniently).

The Simple Perceptron

It was with the model of the simple McCulloch-Pitts neural computational unit that, in the early 1960s, Frank Rosenblatt introduced the concept of the perceptron that is used as a model for learning to recognise and classify patterns [55, 56]. Rosenblatt showed how a neural computational unit with modifiable connections could be trained to classify certain sets of patterns.

The neural unit (see Fig. 7) computes a weighted sum of its inputs from other units, and outputs a one or a zero (a two-class problem), o_i, according to whether this sum is above or below a certain threshold:

$$o_i = f\left(\sum_{n=1}^{Q} w_{in}x_n - \theta_i\right) = f(\sigma) = \begin{cases} 1, & \text{if } \sigma \geq 0; \\ 0, & \text{otherwise}. \end{cases}$$

where $\boldsymbol{x} = \{x_1, x_2, \ldots x_Q\}$ is the input vector. The w_{in} are the connection weights which represent the positive (excitatory) or negative (inhibitory) strength of the synapse connecting any two given neural units, f is the activation or transfer function (a linear or nonlinear function; we consider a binary Heaviside function), and θ_i is the threshold of unit i.

In general, the simple perceptron comprises of a single-layer network of identical neural computational units. The number of output nodes will equal the number of pattern classes required in the output vector. Fig. 8 shows an example simple perceptron.

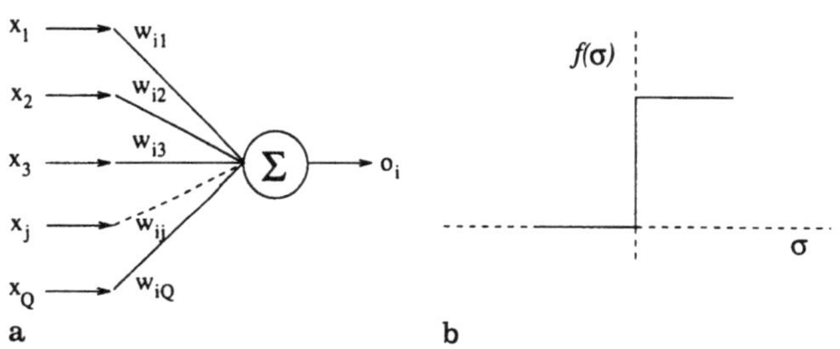

Fig. 7. a Schematic representation of a neural unit, and **b** graphical illustration of a hard-limiting threshold binary unit

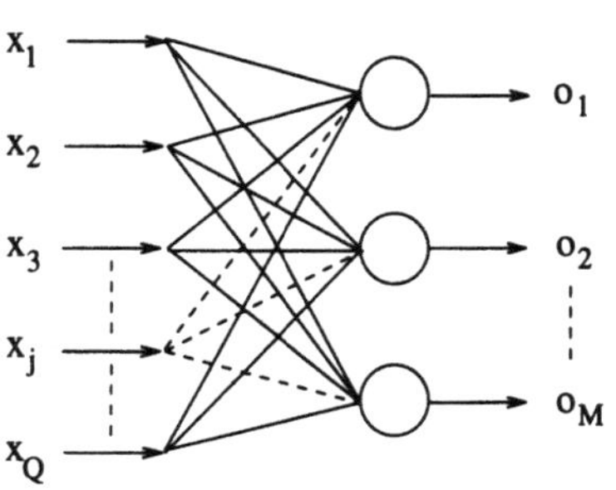

Fig. 8. A single-layer simple perceptron with multiple outputs

The input is propagated through the units in a feed-forward manner and immediately produces the output. As a result of differences among the values of the connection weights, different inputs can produce different outputs.

The classification behaviour of the simple perceptron can best be analysed geometrically, by plotting a map of the decision regions formed in the multi-dimensional space spanned by the Q-dimensional input variables. Consider a single output simple perceptron. Here, the perceptron creates two decision regions (classes) separated by a $(Q - 1)$-dimensional hyper-plane. The equation of the hyper-plane, for a given unit i, is given by

$$\sum_{n=1}^{Q} w_n x_n = \theta \tag{36}$$

or in vector notation

$$\mathbf{w}' \cdot \mathbf{x} = w_0 \tag{37}$$

where it is convenient to implement the threshold as just another weight w_0. If we can draw a hyper-plane that separates the two regions, then the problem is *linearly separable*. If an input vector lies on one side of the hyper-plane, the perceptron will output 1; if it lies on the other side, the output will be a 0. In the general case, if there are several output units (i.e. several classes), we must be able to find one such hyper-plane for *each* output. A two-class example for $N = 2$ is shown in Fig. 9. The linear separability result for a simple perceptron is very similar to the case of linear discriminant analysis as discussed above. However, in the case of the perceptron, the slope and threshold are not known but are learned through an iterative procedure.

The problem of learning is one of locating the hyper-plane to achieve linear separability by changing the connection weights. To train the perceptron we present each example from the training set and use both the output that we want the network to produce (t_i) along with output the network actually does produce (o_i) to generate a difference, $t_i - o_i$. A simple learning rule, based on Hebb's physiological learning rule [57], involves changing the connection weights in a way that is proportional to the input activation to the neural unit making the connection. Small input values will cause small changes in the connection weight,

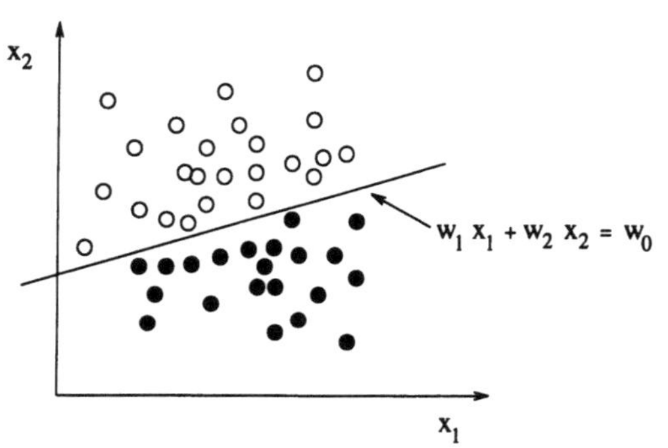

Fig. 9. Linear separability of a simple perceptron, for the case of two classes and two variables

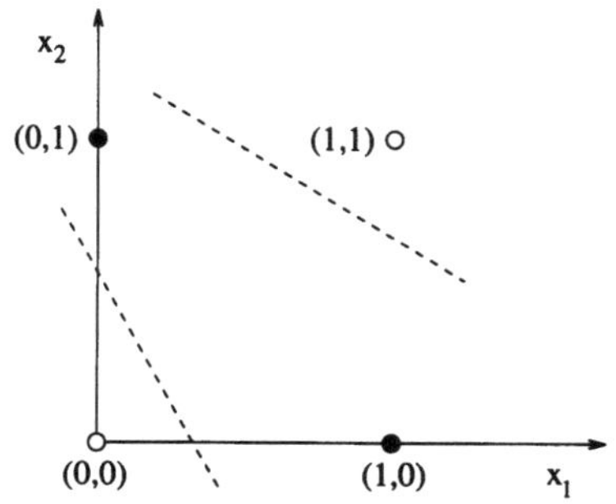

Fig. 10. The exclusive-OR problem is not linearly separable

whereas large input changes will exert large changes in the weight (perhaps "overshooting" the solution vector). Geometrically, changes in the weights correspond to changes in the orientation of the hyper-plane. This learning rule, called the *delta rule*, can be formulated as

$$w_{\text{in}}^{\text{new}} = w_{\text{in}}^{\text{old}} + \Delta w_{\text{in}} \qquad \text{where } \Delta w_{\text{in}} = \eta\,(t_{\text{i}} - o_{\text{i}})x_{\text{n}} \tag{38}$$

where the proportionality parameter η is called the *learning rate*. Initially, the weights are assigned random values, usually between 0 and 1. The training samples are sequentially presented to the perceptron indefinitely until some stopping criterion is satisfied. For the simple perceptron it can be shown that, *if* the problem is linearly separable, the learning rule converges to weights which achieve the desired input-output association in a finite number of steps [58, 59]. The perceptron training algorithm is non-parametric since no assumptions are made about the sample probability density distribution and no means or covariances are evaluated.

At the time of its introduction, the perceptron created much excitement about how future intelligent classification systems could be developed. However, such excitement was short-lived as Minsky and Papert [59] noticed that perceptron convergence procedure could only be guaranteed for linearly separable functions, which most real data did not satisfy. When inputs are not separable or when distributions overlap, the decision boundaries may oscillate continuously and never converge to a stable solution. For example, the disjoint exclusive-OR problem cannot be separated by a single straight line (see Fig. 10).

Multi-Layer Perceptron

To overcome the limitations of the simple perceptron, a multi-layer perceptron (MLP) network with one or more intermediate or hidden layers of identical neural computational units was devised. The input is propagated through the network in a feed-forward manner and immediately produces the output. Layers that are neither input nor output are termed *hidden* layers and make up the MLP's internal

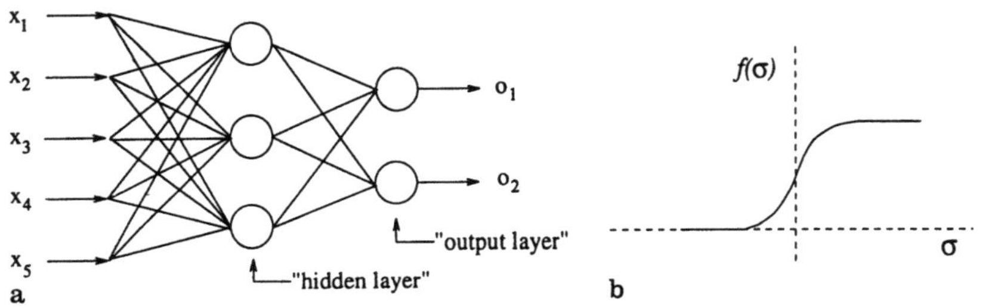

Fig. 11. a A two-layer feed-forward perceptron, and **b** a typical nonlinear (sigmoidal) activation function

representation. Figure 11 shows a two-layer perceptron which includes one hidden layer.

Notice that the activation function in each unit must be nonlinear and differentiable, as a multi-layer linear feed-forward network is equivalent to a single-layer linear network.

Although the discriminant capabilities of multi-layer perceptrons were realised long ago, no effective training algorithms were available. This was only recently overcome with the availability of new learning algorithms. The most well-known is the *error back-propagation* (BP) algorithm or generalised delta rule [60, 61].

In the back-propagation algorithm, the network adjusts its weights each time it sees an input-output pair. Each pair requires two stages a forward pass and a backward pass. The forward pass involves presenting a sample input to the network and letting activations flow until they reach the output layer. During the backward pass, the network's actual output (from the forward pass) is compared with the target (or desired) output, t_j, and error estimates δ_j are computed for each output unit j (for a sigmoidal activation function):

$$\delta_j = o_j(1 - o_j)(t_j - o_j)\,. \tag{39}$$

The error estimates of the output units can then be used to derive error estimates for the units in the hidden layers:

$$\delta_i = o_i(1 - o_i)\sum_n w_{ni}\delta_n \tag{40}$$

where δ_n is the error contributed by each unit in the layer immediately above. This is done until all errors have been propagated back to the input layer. The connection weights can be adjusted by

$$\Delta w_{ij}(\tau) = w_{ij}(\tau + 1) - w_{ij}(\tau) = \eta\delta_i o_j \tag{41}$$

where $\Delta w_{ij}(\tau)$ is the weight change at iteration τ and, as before, η is the learning rate. The choice of η affects the convergence rate of the system. The forward-backward procedure is repeated for each input-output pair presented to the network. After the network has seen all input-output pairs, one *epoch* has been completed. Training such a network often requires many epochs.

Normally, during the process of classification, the class of the input pattern is determined by the output neuron with the maximum signal level (above a predetermined threshold). Also, the output neuron's signal level, normalised over all output neurons, is intuitively used as a measure of the degree of confidence in the classification.

The BP algorithm inherently computes an energy function which the algorithm attempts to minimise. This energy function, E, represents the amount by which the output of the net differs from the required output, or the mean-square error function:

$$E = -\frac{1}{2}\sum_{m}\sum_{j}(t_{mj} - o_{mj})^2 . \tag{42}$$

Note that the energy minimisation is calculated for all patterns of m (where $m = 1, 2, \ldots, N$). The dynamics of the minimisation process corresponds to a series of weight adjustments which can be visualised as a gradient descent in the multi-dimensional weight space. The weight space is generally not smooth, but very "hilly", and can give rise to the BP algorithm getting "stuck" in local minima thereby leading to suboptimal solutions. Various techniques can be used to reduce the possibility of such a scenario occurring. One technique frequently used is to perturb the weights by adding random noise. Also, the gradient descent minimisation process can be very slow, particularly if η is small. This problem has been addressed by various authors [62, 63]. A simple, but efficient, technique involves introducing a momentum term, μ, to "push" the weight changes over local increases in the energy function and increase the convergence along shallow gradients. The weight-update equation is modified to include a contribution from the previous time step:

$$\Delta w_{ij}(\tau + 1) = \eta\delta_i o_j + \mu\Delta w_{ij}(\tau - 1) . \tag{43}$$

Little is known about the convergence behaviour of the BP algorithm, and many practitioners stop the iterative process when some minimum mean-square error value is obtained or when a certain proportion of training set patterns are sufficiently well classified. Training times are typically longer when complex decision regions are required and when networks have a larger number of hidden layers [64].

Some of the important design issues in building an ANN classifier are to find an appropriate network topology (number of hidden layers and number of neurons in each layer) and to evaluate the influence of dimensionality and training sample size.

As we stated previously, a single-layer (no hidden layers) MLP-BP with a hard-limiting activation function gives rise to a linear classification function, i.e. the resulting decision surface is a hyper-plane. With multiple layers and soft-limiting activation functions, arbitrary complex decision surfaces can be realised [65]. Lowe and Webb [66] show that the hidden layers perform feature extraction which maximises class separation, whilst the output layer performs an optimum mapping on to the targets. This enables MLP-BP networks to combine feature extraction and classification simultaneously, this being one possible explanation why such networks have demonstrated to produce good classification over a range of problems. Cybenko [67] formally shows that one hidden layer and any continuous sigmoidal activation function is sufficient to compute arbitrary decision boundaries for the outputs. Unfortunately, this result does not tell us how many units are necessary nor whether it is possible to learn the weights. Furthermore, it has been shown that a one hidden-layer MLP with an infinite number of hidden units is sufficient to approximate any posterior probability density function to any degree of accuracy. Thus, if convergence in a least-squares sense is achieved, the outputs of a MLP-BP network are direct estimations of posterior probabilities and the network has the same computational power as the Bayes decision rule [68, 69].

The choice of the exact number of hidden nodes in a MLP-BP network is a more complex task. A small number of hidden nodes (significantly less than the number of patterns in the training sample) reduces the computational time for training, but not too small as it may be difficult to obtain convergence during training and to create adequate decision surfaces. A MLP-BP with a large number of hidden nodes will have more degrees of freedom and will therefore require a reduced accuracy in the values of the weights to achieve the desired classification accuracy. However, such a large number of parameters (for a limited training set size) may result in the network interpolating the data and producing an increased true classification error rate attributable to over-fitting, and therefore exhibiting poor generalisation capabilities [70]. Very large training set sizes are then required to achieve acceptable performance [71]. Learning theory has begun to establish what is possible for the MLP. In particular it has been shown that, for a single hidden layer fully connected MLP with W weights, one needs in the order W/ε patterns in the training set to expect a generalisation error of less than ε [72]. Empirical results indicate that, for the case of a single hidden layer, the maximum number of hidden nodes should be of the order $M \times (N + 1)$ [73], where N and M are the number of input and output units respectively. Also, Widrow [74] suggests a rule of thumb that the training sample size should be of the order of ten times the number of weights in a network.

Increasing the number of features never increases the classification error rate of the optimal Bayes classifier (infinite number of training samples). However, in the context of finite training sample sizes, a peaking in the classification performance is observed. The additional discriminatory information that is conveyed by the additional features is outweighed by the increased true classification error due to the poor generalisation ability in the higher dimensional space. This is also observed in the design of MLP-BP networks [70].

We now consider two example applications of MLP-BP: i) an infrared spectral peak verification system and ii) a simulation of the relative contribution of various pollution sources to the pollution of a given ecosystem. Much of what follows is drawn from the papers by Wythoff et al. [75] and Karayiannis and Venetsanopoulos [76].

There have been many reports on the use of neural networks for the interpretation of spectra. For example, Thomsen and Meyer [77] employed a single hidden layer MLP to recognise the NMR spectra of sugar alditols; Borggaard et al. [78] implemented various MLP meso-architectures to determine the fat content of pork meat using NIR spectra and compared the results with techniques such as partial least-squares and principal components analysis. Of particular interest in spectral interpretation is that of peak verification and peak recognition. Chemical applications include spectroscopic and chromatographic methods, as well as flow injection analysis. The feasibility of exploiting neural networks to verify and recognise peak-shaped signals in analytical data was undertaken [75]. Here data from the 2 cm^{-1} resolution IR spectrum of a vapour-phase mixture of tetrahydrofuran, 1,1-dichloroethane, benzene, ethyl-benzene, methylene chloride and 1,1,1-trichloroethane at concentrations of 3 ppm were used as inputs to a single hidden layer MLP. The MLP consisted of 14 input layer nodes representing the absorbance values and noise magnitude from the spectrum and a number of hidden layer nodes varying from one to nine. Peak verification was undertaken by using a window of data points obtained from the peak table produced by the IR spectrum workstation. A total of 132 patterns were used for training the MLP, and an independent set of 189 test patterns were presented to the MLP for evaluation. The minimum true mean absolute error between the desired and actual MLP outputs of 0.13 was observed for the case of two hidden layer nodes.

In the ecosystem simulation application a simplified river ecosystem situated in an industrial environment is assumed, where various potential chemical sources are responsible for polluting the ecosystem. Each day, a number (n_p) of chemical pollutants exceeding a normal threshold are detected in the ecosystem and a number of wastes from pollution sources (n_s) are released into the ecosystem. The presence or absence of a total of $n_p = 10$ chemical pollutants, including ammonia, chlorine, cyanides etc., and $n_s = 6$ pollution sources such as, domestic sewage, a pulp and paper mill, a metal plating plant etc., measured for a period of 15 days are used in the single-layer MLP-BP model. An additional source waste pattern is included for the case when there are no chemicals detected in a certain pollution source of the ecosystem. The inputs to the MLP-BP consist of the pollution pattern of the previous day together with the waste pattern of all the pollution sources of the next day (a total of $n_p(n_s + 1)$ inputs). The output of the network is the resulting pollution pattern (n_p outputs). A total of $n_p(n_s + 1)$ hidden units are employed. After training the network, the appearance of chemical pollutants exceeding a normal threshold every day can then be simulated. The relative effect of the pollution sources can be evaluated by testing the performance of the network when each one of the pollution sources, or any combination of sources, is considered to be active. The network was able to identify the relative impact of the pollution sources which were the major contributors to the pollution of the river.

In summary, in the context of classification, neural networks seem to work well and tend to be more robust in a variety of applications. They are able to handle noisy or incomplete inputs and make no statistical assumptions on the behaviour of the data. The ANN approach is not necessarily the best one among all approaches covered here, it just performs well quite often. The major disadvantages of neural networks are the long training times required and the difficulty in interpreting the connection weights in terms of the classification performance.

Classification Trees

The primary goal of supervised learning systems is to acquire classification or decision rules from a set of training samples, each sample belonging to a particular class, and then to assign new unseen samples to these classes. So far, the classification techniques that we have presented needed some quite elaborate and computationally intensive mathematical procedures to produce the decision rules. Also, once training has been accomplished, invoking any of these rules to classify an unseen sample will also require some sort of computational facility. However, such techniques may, in some cases, be incompatible with the human user for various reasons. Firstly, the human use and interpretation of the results of such techniques can be problematic because the explanation of the results relies on the mathematical understanding of the techniques. Secondly, it is generally accepted that the underlying nature of human reasoning is not a numerical process but rather a symbolic process. Human beings understand better simple condition-action rules of the form "If the concentration of cadmium exceeds 10 ppm, then conclude H" or, more generally, in disjunctive normal form of the type "If both X and Y are true or Z is true". Finally, in some applications, the pattern feature values are categorical. For example, the colour of some object can be red, green, or blue. We can always invent ways of converting non-numeric feature values into some numeric equivalent. However, in some applications, this may not be possible due to the lack of knowledge of how best to specify the conversion or due to the large number of categorical features. To address these issues, other techniques that are more compatible with human reasoning have been developed. These are generally referred to as *machine learning* methods. Comparable techniques have also been developed by the statistics community [79].

One of the most popular machine learning algorithms involves learning a *decision tree* (ID3 and its successor C4 and C4.5) [80] – also referred to as a *classification tree* (CART) [79] or *discrimination tree* [81]. The basis to the construction of classification/decision trees has been approached in many different ways [82–84]. A significant advantage of the decision tree is the ability to express the tree as a set of rules and thus provide a procedure for inductive inference and the acquisition of rules, which humans can easily understand. The tree can determine the class membership of any pattern from its feature values independently of any a priori information about the functional form of the distribution of pattern vectors.

A decision tree is generated by hierarchically partitioning the feature space by using a series of local searches for good partitions or splits. The construction of the tree is formulated in a top-down fashion, starting at the root node, and recursively splitting the decision region into two half spaces. This creates a binary decision tree (two branches per internal node), although more general decision trees can be constructed by using multiple partitions at each node. Note that once a node is split no backtracking is permitted, i.e. the decision is never revised. A node decision function defines the dividing boundary at a given tree node, where each decision function uses only a subset of features (usually a single feature) as its argument. In general, the decision function is implemented as a simple hard threshold function. To improve the overall decision performance of the tree, a heuristic measure of the "goodness" of splitting each node is made in terms of a "mutual information gain" criterion or "node impurity" measure. This heuristic is implemented as an evaluation function (see later). Once a node is designated as a terminal node a decision rule is invoked to assign classification labels to each terminal node, usually with a class label that is in a majority over the training patterns.

From a geometric point of view, the result of the tree-growing process is a partitioning of the feature space by hyper-planes parallel to the axes of the feature space so that the feature space is covered with hyper-rectangular regions (for univariate splits). The partitioning of the feature space may also be undertaken along hyper-surfaces which are not necessarily parallel to the feature axes; by the linear [79, 85] or non-linear [86] combinations of features. A simple perceptron [87] or MLP [86] can be used in each tree node for implementing the hyper-surface decision test. How efficiently these regions cover the data will determine the classification performance of the decision tree.

Classification trees are also related to the multi-layer feed-forward perceptron [88, 89]. A binary classification tree is initially constructed from the training and, if applicable, test sets. The tree is then "mapped" into a MLP with two hidden layers and a single output layer – the first hidden layer relating to the number t of non-terminal nodes of the classification tree and the second hidden layer corresponding to the $t + 1$ paths from the root of the tree to each terminal/leaf node. This enables a decision tree to be simulated with a MLP, thereby avoiding the problem of specifying the number of hidden neural units in advance and the problem of slow convergence rates obtained with the back-propagation algorithm.

Lead/Cadmium Pollution. As an illustration consider the following example data set. Lead and cadmium levels are measured in the teeth of young school age children in two regions in North-West Belgium, one data set obtained in an urban heavy industrial area 2 Km from a non-ferrous smelting plant (Hoboken) and another data set in a seaside rural-urban area (De Haan) [90]. A graph of the distribution of a subset of the two data sets for the two features lead (Pb) and cadmium (Cd) concentrations, in parts per million (ppm), is shown in Fig. 12. Also shown is the resulting simple binary decision tree with two classes –

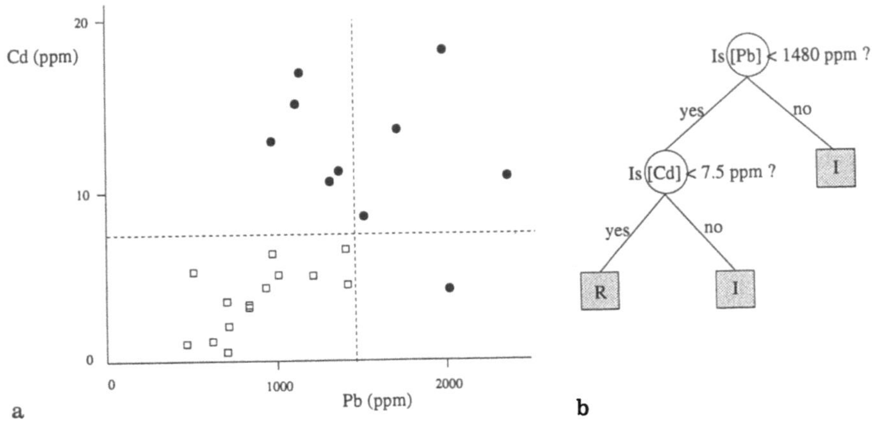

Fig. 12. a Example feature space, and **b** the decision tree

industrial (I) and rural (R). The terminal (leaf) nodes of the tree correspond to the classes and all paths leading to a terminal node is a conjunction of the various tests along the path. If there are multiple paths for a given class (i.e. a class has more than one terminal node), then the paths represent disjunctions. In the above example we have acquired the following rules from the data set:

if [Pb] < 1480 ppm AND [Cd] < 7.5 ppm then "rural area"

and

if [Pb] ≥ 1480 ppm OR [Cd] ≥ 7.5 ppm then "industrial area".

As mentioned previously, to obtain the best tree an evaluation function is required at each stage of the recursive process to measure the "goodness" of splitting a given frontier node, t. The evaluation function tries to reduce the degree of randomness (or "impurity") in the selection of features in the current node and future nodes. There exists a variety of splitting functions; two of the most popular ones are the *Gini* function [79]

$$i(t) = \sum_{k \neq 1}^{K} P(\omega_k|t)P(\omega_1|t) \tag{44}$$

and the *entropy* function (i.e. information content) [80]

$$i(t) = -\sum_{k}^{K} P(\omega_k|t) \log P(\omega_k|t) \tag{45}$$

where $P(\omega_k|t)$ is the probability of class $k (k = 1, 2, \ldots, K)$ estimated from the case frequencies at the tree node t to be split.

To decide on which feature to split at a given tree node, the function evaluates the degree of randomness prior to splitting on a given feature and compares it with the randomness after splitting. That is, we calculate (assuming V_F values for the given feature F) the *gain criterion*

$$\Delta i(t;F) = i(t) - \sum_{v_F=1}^{V_F} P(S_{v_F}|t)i(t_{v_F}) \tag{46}$$

where $P(S_{v_F}|t)$ is the probability distribution of data points for a given feature value v_F, estimated from the case frequencies for a given v_F at the given tree node t. This is done for all possible features at node t. The feature which provides, for the case of the entropy measure, the maximum gain in information content is the feature chosen for splitting. Detailed examples are given in [79–81].

We have assumed that the splitting is performed in the case of categorical (unordered) features. However, for continuous (ordered) features, splits for all values found in the training sample will need to be considered [91]. For ordinal (discrete, ordered) values the discrete values (e.g. very cold, cold etc.) are mapped onto integers and then treated as continuous features.

For a given set of training patterns a tree can grow such that every terminal node will contain members of only one class. This will yield a 100% classification rate on the training patterns and corresponds to "noise fitting". This 0% apparent error rate is not the true error rate, which is most likely to be larger. That is, the predictive classification power of the tree is lower than is apparent, corresponding to the situation of "over-fitting". Although the more complex (larger) tree reflects the true structural relationships that exist in the training data, it will also reflect the noisy patterns that may be present in the training data set. Therefore, the larger tree will most often perform worse on (new) test data. A smaller tree will typically have a larger apparent error rate but a smaller true error rate than a larger fully grown tree. Also, the true error rate of the smaller tree is no less than half the true error rate of the fully grown tree [16]. It is therefore important to decide when to stop splitting any node to avoid over-fitting (see, however, [92] for a counter-argument). In general, the problem of growing the smallest decision tree that correctly classifies the training data is known to be computationally intractable (see, for example, [93]). We therefore rely on heuristic approaches to approximate this. There are various ways for determining when node splitting is no longer significant and should be stopped, such as, the standard statistical χ^2 test and error reduction. In general, such single node, lookahead threshold-based tests are not the best techniques to avoid over-fitting because determining the optimal threshold is difficult – a threshold that is too low may result in little improvement in tree size and a threshold that is too high may terminate splitting too early, risking under-fitting, which is potentially worse than over-fitting.

A superior approach involves *pruning* the tree given a fully expanded tree grown (grown with a given set of training samples) [79]. The tree can be selectively pruned upwards from the terminal nodes to find a best sub-tree having the lowest true error estimate. There are several methods to determine the best pruned sub-tree. One simple technique, called *reduced-error pruning* [94], involves dividing the samples into a training and test set, growing the full tree on the training set (as before), and pruning using the test set. A branch of the tree is pruned if the misclassification error rate on the test set is reduced. From all the nodes, choose the one with the largest reduction in the misclassification error rate as the sub-tree to prune. This process continues until no improvement in the error rate is observed (or when the error rate actually increases). The pruning process then stops. This works reasonably well if the sample size is large, the main problem being that the final pruned tree is designed based on both training and test sample sets. Also, it is possible to obtain a number of sub-trees with the same reduction in error rate and a heuristic is normally needed to choose the most appropriate sub-tree to prune. Another technique, called *cost-complexity* or *error-complexity pruning* [79], takes into account both the error rate and the complexity (size) of the tree. It is based on generating a finite number of sub-trees with progressively fewer terminal nodes by finding the weakest link in the tree. The weakest link in a tree is the node t (in the sub-branch T_t of the fully grown tree T_0 formed by node t) that can be deleted with the minimum value of $g(t)$:

$$g(t) = (R(t) - R(T_t))/(|T_t| - 1) \tag{47}$$

for all nodes $t \in T_0$, where $|T_t|$ is the number of terminal nodes in T_t, and where $R(t)$ and $R(T_t)$ are the apparent error rates of the node t and sub-branch T_t, respectively. The weakest link in T_0 is pruned and the process is repeated. The result is a finite sequence of sub-trees $T_0, T_1, T_2, \ldots, r_{T_0}$ (r_{T_0} is the root of T_0). The final step is to select the best sub-tree. In the case of large samples, the best sub-tree can be determined by evaluating the set of trees using a set of test samples and choosing the sub-tree with the minimum error rate. For smaller sample sets, re-sampling is more appropriate [79]. Mingers [95] presents empirical comparisons of several pruning methods.

In many applications such as environmental sciences, one or more features could be missing or unknown in a data set. This could occur in situations where specific values were not recorded in a particular measurement, or were not relevant etc. This can be a problem with decision trees as the missing feature may be involved in a test in one of the nodes, with the result that a single sample can follow multiple branches in the tree. Usually missing feature values can be filled in simply by using the most probable or median value. Breiman et al. [79] use an additional feature evaluation splitting test at each node in the tree as a substitute test, called a *surrogate split*, which is most similar to the original split test.

Using the two-class Pb/Cd pollution example presented above (with a total of 39 samples), we applied the C4.5 decision tree method and obtained an overall misclassification error rate of 15.8% (based on ten-way cross-validation). The rules obtained were as follows.

If [Pb] > 1429 ppm then "industrial area"

and

$$\text{If } [\text{Pb}] \leq 1429 \text{ ppm and}$$
$$[\text{Cd}] \leq 12.16 \text{ ppm then "rural area"}$$
$$[\text{Cd}] > 12.16 \text{ ppm then "industrial area"}.$$

As we have seen, decision trees have the advantage of being more compatible with human reasoning, have the ability to handle categorical data, are able to classify new data efficiently, demonstrate good generalisation capabilities, and provide a means of constructing a tree that avoids under-fitting or over-fitting. They are also robust in the presence of outliers and misclassified points [79], are able to handle multivariate splits (by a linear combination of features), and can cope with continuous classes.

However, one disadvantage includes the difficulty of performing incremental updates to the decision tree (at present all training samples must be available during the growth/pruning stages of the tree).

Performance Comparison of Multi-Layer Perceptrons and Classification Trees

A number of works have been published comparing the performance of decision trees and other classification methods including, multi-layer perceptrons, linear and quadratic classifiers, nearest-neighbour etc. [16, 96–98].

Atlas et al. [96] compared empirically the MLP-BP and CART methods using three different problems in power systems load forecasting and power system security, and in vowel identification. MLP-BPs were found to have a better classification performance (using an independent test set) than CART, although the differences were not statistically significant except in the case of the power security problem. CART's performance improved when multivariate splits (linear combinations of features) were used.

Brown et al. [97] also compared the two methods on three problems in radar emitter data (with few features but a large sample size), modem/transceiver data (many features but a limited training sample size), and digit recognition data with binary features. The MLP-BP performed well on large data sets with a selected subset of features, whereas CART effectively handled problems with a large number of (possibly not all relevant) features and a small sample size. This points to the advantage of CART for variable selection.

Weiss and Kulikowski [16] provided an extensive comparative classification performance analysis of decision trees, MLP-BP and statistical classifiers. A total of four data sets (three of which involved medical diagnosis) were used. Two of the medical data sets were average-to-high dimensional and consisted of a large number of training samples. An optimal number of hidden units based on the minimal classification error rate was calculated for the application of the MLP-BP to each data set. Classification error rate was obtained using either independent test sets or, in the case of small sample sizes, cross-validation. The decision tree methods (CART and C4) performed the best, followed by the MLP-BP method. Overall, the quadratic classifier performed the worse.

In summary, decision tree methods and MLP-BP produce comparable error rates. However, decision trees do well with problems that do not have a small ratio of the number of features to the training sample size, whereas MLP-BP performs well on problems with large amounts of data and a small number of features. Both CART and MLP-BP are able to approximate nonlinear decision boundaries but CART, with univariate splits, constructs classification regions with "ragged" boundaries whereas MLP-BP, owing to the fact that a single point in input space can affect all units, constructs smooth boundaries. Both decision tree methods and MLP-BP are capable of handling over-fitting/under-fitting by pruning (in the case of decision trees) or by cross-validation techniques. Decision trees are fast to train, as compared with MLP-BP networks, and are easier to interpret (with rule acquisition) than MLP-BP. MLP-BP networks are good for generalisation (i.e. high predictive power) [98], are generally less susceptible to missing values, and are fault-tolerant.

Software for Pattern Classification

Many software packages that run under different operating systems are available for the purposes of pattern classification. These include SPSS, S-Plus, SAS, etc. Most of these packages run under MS-DOS, Unix and, more recently, under MS-Windows. Each package will provide an underlying operating environment (menu-driven interface, file input/output, output display etc.), some pre-processing functions and a selection of classifiers to choose from. The larger packages tend to be generic in nature (i.e. provide a basic functionality which would satisfy many applications). Specific software modules can be incorporated as the user requires them. Many applications, however, would only use a small subset of the package and, in many cases, would not provide all of the application-specific functionality. Several packages specifically designed for chemometrics also include one or more classifiers e.g. PARVUS [99], ARTHUR [32], SIRIUS [100], SIMCA 4R [101], and more recently SCAN [102]. Also, there is often a large variation in the quality of packages (here measured in terms of the user-interface, output graphics etc.). In many cases, the data display and user-interface can be quite primitive (e.g. 2-D and 3-D interactive graphics facilities are rarely provided). Some new software packages, offering better user-oriented facilities and making available applications-specific software modules are being developed (e.g. PARIS [50]).

Conclusions

Various pattern classification techniques used in the context of environmental chemistry have been presented. Both parametric and non-parametric Bayesian classifiers have been outlined, as well as artificial intelligence-based techniques such as artificial neural networks and classification trees. Environmental applications were used to illustrate the various classifiers. The implementation of a given classifier is very much dependent on the application under consideration – a classifier may perform well in the context of one application but may give poor results in another. Therefore, it is important that the environmental scientist makes a judicious selection of classifiers (usually made on the basis of results obtained from data pre-processing) for a given application. The choice of a particular classifier depends on many factors including the dimensionality, amount of noise, class distribution, number of training and test samples, presence of outliers etc.

References

1. Kawata S (1989) Imaging spectroscopy and pattern recognition Bunseki 5:345–349
2. Pontaul A, Cuff C (1944) Order-disorder effects on the short wavelength infrared spectra of kaolinites Clay and Clay Minerals – submitted for publication
3. Raudys S, Jain A (1991) Small sample size effects in statistical pattern recognition: Recommendations for practitioners IEEE Trans on Pattern Analysis and Machine Intelligence pages 252–264
4. Cybenko G (1988) Continuous valued neural networks: approximation theoretic results In Proc Interface88, The Symposium on the Interface: Computer Science and Statistics pages 174–183
5. Jain A (1987) Advances in statistical pattern recognition In P. Devijver and J. Kittler, editors, Pattern recognition theory and applications pages 1–19 NATO ASI Series, Springer-Verlag
6. Ismail S, Grass F, Varmuza K (1988) Pattern recognition techniques as a tool for identifying sources of contamination in environmental sample analysis J Trace Microprobe Tech 6:563–573
7. Tomas X, Rius J, Obiols J, Sol A (1988) Application of pattern recognition to speciation data of heavy metals in suspended particulates of urban air J Chemometrics 3:139–150
8. Higashi K, Hagira K (1988) Identification of oil spilled at sea by high performance gel permeation chromatography pattern recognition Water Science Technology 20:55–62
9. Scott D (1988) Classification of binary mass spectra of toxic compounds with an inductive expert system and comparison with simca class modeling Analytica Chimica Acta, 211:11–29
10. Scott D, Dunn W, Emery S (1988) Pattern recognition classification and identification of trace organic pollutants in ambient air from mass spectra J Res Natl Bur Stand (US) 93:281–283
11. Dunn W, Koehler M, Emery S, Scott D (1987) Application of pattern recognition to mass spectral data of toxic organic compounds in ambient air Chemometrics and Intelligent Laboratory Systems 1:321–334
12. Pontash K, Smith E, Cairns J (1989) Diversity indices, community comparison indices and canonical discriminant analysis: interpreting the results of multispecies toxicity tests Water Res 23:1229–1238
13. Davis B (1989) Data handling and pattern recognition for metal-contaminated soils Environ Geochem Health 11:137–143
14. Saaksjarvi E, Khalighi M, Minkkinen P (1989) Waste water pollution modeling in the southern eare of Lake Saimaa, Finland, by the simca pattern recognition method Chemometrics and Intelligent Laboratory Systems 7:171–180
15. Boey K, Jeyaratnam J (1988) A discriminant analysis of neuropsychological effect of low lead exposure Toxicology 49:309–314
16. Weiss S, Kulikowski C (1991) Computer systems that learn Morgan Kaufmann Publishers San Mateo, CA

17. Duda R, Hart P (1973) Pattern classification and scene analysis John Wiley, New York
18. Friedman JH, Stuetzle W (1981) Projection pursuit regression J American Statistical Association 76:817–823
19. Friedman JH (1991) Multivariate adaptive regression splines Annals of Statistics 19:1–141
20. Anderson T (1958) An introduction to multivariate statistical analysis. John Wiley, New York
21. Ripley BD (1994) Neural networks and related methods for classification J Royal Statist Soc B 56:409–456
22. Hastie T, Tibshirani R, Buja A (1993) Flexible discriminant analysis by optimal scoring – obtained via ftp
23. Hastie T, Buja A, Tibshirani R (1993) Penalized discriminant analysis – obtained via ftp
24. Tatsuoka M (1971) Multivariate analysis John Wiley, New York
25. Rao C (1948) The utilization of multiple measurements in problems of biological classification J Royal Statist Soc (Ser B) 10:159–203
26. McLachlan G (1992) Discriminant analysis and statistical pattern recognition John Wiley, New York
27. Brereton R (1992) Multivariate pattern recognition in chemometrics Elsevier, Amsterdam
28. Sharaf M, Illman D, Kowalski B (1986) Chemometrics John Wiley, New York
29. Massard D, Vandeginste B, Deming S, Michotte Y (1988) Chemometrics: A textbook. Elsevier, Amsterdam
30. Nilsson N (1965) Learning machines McGraw-Hill, New York
31. Fisher R (1936) The use of multiple measurements in taxonomic problems Annals of Eugenics 7:179–188
32. Harper A, Duewer D, Kowalski B (1977) ARTHUR and experimental data analysis: the heuristic use of a polyalgorithm In B. Kowalski, editor, Chemometrics: Theory and practice Am Chem Soc Symp Ser 52
33. Australian Institute of Marine Science (1992) Long term monitoring of the Great Barrier Reef: dissolved and particulate nutrients Australian Institute of Marine Science, Townsville, Australia
34. Norusis M (1993) SPSS for Windows Release 6.00 SPSS Inc
35. Anderson J (1972) Separate sample logistic discrimination Biometrika 59:19–36
36. Breiman L, Ihaka R (1984) Nonlinear discriminant analysis via scaling and ace Technical report, Univ of California at Berkeley
37. Hastie T, Tibshirani R (1990) Generalised additive models Chapman and Hall, London, England
38. Frank I (1990) A nonlinear PLS model Chemometrics and Intelligent Laboratory Systems 8:109–119
39. Defrise-Gussenhoven E (1955) Mesure de divergence λ^2 entre un sujet determiné et une population multi-variée normale Bulletin de l'Institut Royal des Sciences Naturelles de Belgique 31:1–15
40. Coomans D, Broeckaert I, Derde M, Wold S, Massart D (1984) Use of a microcomputer for the definition of multivariate confidence regions in medical diagnosis based on clinical laboratory profiles Comp Biomed Res 17:1–14
41. Wold S (1978) Cross-validatory estimation of the number of components in factor and principal component analysis Technometrics 20:397–411
42. Derde M, Coomans D, Massart D (1982) Effect of scaling on class modeling with the SIMCA method Analytica Chimica Acta 141:187–192
43. Wold S (1976) Pattern recognition by means of disjoint principal components models Pattern Recognition 8:127–139
44. Frank I (1988) DASCO: A new classification method Chemometrics and Intelligent Laboratory Systems 4:215–222
45. Hand DJ (1981) Discrimination and classification John Wiley, New York
46. Aeberhard S, Coomans D, de Vel O (1994) Comparative analysis of pattern classifiers in a high dimensional setting Pattern Recognition – accepted for publication
47. Fukunaga K, Hayes R (1989) Effects of sample size in classifier design IEEE Trans on Pattern Analysis and Machine Intelligence 11:873
48. Gnanadesikan R et al. (1989) Discriminant analysis and clustering Statistical Science 4:34–69
49. Friedman JH (1989) Regularized discriminant analysis J American Statistical Association 84:165–175
50. Aeberhard S, Coomans D, de Vel O (1993) Improvements to the classification performance of regularized discriminant analysis J Chemometrics 7:99–115

51. Longstaff ID (1987) On extensions to Fishers linear discriminant function IEEE Trans on Pattern Analysis and Machine Intelligence 9:321–325
52. Coomans D, Broeckaert I (1986) Potential pattern recognition in chemical and medical decision making Research Studies Press John Wiley and Sons, England
53. Loftsgaarden D, Quesenberry C (1965) A nonparametric estimate of a multivariate density function Annals Math Statistics 36:1049–1051
54. McCulloch WS, Pitts W (1943) A logical calculus of ideas immanent in nervous activity Bulletin of Mathematical Biophysics 5:115–133
55. Rosenblatt F (1958) The Perceptron: a probabilistic model for information storage and organization in the brain Psychological Review 65:386–408
56. Rosenblatt F (1962) Principles of neurodynamics: Perceptrons and the theory of brain mechanisms. Spartan, New York
57. Hebb DO (1949) The organization of behavior. Wiley, New York
58. Block HD (1962) The Perceptron: a model for brain functioning Reviews of Modern Physics 34:123–135
59. Minsky M, Papert S (1969) Perceptrons MIT Press, Cambridge
60. Werbos P (1974) Beyond regression: new tools for prediction and analysis in the behavioral sciences. PhD thesis, Committee on Applied Mathematics, Harvard University
61. Rumelhart D, Hinton G, Williams R (1986) Learning internal representations by error propagation In D. Rumelhart and J. McClelland, editors, Parallel distributed processing pages 318–364 MIT Press
62. Hertz J, Krogh A, Palmer R (1991) Introduction to the theory of neural computation Addisson-Wesley, Redwood City, CA
63. Fahlmann S (1989) Faster learning variations on back-propagation: an empirical study In Proc of the 1988 Connectionist Models Summer School pages 38–51, Eds D. Touretzsky et al.
64. Lippmann R (1989) Pattern classification using neural networks IEEE Communications Magazine pages 47–64
65. Lippmann R (1987) An introduction to computing with neural networks IEEE ASSP Magazine 4:4–22
66. Lowe D, Webb A (1991) Optimized feature extraction and the Bayes decision in feed-forward classifier networks IEEE Trans on Pattern Analysis and Machine Intelligence 13:355–364
67. Cybenko G (1989) Approximation by superpositions of a sigmoidal function Mathematics of Control, Signals, and Systems 2:303–314
68. Lee D, Srihari S, Gaborski R (1991) Bayesian and neural network pattern recognition: a theoretical connection and empirical results with handwritten characters In I.K. Sethi and A.K. Jain, editors, Artificial neural networks and statistical pattern recognition pages 89–108 Elsevier Science Publishers
69. Wan E (1990) Neural network classification: a Bayesian interpretation IEEE Trans on Neural Networks 1:303–305
70. Raudys S, Jain A (1991) Small sample size problems in designing artificial neural networks In I.K. Sethi and A.K. Jain, editors, Artificial neural networks and statistical pattern recognition pages 33–50. Elsevier Science Publishers
71. Geman S, Bienenstock E, Doursat R (1992) Neural networks and the bias/variance dilemma Neural Computation 4:1–58
72. Baum E, Haussler D (1989) What size net gives valid generalization? Neural Computation 1:151–160
73. Maren A, Harston C, Pap R, (1990) Handbook of neural computing applications Academic Press, San Diego
74. Widrow B (1987) Adaline and Madaline In Proc IEEE 1st Intl Conf on Neural Networks pages 143–158
75. Wythoff BJ, Levine SP, Tomellini SA (1990) Spectral peak verification and recognition using a multilayered neural network Anal Chem 62:2702–2709
76. Karayiannis N, Venetsanopoulos A (1990) Applications of neural networks to environmental protection In Proc Inter Conf Neural Networks pages 334–337 Paris
77. Thomsen J, Meyer B (1989) Pattern recognition of the ^{1}H NMR spectra of sugar alditols using a neural network J Magnetic Resonance 84:212–217
78. Borggaard C, Thodberg H (1992) Optimal minimal neural interpretation of spectra Anal Chem 64:545–551

79. Breiman L, Friedman J, Olshen R, Stone C (1984) Classification and regression trees Wadsworth, Belmont, California
80. Quinlan J (1983) Learning efficient classification procedures and their application to chess end-games In R. Michalski, J. Carbonell, and T. Mitchell, editors, Machine learning Tioga Publishing Company
81. Pao Y (1989) Adaptive pattern recognition and neural networks Addison-Wesley, Reading, Mass
82. Friedman J (1977) A recursive partitioning decision rule for nonparametric classification IEEE Trans on Computers 26:404–408
83. Payne H, Meisel W (1977) An algorithm for constructing optimal binary decision trees IEEE Trans on Computers 26:905–916
84. Hunt E, Marin J, Stone P (1966) Experiments in induction Academic Press, New York
85. Murthy S, Kasif S, Salzberg S, Beigel R (1993) OC1: Randomised induction of oblique decision trees – obtained via ftp
86. Guo H, Gelfand S (1992) Classification trees with neural network feature extraction IEEE Trans on Neural Networks 3:923–933
87. Sankar A, Mammone R (1990) A fast learning algorithm for tree neural networks In Proc 1990 Conf on Information Sciences and Systems pages 638–642 Princeton, NJ
88. Brent R (1991) Fast training algorithms for multi-layer neural nets IEEE Trans on Neural Networks 2:346–354
89. Bigot P, Cosnard M (1993) Probabilistic decision trees and multilayered perceptrons In Proc of the European Symp on Artificial Neural Networks pages 91–96 Brussels, Belgium
90. Coomans D, Slop D, Cleymaet R (1991) Lead and Cadmium content in tooth surface enamel of Belgium school children from different geographic areas Technical report, Dept of Mathematics and Statistics (James Cook University, Australia) and Eenheid Prothetische Tandheelkunde (Vrije Universiteit Brussel, Belguim)
91. Quinlan J (1993) C4.5: Programs for machine learning Morgan Kaufmann, San Mateo
92. Schaffer C (1993) Overfitting avoidance as bias Machine Learning 10:153–178
93. Hyafil R, Rivest R (1976) Constructing optimal binary trees is NP-complete Information Processing Letters 5:15–17
94. Quinlan J (1987) Simplifying decision trees Int J Man-Machine Studies 27:221–234
95. Mingers J (1989) An empirical comparison of pruning methods for decision tree induction Machine Learning 4:227–243
96. Atlas L, Cole R, Muthusamy Y, Lippman A, Connor J, Park D, El-Sharkawi M, Marks R (1990) A performance comparison of trained multilayer perceptrons and trained classification trees Proc IEEE 78:1614–1619
97. Brown D, Corruble V, Pittard C (1993) A comparison of decision tree classifiers with backpropagation neural networks for multimodal classification problems Pattern Recognition 26:953–961
98. Tsoi A, Pearson R (1991) Comparison of three classification techniques, CART, C4.5 and multi-layer perceptrons In R. Lippmann, J. Moody, and D. Touretzky, editors, Advances in neural information processing systems volume 3 Morgan Kaufmann Publishers
99. Forina M, Armanino C, Lanteri S, Leardi R (1989) PARVUS Elsevier Scientific Software, Amsterdam
100. Kvalheim O, Karstang T (1992) SIRIUS – A program for analysis and display of multivariate data In R. Brereton, editor, Multivariate pattern recognition in chemometrics pages 303–320 Elsevier
101. Nilsson-Lindgren A (1991) SIMCA 4R – Soft independent modeling of class analogy Umetri AB, Umea, Sweden
102. Todeschini R, Cosentino U, Frank I, Moro G (1992) SCAN – Software for chemometrics analysis Jeril Inc Stanford, USA
103. Ayre J, Aeberhard S, Coomans D, de Vel O (1994) PARIS – A Pattern Analysis/Recognition Interactive Software, Quality Interactive Software Pty Ltd, Townsville, Australia

Subject Index

Springer-Verlag and the Environment

We at Springer-Verlag firmly believe that an international science publisher has a special obligation to the environment, and our corporate policies consistently reflect this conviction.

We also expect our business partners – paper mills, printers, packaging manufacturers, etc. – to commit themselves to using environmentally friendly materials and production processes.

The paper in this book is made from low- or no-chlorine pulp and is acid free, in conformance with international standards for paper permanency.

MIX
Papier aus verantwortungsvollen Quellen
Paper from responsible sources
FSC® C105338

If you have any concerns about our products,
you can contact us on
ProductSafety@springernature.com

In case Publisher is established outside the EU,
the EU authorized representative is:
Springer Nature Customer Service Center GmbH
Europaplatz 3, 69115 Heidelberg, Germany

Printed by Libri Plureos GmbH
in Hamburg, Germany